VETERINARY TOXICOLOG

VETERINARY TOXICOLOGY

HARIKRISHNAN RAO
VIVEKANAND SHASTRI
SANTHOSH KUMAR

KNOWLEDGE BAKERS

ISBN: 978-93-49964-86-0

Published by : **Knowledge Bakers**
Pune, India
Email: knowledgebakers2019@gmail.com

PREFACE

Veterinary Toxicology is a comprehensive guide that delves into the study of toxic substances and their adverse effects on animals. This book is designed to serve as an essential resource for veterinary students, professionals, toxicologists, and researchers involved in understanding the impacts of toxicants on animal health. It provides both fundamental and advanced knowledge of toxicology, integrating theoretical insights with practical applications to ensure a thorough understanding of the field.

The book begins by laying the foundation of veterinary toxicology, introducing the basic principles of toxicology, such as dose-response relationships, mechanisms of toxicity, and the factors influencing the toxicity of substances in animals. It provides an in-depth analysis of how toxins enter the body, how they are metabolized, and their effects on various organ systems. These concepts are essential for comprehending the complexities of veterinary toxicology and for diagnosing and treating poisoning in animals.

A significant portion of the book is dedicated to discussing the diverse range of toxicants that can affect animals, including chemical, biological, and environmental toxins. It examines commonly encountered poisons in veterinary practice, such as plant toxins, pesticides, heavy metals, and industrial chemicals, providing detailed information on their sources, modes of action, clinical signs, diagnosis, and management. Additionally, the book discusses emerging toxicological concerns, such as the impact of pharmaceutical contaminants, endocrine-disrupting chemicals, and other pollutants that are increasingly relevant in modern veterinary practice.

The text also covers the methods used to assess toxicity in animals, including diagnostic tools and laboratory techniques that enable veterinarians to identify toxic substances, evaluate the severity of poisoning, and determine appropriate treatment options. Special emphasis is placed on the importance of early diagnosis and intervention in mitigating the harmful effects of toxic exposure. The role of veterinary toxicologists in collaborating with clinicians, researchers, and public health officials in managing animal poisonings and preventing outbreaks of toxicological incidents is also discussed.

One of the critical aspects of Veterinary Toxicology is its focus on the prevention of toxic exposures in animals. The book provides guidelines on risk assessment, including environmental factors, management practices, and preventive measures that can reduce the likelihood of poisoning incidents. It emphasizes the importance of public awareness, education, and proper disposal of toxic materials in safeguarding animal health. Additionally, the book explores the ethical and regulatory frameworks that govern the use of chemicals in animal husbandry, agriculture, and the environment.

The management of toxicological emergencies is another key theme in the book. It provides practical advice on the acute treatment of poisoned animals, including the use of antidotes, supportive care, and decontamination techniques. Case studies and clinical examples illustrate the various treatment protocols for different types of poisoning, offering readers valuable insights into the practical aspects of veterinary toxicology.

The book highlights the role of research and the importance of evidence-based approaches in advancing veterinary toxicology. It discusses the latest advancements in toxicological research, including the development of new diagnostic techniques, better understanding of molecular mechanisms of toxicity, and the application of genomics and proteomics in identifying susceptibility to toxicants. This is particularly relevant in an era of rapidly advancing technology and knowledge, where new challenges in toxicology constantly emerge.

Veterinary Toxicology also addresses the growing need for veterinarians to be aware of the legal and regulatory aspects of toxicology, including legislation regarding the use of hazardous substances and animal welfare. It outlines the responsibilities of veterinary professionals in preventing and responding to toxicological incidents, as well as the need for effective communication with pet owners, farmers, and the general public regarding the risks of toxic exposure.

Finally, the book explores the future directions of veterinary toxicology, including the increasing need for veterinary professionals to engage in interdisciplinary collaborations with toxicologists, environmental scientists, and public health experts. It emphasizes the role of veterinarians in preventing, diagnosing, and treating toxicological diseases and in advancing research that can improve the health and well-being of both domestic and wildlife populations.

Veterinary Toxicology is an indispensable resource that provides comprehensive knowledge and practical guidance on managing toxicological risks in veterinary practice. It is designed to help readers understand the complexities of toxic substances and their effects on animal health, enabling them to make informed decisions in clinical, research, and public health settings.

Harikrishnan Rao

Vivekanand Shastri

Santhosh Kumar

CONTENTS

1

INTRODUCTION

VETERINARY TOXICOLOGY AIMS AND TASKS

Toxicology is the study of poisons and their effects on living organisms. In veterinary medicine, this has come to mean an understanding of sources of poisons, circumstances of exposure, diagnosis of the type of poisoning, treatment, and application of management or educational strategies to prevent poisoning. More so than many of the specialties in veterinary medicine, toxicology is based on the important principle of dose and response. That is, there is a graded and possibly predictable response based on increasing exposure to the toxicant.

Although alchemy has long since been abandoned, Paracelsus' principle of what makes a poison is still true and relevant in the daily practice of nutrition, therapeutics, and toxicology. Today, with emphasis on synthetic drugs, natural or alternative therapies, and the rapidly growing field of nutraceuticals, there is increasing need to be aware of the dosage and response principle for both beneficial and detrimental effects in the daily practice of veterinary medicine.

Determinants of exposure that affect dosage may be more than simply the gross amount of material ingested or applied to the skin. Rather, the effective dosage at a susceptible receptor site determines the ultimate response. Thus species differences in metabolism, vehicle differences that promote skin penetration, specific drug or chemical interactions that potentiate response and organ dysfunction that limits elimination can all influence the ultimate dosage. Clinicians must consider all of these possibilities when working to diagnose a potential toxicosis or apply therapeutic agents to their patients.

Toxicology involves the knowledge of poisons, including their chemical properties, identification, and biologic effects, and the treatment of disease conditions caused by poisons. Toxicology shares many principles with pharmacology, including the dynamics of absorption, distribution, storage, metabolism, and elimination; mechanisms of action; principles of treatment; and dose-response relationships.

Toxicology literature is best understood if some basic terminology is remembered. A poison or toxicant is usually considered any solid, liquid, or gas that when introduced into or applied to the body can interfere with homeostasis of the organism or life processes of its cells by its own inherent qualities, without acting mechanically and irrespective of temperature.

The term toxin is used to describe poisons that originate from biological sources and are generally classified as biotoxins. Biotoxins are further classified according to origin as zootoxins (animal origin), bacterial toxins (which include endotoxins and exotoxins), phytotoxins (plant origin), and mycotoxins (fungal origin).

Poisons may be categorized as organic, inorganic, metallic, or biological. A further distinction is made by some between synthetic and natural agents. Synthetic agents may have been designed specifically as toxicants that may have a very broad or very narrow range of toxicity and/or may produce effects in very specific targets. Natural products used in nutrition, medicine, or commerce are sometimes believed to be less hazardous than synthetic products. However, natural products are not inherently more or less toxic than synthetic molecules. Indeed, some of the most toxic agents known (e.g., botulinum toxin, tetrodotoxin) are of natural origin. Knowledge of the chemical nature and specific effects of toxicants is the only certain way to assess hazard from exposure.

The terms toxic, toxicity, and toxicosis are often misunderstood or misused. The word toxic is used to describe the effects of a toxicant (e.g., the "toxic" effects of organophosphate insecticides may be described as cholinesterase inhibition; vomiting, salivation, dyspnea, and diarrhea).

However, toxicity is used to describe the quantitative amount or dosage of a poison that will produce a defined effect. For example, the acute lethal dosage to cats of ethylene glycol would be described as 2 to 5 ml/kg body weight. The toxic effects of ethylene glycol are acidosis and oxalate nephrosis. Finally the state of being poisoned by a toxicant, such as ethylene glycol, is toxicosis.

Toxicology is a science of poisons. Fogleman refers to toxicology as a "many splendored thing." Although its splendour may not be obvious to the physician dealing with a hopeless case of heroin addiction or to the veterinarian faced with an outbreak of malicious poisoning in domestic pets, the many facets of toxicology are beyond dispute.

It has frontiers with pharmacology, physiology and pathology; with chemistry, biochemistry and biology; with agriculture, industry and economics; with forensic science and clinical medicine; with ecology, with pollution, and hence with the whole future of life on this planet. To understand it properly calls for knowledge of all these subjects, a breadth of learning impossible of achievement. Hence the discipline is usually divided into several branches such as clinical toxicology, chemical toxicology, forensic toxicology, industrial toxicology and veterinary toxicology.

Mathieu Orfila is considered to be the modem father of toxicology, having given the subject its first formal treatment in 1813 in his Traite des poisons, also called Toxicologie generale.

Theophrastus Phillipus Auroleus Bombastus von Hohenheim (1493-1541) (also referred to as Paracelsus, from his belief that his studies were above or beyond the work of Celsus - a Roman physician from the first century) is also considered "the father" of toxicology. He is credited with the classic toxicology maxim "All things are poison and nothing is without poison; only the dose makes a thing not a poison." This is often condensed to: "The dose makes the poison".

In the last century, our society has become chemically oriented. Along with benefits, problems related to health hazards have arisen. Toxicology, a science that deals with the harmful effects of chemical substances on biological mechanisms, is attempting to study these problems scientifically and to find practical solutions. The research techniques of toxicology are those of classical medical sciences: biochemistry, pharmacology, pathology, etc. Modern toxicology is a multidisciplinary field and borrows freely, in addition to the aforementioned areas, from veterinary sciences, environmental sciences and several others. No matter which has provided his major basic training, a toxicologist should always be interested in the dynamic aspects of chemico-biological interactions. On the occasion, he will also have some interest in their forensic implications.

Veterinary toxicology faces problems related to the general increase in the use of chemicals and especially to their increased use in livestock husbandry and agriculture. Since these two industries supply mankind with much of what he eats, careful considerations must be given to the possible effects on the human population of the use of chemicals on animals and crops. Toxicologists must bear in mind possible public health hazards when evaluating the toxicity of chemical agents.

Veterinary toxicologists are also involved in the evaluation of the toxic effects of chemicals, which are primarily intended for use in or on humans, since these compounds must be evaluated in animal system prior to their limited or widespread use in man.

The principles of the preclinical evaluation of the safety of new compounds for their proposed use in veterinary medicine and in human medicine remain the same. Thus, it is quite evident that veterinary toxicology and human toxicology are so closely interrelated that they are indeed inseparable.

Definition of a poison. It is difficult to arrive at a completely satisfactory definition of a poison. A poison may be defined as any substance or matter (solid, liquid, or gaseous) which, when applied to the body outwardly, or in any way introduced into it in a very small dose, can destroy life by its own inherent qualities, without acting mechanically, and irrespective of temperature.

It is doubtful if anymore modern definition describes it better, because a poison is almost impossible to define, and any attempt to do so is a exercise in semantics rather than in toxicology. Whether or not a substance is poisonous depends on the quantity taken, the species to which it is given, and the route by which it enters the body. A small intake of vitamin A is essential to prevent night blindness, but excess may lead to serious gastrointestinal disorders, as arctic explorers who have eaten polar bear liver have found to their cost. To the layman, sugar and salt are the epitome of the nontoxic, but cattle have been poisoned by sugar, and salt poisoning in pigs is well known. Cobra venom may be drunk with impurity, but is lethal if administered parenterally. It would be difficult to find a definition adequately to cover retinal, sucrose, sodium chloride and cobra toxin.

EC and Lithuanian legal documents for hazardous substances and pollution

REACH is the European Community Regulation on chemicals and their safe use (EC 1907/2006). It deals with the Registration, Evaluation, Authorisation and Restriction of Chemical substances. The law entered into force on 1 June 2007.

The aim of REACH is to improve the protection of human health and the environment through the better and earlier identification of the intrinsic properties of chemical substances. At the same time, REACH aims to enhance innovation and competitiveness of the EU chemicals industry. The benefits of the REACH system will come gradually, as more and more substances are phased into REACH.

The REACH Regulation places greater responsibility on industry to manage the risks from chemicals and to provide safety information on the substances. Manufacturers and importers are required to gather information on the properties of their chemical substances, which will allow their safe handling, and to register the information in a central database run by the European Chemicals Agency (ECHA) in Helsinki. The Agency acts as the central point in the REACH system: it manages the databases necessary to operate the system, co-ordinates the in-depth evaluation of suspicious chemicals and is building up a public database in which consumers and professionals can find hazard information.

The Regulation also calls for the progressive substitution of the most dangerous chemicals when suitable alternatives have been identified.

One of the main reasons for developing and adopting the REACH Regulation was that a large number of substances have been manufactured and placed on the market in Europe for many years, sometimes in very high amounts, and yet there is insufficient information on the hazards that they pose to human health and the environment. There is a need to fill these information gaps to ensure that industry is able to assess hazards and risks of the substances, and to identify and implement the risk management measures to protect humans and the environment.

Classification of poisons

The graduate veterinarian's ability to serve his clientele is no better than his ability to accurately differentiate between the various possible causes of the illnesses presented to him. It should be noted that mere memorization of clinical signs is not the only way to recall a certain toxicological problem. Perhaps, a better method of recognition is to classify the common toxicoses according to their specific lesions. Since there are many plants, feed additives, insecticides, herbicides, fungicides, and heavy metals that may fall under a specific lesion, each lesion could be subdivided and indexed to help with differential diagnosis. The following classification based on toxic effects of poisons is far from complete, but it may act as a starting point for future indices of diagnostic classifications.

Gastrointestinal Syndrome. This is seen in poisoning from oak, acute copper, E. coli, Salmonella, carbamate fungicides, ANTU, acute crotalaria, paraquat herbicides, thallium, bracken fern, warfarin, etc. The important clinical signs include: bloody diarrhea, hematuria, enteritis, sometimes GI irritation. Death may occur within one to a few days. In case of wild indigo, staphylococcus, peptides and amides, and borates deaths seldom occur. Severe gastrointestinal syndrome is caused by pokeweed, sneezweed, inorganic arsenic herbicides. Death may occur within 1-2 days with bloody diarrhea, hemolytic crisis (pokeweed), and excess salivation (sneezweed).

Neuromuscular (paresis) effects. Some poisons causing these effects act rapidly, some act slowly. Oleander, buttercup, buckeye, lead (in poultry) act rapidly causing death in 1-2 days. Others include: death comas, poison hemlock, organophosphates. All the above cause in coordination plus GI syndrome where as jimsonweed, 2,4-D, organic tin, ticks cause only in coordination. Other examples of toxicants causing CNS effects include botulism, horse tail, bracken fern (in horses), pigweed, etc.

Bone-Teeth-Hoof-Hair Deformities. Ergot-sloughing of tips of tail, ears, teats, etc. Fluoride- bone and teeth lesions Selenium (chronic), poisonvetch, copper deficiency, chronic thallium, lead nitrate, chlorinated naphthalene, are all associated with hair problem.

Kidney Lesions. Depending on the types of renal damage caused by toxicants kidney lesions are categorized as follows:

a. Hematuria: This may be accompanied by fever as in the case of bracken fern or there may be no fever as seen in poisoning from oak, cassia, inorganic: mercury or cadmium.

b. Hemoglobinuria: This is caused by crucifers (mustard) with photosensitization. Similar condition is seen in chronic copper toxicity.

c. Oxalates: These are formed in poisoning from amaranthus (which also causes perirenal edema), halogeton, black greasewood, rhubarb, and ethylene glycol.

d. Perirenal Edema: Amaranthus and oak (with as cites); nightshade (no as cites).

e. Degenerative Changes in Kidneys: These can be caused by chronic organic mercury compounds, chronic thallium, sulfonamides and so forth.

Liver Lesions. These are caused by several toxicants such as selenium (acute,) aflatoxins, crotalaria, tannic acid, carbon tetrachloride, phenothiazines, gossypol, tarweed, etc.

Photosensitization. Phenolic fungicides (necrosis of contact tissue); vehicles for insecticides; blue-green algae (death within 1-2 days); crucifers (with hemoglobinuria), snow-on-the-mountain (without hemoglobinuria); phenothiazines, alfalfa; St. Johnswort, horse brush.

Hemorrhagic Syndrome. Bracken fern (with fever); those that do not have fevers include: sweet clover, warfarin, pindone, radiation, mycotoxins, crotalaria.

Fevers. Castorbean, oleander, have rapidly acting toxic materials and death may occur within 1-2 days. On the other hand, milkweed, bracken fern, buckeye, locust, etc. are known to be slow- acting.

Abortions and/or Anomalies. Fusarium fungal metabolities, poisonvetch, lupine, broomweed, ergot, 2, 4, 5-T, nitrates. The following cause birth defects: chronic selenium, veratrum, oak, locoweed, jimsonweed, hem-lock, lead, and mercury.

The classification of poisons made in the previous section is based on the organ or system that is the 'target' site for the effect of the toxic chemical. Such classification that is based on the toxic effects of poisons on the body may be unsatisfactory because the same substance can have different effects on different organs of the body. It can also vary in its action between one species and another. Therefore a relatively simple classification can be made on the basis of the structural features of the toxic chemicals that are responsible for their toxic properties and their affinity for 'target' sites in the animal body. Various diverse chemical compounds can be thus divided into two groups, namely, inorganic and organic compounds.

Inorganic compounds include metals, metalloids, their salts and acids and alkalis.

Organic compounds on the other hand include all carbon compounds other than carbonates, and the metallic carbides and cyanides.

Furthermore, an analytical toxicologist endeavours to separate poisons into characteristic groups and several classifications can be made according to the analytical procedures involved. A typical subdivision is: (1) volatile poisons, (2) metallic poisons, (3) toxic anions, (4) non-volatile organic poisons isolated by 'solvent extractions, (5) miscellaneous poisons. Finally, poisons may often be classified conveniently by their origin (plant poisons) or use (pesticides).

CHEMICALS CLASSIFICATION AND LABELLING

The CLP Regulation (for "Classification, Labelling and Packaging") is a European Union regulation which aligns the European Union system of classification, labelling and packaging chemical substances and mixtures to the Globally Harmonised System (GHS). It is expected to facilitate global trade and the harmonised communication of hazard information of chemicals and to promote regulatory efficiency. It complements the Registration, Evaluation, Authorisation and Restriction of Chemicals (REACH) Regulation (EC No 1907/2006) and replaces the current system contained in the Dangerous Substances Directive (67/548/EEC) and the Dangerous Preparations Directive (1999/45/EC).

The regulation incorporates the classification criteria and labelling rules agreed at UN level, the so called Globally Harmonised System of Classification and Labelling of Chemicals (GHS). It introduces new classification criteria, hazard symbols (pictograms) and labelling phrases, while taking account of elements which are part of the current EU legislation.

The regulation requires companies to appropriately classify, label and package their substances and mixtures before placing them on the market. It aims to protect workers, consumers and the environment by means of labelling which reflects possible hazardous effects of a particular chemical. It also takes over provisions of the REACH Regulation regarding the notification of classifications, the establishment of a list of harmonised classifications and the creation of a classification and labelling inventory.

2

TOXICOKINETICS

MIGRATION OF SUBSTANCES THROUGH BIOLOGICAL MEMBRANES

The passage of xenobiotics through cellular membranes can be either energy-independent ("passive" transport) or can require the expenditure of energy through "specialized" or "active" transport systems. Passive transport of xenobiotics can be accomplished through simple diffusion or filtration. Specialized, energy-dependent, cellular transport systems include the process specifically referred to as "active transport," along with facilitated transport and pinocytosis.

Passive transport of xenobiotics. Simple diffusion and filtration are nonsaturable processes, which do not require the expenditure of energy to transport xenobiotics across cellular membranes. Both of these mechanisms of passive transport are dependent on the concentration gradient for a given xenobiotic, with the rate of transport being proportional to the difference in that chemical's concentration between the two sides of a particular membrane (Fick's law).

Simple diffusion is the most common mechanism by which xenobiotics cross cellular membranes. Uncharged (nonionized), lipid-soluble molecules, especially those which are small, are more readily diffusible across the phospholipid bilayers of biological membranes than charged (ionized) molecules, which are generally less lipid-soluble. The Henderson-Hasselbalch equation can be used to predict whether a particular xenobiotic will be in the nonionized or ionized state in a particular biological matrix. In this equation, the difference between the association constant (pKa), which is equivalent to the pH at which equal amounts of a xenobiotic are in the nonionized and ionized states, and the pH of the biological matrix in which the xenobiotic will exist (i.e., pKa " pH) is equal to the common log of the quotient of nonionized xenobiotic divided by ionized xenobiotic for weak acids and the log of the reciprocal quotient (ionized xenobiotic divided by nonionized xenobiotic) for weak bases. Filtration involves the passage of xenobiotics through patencies or pores within cellular membranes and is determined, in large part, by the size of the xenobiotic molecule and pore-size, which varies in different organs and tissues.

Specialized transport of xenobiotics. Active transport is an energy-dependent, saturable process by which xenobiotics are transported across biological membranes against electrochemical or concentration gradients.2-4 Specific examples of active transport systems include the multidrug- resistant proteins (P-glycoproteins) and members of the organic cation transporter family.3 Facilitated or carrier mediated transport can require the expenditure of energy, but, in contrast to active transport, xenobiotic transport by this mechanism is not against a concentration gradient.

Pinocytotic transport involves cellular engulfment of small amounts of xenobiotic and the transfer of this amount of chemical through the cellular membrane.

ADME notion

The basic concepts regarding the toxicokinetics and toxicodynamics of xenobiotics are clinically relevant to veterinary toxicology and need to be understood by veterinary practitioners, professional students, and other personnel who will be participating in the diagnosis and treatment of animal intoxications. In discussing the aspects of toxicokinetics and toxicodynamics most pertinent to animal toxicoses, it is first necessary to define several terms. "Xenobiotic" is a general term referring to any chemical foreign to an organism or, in other words, any compound not occurring within the normal metabolic pathways of a biological system.

Depending on the compound and the level of exposure, interactions between xenobiotics and animals can be benign, therapeutic, or toxic in nature. The pharmacokinetics and pharmacodynamics of a therapeutic xenobiotic influence the time course and efficacy of that compound in a pharmacological setting. Likewise, the toxicokinetics and toxicodynamics of a toxic xenobiotic determine the "when," "how long," "what," and "why" for the adverse effects of that toxicant.

The "disposition" of a xenobiotic is what the animal's body does to that compound following exposure. The disposition or fate of a xenobiotic within the body consists of the chemical's absorption, distribution, biotransformation and excretion characteristics, which are collectively abbreviated as ADME. "Toxicokinetics" refers to the quantitation and determination of the time course of the disposition or ADME for a given toxic xenobiotic.

- There are a variety of specialized toxicokinetic terms, including bioavailability, volume of distribution, clearance, half-life, one-compartment model and first-

and zero-order kinetics, which will be discussed later in this chapter under the separate components of ADME.

- The term "toxicodynamics" describes what a toxicant does physiologically, biochemically, and molecularly to an animal's body following exposure. The toxicodynamics of a given toxic xenobiotic are dependent on the mechanism of action of that toxicant and the relationship between toxicant concentration and the observed effects of the toxicant on biological processes in the animal (i.e., the dose-response relationship).
- The disposition and/or toxicokinetics of a particular xenobiotic also play a role in determining the organs or tissues affected by a toxicant, and the clinical presentation and time course of a toxicosis resulting from excessive exposure to that compound.
- Toxicokinetics involves the relationship between tissue concentration of a toxicant and time. There are two types of kinetics:

1. Zero-order kinetics and
2. First order kinetics.

In zero-order kinetics, a toxicant is eliminated from the body at a fixed amount per unit time. This occurs when the toxicant elimination process is saturated and a fixed amount of toxicant (e.g., mg or g) is excreted per unit time.

Many toxicants are eliminated from animal tissues in a first-order manner. Thus, in first-order kinetics a constant fraction of a toxicant/drug is eliminated per unit time. For example, if the initial level were 500 mg and it took ten days to reach 250 mg, then it will take additional 10 days to reach 125 mg. During the first 10 days a total of 250 mg toxicant was excreted and during the next 10 days only 125 mg. In each case, the amount excreted was a constant fraction (i.e., 50%) of the original amount. The ten-day period is called half-life of the toxicant. Various species handle excretion of a toxicant in different manner and therefore a given toxicant will have different half- lives in different species.

Possibilities of poisons entering into an animal body and methods of absorption

Absorption. In order to exert their toxic effects most poisons must be absorbed into the blood stream. Under natural conditions toxicants may enter the body through the lungs, gut and skin. Substances may be injected into the body subcutaneously, intramuscularly, intravenously and intraperitoneally.

1. Respiratory tract: The very extensive highly vascular pulmonary mucous membrane affords an excellent channel of absorption for gases and for solids and liquids, particularly when in a fine state of dispersal (as aerosols or dusts).

2. Alimentary tract: It is the most usual route of entry of a poison. In all species, much absorption occurs in the small intestine; in the dog it can also take place from the stomach, in the ruminant from the rumen and reticulum, and in all species, especially in non-ruminant herbivora, from the large gut. The contents of the alimentary tract can modify the action of poison. The hydrochloric acid present in gastric juice may aid the solution of originally insoluble materials. A full stomach or rumen may delay symptoms of poisoning or may dilute the poison to such an extent that it is relatively harmless.
3. Dermal exposure: The unbroken skin does not offer a favourable channel of absorption to most compounds. Nicotine is one of the exceptions which in aqueous solutions, is very efficiently absorbed through intact skin. Absorption takes place more readily from oily solutions or emulsions.

Absorption through damaged or abraded skin or a wound occurs as through moist mucous membranes. Subcutaneous or intramuscular injection is equally effective, while intravenous administration is the fastest way of introducing a poison into the blood stream.

With the exception of caustic and corrosive toxicants that cause adverse effects at the site of exposure, a toxic xenobiotic is generally first "absorbed" or taken up into the body. Absorption involves crossing cellular membranes, which are typically composed of phospholipid bilayers containing various sized pores and embedded proteins. The route of exposure and physiochemical properties of a toxicant, such as its resemblance to endogenous compounds, its molecular size and relative lipid and water solubilities, the magnitude of a molecule's association constant, and whether a compound can be classified as a weak acid or as a weak base, all determine the manner and quantities in which a xenobiotic is absorbed across cell membranes.

The most common routes of exposure for xenobiotics in animal toxicology are oral (gastrointestinal), dermal (percutaneous), and inhalation (pulmonary). In rare instances of iatrogenic intoxications, xenobiotics can be injected subcutaneously, intramuscularly, intraperitoneally, or even intravenously. There are unique aspects to the absorption of xenobiotics associated with each route of exposure, especially with regard to the bioavailability of potential toxicants.

"Bioavailability" (often represented by "F" in toxicokinetic equations) represents the fraction of the total dose of a toxic xenobiotic that is actually absorbed by an animal. In intravenous exposures, the bioavailabilityof a toxic xenobiotic is 100% since the entire dose of the toxicant reaches the peripheral circulation. The absorption of gases and vapors in the respiratory tract is largely dependent on the ratio (blood-to-gas partition coefficient) between the equilibrium concentrations of the toxicant dissolved in the blood and the gaseous phase of the toxicant in the alveolar spaces. The size of aerosolized particles will determine to a large degree whether a xenobiotic is deposited in the nasopharyngeal region (particles >5 μm) or within the alveoli of the lungs (<1 μm). The stratum corneum and its associated keratinized structures often impede the percutaneous absorption of xenobiotics, and there are variations in the absorptive ability of skin in different anatomical locations. Dermal absorption is frequently dependent on the vehicle in which a toxicant is dissolved and is generally greater for lipid soluble compounds as compared with chemicals that are highly soluble in water. The bioavailability of toxic xenobiotics that are ingested can be negatively impacted by acidic degradation in the stomach and/or enzymatic breakdown in the small intestine. Decreased gastrointestinal transit time can diminish xenobiotic bioavailability by limiting the access of toxicants to those regions of the digestive tract where rates of absorption are greatest. Some potential toxicants, especially certain heavy metals (e.g., lead and cadmium), resemble essential minerals such as calcium and zinc, respectively. The gastrointestinal absorption of these toxic nonessential metals involves interactions with dietary levels of the corresponding essential metals and regulated mechanisms of gastrointestinal uptake designed for these required minerals.

Hepatic biotransformation of xenobiotics can also influence the aparent bioavailability of ingested toxicants. Following oral exposure, xenobiotics absorbed from the gastrointestinal tract are transported to the liver via the hepatic portal circulation. For some xenobiotics, rapid hepatic degradation (and in some instances prior biotransformation in gastrointestinal cells) prevents access of the compound to the systemic circulation, resulting in an apparently decreased bioavailability from what is termed the "first-pass effect" or "presystemic elimination." In contrast, the bioavailability f some chemicals is enhanced by a cycle of biliary excretion and subsequent reuptake from the intestines referred to as "enterohepatic recirculation.

POISON DISTRIBUTION

Distribution and Accumulation. Hepatic storage: All foreign compounds entering the body pass to the liver which is the major detoxifying organ in the body and by virtue of this fact many poisons accumulate in this organ. It is therefore not unexpected to find hepatic lesions as a consequence of exposure to many toxicants.

Extra hepatic storage: Some poisons are selectively deposited in certain organs or tissues. Iodine is largely taken up by the thyroid glands; strontium, fluorine and lead are deposited in the bones. A knowledge of the distribution of a particular poison is of great help in the selection of organs for chemical analysis.

"Distribution" refers to the translocation of a xenobiotic from the site of absorption to various body organs and tissues and involves both transport of the chemical within the circulation and cellular uptake of the xenobiotic. The rate of xenobiotic transfer into a particular organ or tissue is determined by the physiochemical properties of the specific xenobiotic (e.g., lipid solubility and molecular weight), the blood flow to the organs or tissues in question, and the rate of diffusion of the xenobiotic across the endothelial walls of the capillary bed into cells within a particular organ or tissue.

The "volume of distribution" (Vd) for a given xenobiotic represents the quotient of the total amount of that chemical in the body divided by the concentration of the xenobiotic within the blood and is used to describe the extent to which a xenobiotic is distributed within the body. The Vd is a clinically relevant indicator as to whether a chemical is primarily contained within the plasma compartment (relatively low Vd) or whether a compound is widely distributed throughout the body within the interstitial and/or intracellular compartments of various organs and tissues (relatively high Vd).

Xenobiotic storage depots. Xenobiotics can be stored within a variety of different body organs and tissues. Depending on the anatomical and physiological relationships between the storage depot and the target organ(s) and/or tissue(s) for a specific toxicant, storage of toxic xenobiotics can function as either a protective mechanism or as a means by which the toxic effects of a xenobiotic are potentiated. An understanding of the storage sites of toxic xenobiotics can provide additional insight about circumstances that would be expected to exacerbate a particular toxicosis along with indicating which organs or tissues would be expected to have the highest concentrations for diagnostic sampling. Plasma proteins represent a storage site for many xenobiotics (e.g., salicylates, barbiturates, cardiac glycosides) and important physiological constituents, including steroid hormones, vitamins, and various essential minerals. Displacement of toxic xenobiotics from plasma proteins can greatly increase the amount of unbound toxicant distributed to target organs or tissue. A wide variety of xenobiotics accumulate in the liver and kidneys, making these organs ideal sites for postmortem sample collection in cases of suspected toxicoses. Some toxic metals, such as cadmium, accumulate in the liver and kidneys because of the high endogenous concentrations and induction of metallothionein in these organs. Fat and bone are storage depots for a variety of different xenobiotics, and rapid depletion of body fat stores (weight loss) or increased remodeling of bone during growth or pregnancy have the potential to increase the exposure of target organs or tissue to previously stored toxicants.

Potential tissue barriers to xenobiotic distribution. The blood-brain barrier is frequently mentioned in the current literature with regard to its ability to limit exposure of the central nervous system (CNS) to toxic xenobiotics. Other potential barriers to chemical uptake also occur in the eyes, testes, prostate, joints, and placenta. In these instances only small, nonionized, lipid- soluble molecules are able to cross the membranes and gain access to potential target tissues. The "blood-brain barrier" to xenobiotic uptake consists of the relatively nonporous CNS capillary endothelium, which contains multidrugresistant protein and is surrounded for the most part by glial cells. The extremely low protein content of the interstitial fluid within the CNS also contributes to the apparent inability of many protein-bound, toxic xenobiotics to reach clinically relevant concentrations in the brain. Since the blood-brain barrier is not fully formed at birth and is less well-developed in some breeds of dogs (e.g., collies and collie crosses), immature animals and collie-related breeds are more susceptible to the adverse effects of compounds normally "blocked" by the blood-brain barrier.

Poison biotransformation

Biotransformation and Detoxication. "Metabolism" can be used to refer to the fate or disposition of a xenobiotic or the sum total of the chemical transformations of normal body constituents, which occur in living organisms. Biotransformation, on the other hand, is a general term referring to the metabolic conversion of both endogenous and xenobiotic chemicals into more water-soluble forms. For the purposes of this chapter, xenobiotic "metabolism" and "biotransformation" are synonymous and refer to the generally two-phase process by which chemicals are converted to more water-soluble forms for excretion from the body. In xenobiotic metabolism/biotransformation, the lipophilic (lipid-soluble) properties of xenobiotics that favour absorption are biotransformed into physiochemical characteristics (hydrophilicity or water solubility) that predispose compounds to excretion in the urine or faeces. Although multiple organs within the body have biotransformation capabilities, most xenobiotics are biotransformed in the liver.

Phase I and phase II xenobiotic biotransformation. Xenobiotics are usually biotransformed in two phases (I and II), which involve enzymes having broad substrate specificity. Phase I reactions generally involve oxidation, hydrolysis, or reduction, and convert apolar, lipophilic xenobiotics into metabolites, which have greater polarity and hydrophilicity. In these instances, hydroxyl, amino, carboxyl, or thiol moieties are usually either exposed or added to increase water solubility. Oxidation reactions, especially those catalyzed by cytochrome P450 enzymes, are the phase I biotransformations most commonly involved in xenobiotic metabolism, and many xenobiotics are able to induce cytochrome P450 activity.2,5,6 During phase II biotransformation, the xenobiotic or its metabolites are conjugated with a functional group (e.g., glucuronide, sulfate, amino acids, glutathione, or acyl or methyl groups), resulting in a compound with dramatically increased water solubility.

Not all mammalian species have equal phase II biotransformation capabilities, and the inability of domestic cats to glucuronidate xenobiotics is especially clinically relevant to veterinary toxicologists. Most xenobiotic biotransformations result in less toxic metabolites. However, there are xenobiotics (e.g., acetaminophen and aflatoxin B1) for which the products of hepatic phase I metabolism are actually more toxic than the parent xenobiotic. In these instances of "metabolic activation," "bioactivation," "toxication," or "lethal synthesis," any factors that increase hepatic biotransformation of the parent compound will enhance the amount of toxic metabolite to which the animal is exposed.

Many foreign compounds that are introduced into the body undergo chemical transformation, and this process is generally referred to as metabolic transformation or biotransformation. Very frequently this process is referred to as the detoxication mechanism. However, this term is misleading because the metabolic transformation of a foreign compound may result into increased or decreased toxicity of the product. Two categories of enzyme systems are known to exist in mammals that cause chemical transformation of foreign compounds.

One category consists of enzymes that normally occur in the tissues and are responsible for transformation of normal endogenous substances in tissues as well as foreign chemical e.g., alcohol dehydrogenase. Second category consists of an enzymes system that is very important in toxicology. These are known as drug metabolizing enzymes, which are present in many tissues but are particularly in liver cells. These enzymes are located in small particles called microsomes which in turn are located in the smooth surface of endoplasmic reticulum. These microsomal enzymes are capable of catalyzing a variety of biotransformation reactions, the major ones being oxidation, reduction, and hydrolysis.

- The drug metabolizing enzymes system is also called as mixed function oxidase (MFO) enzyme system for the obvious reason that is capable of catalyzing various types of reactions. The oxidase system requires the presence of a cofactor NADPH, and molecular oxygen for its activity.
- NADPH reduces a component of the microsomes which reacts with molecular oxygen to form an
- active oxygen intermediate which oxidizes the drug or the foreign chemical. The components of microsomes which is reduced by NADPH is a heme protein called cytochrome P-450 (CYP450).
- R.T. Williams divided the biotransformation mechanism of foreign chemicals into two major types:

(1) the non synthetic reactions involving oxidation, reduction and hydrolysis, and

(2) the synthetic or conjugations involving production of a product that is biosynthesized from the chemical (or its metabolite) plus an endogenous metabolite such as glycine.

If we call the non synthetic reactions as phase I of biotransformation mechanism, then conjugation reactions will be phase II, which are described below.

Conjugation Reactions: These are reactions in which a poison metabolite is combined with some compound provided by the animal body. Following is the list of some of these processes.

Glycine Conjugations: Benzoic acid combines with glycine to form hippuric acid.

Glucuronic Acid: Conjugation with this occurs in many cases. e.g., dinitrophenol, chloral hydrate.

Sulfate Conjugation: Some phenols conjugate with SO4 to form ethereal sulfates.

Cysteine Conjugation: Arsenic trioxide and mainly benzene, polycyclic hydrocarbons, and certain halogenated hydrocarbons conjugate with cysteine to form mercapturic acids.

Acetylation Reactions: This takes place between CoASH and amino groups of aromatic compounds, e.g., sulfanilamide derivatives. Dog is deficient to acetylate aromatic amine groups. Methylation Reactions: It is relatively uncommon form of detoxication and is confined to the heterocyclic nitrogen atoms in compound of the pyridine and quinoline type.

Thiocyanate Formations: Inorganic cyanides are converted by the enzymerhodanese to thiocyanate. It is a true detoxication, for sodium thiocyanate is nearly 200 times less toxic than sodium cyanide and is slowly-excreted in the urine.

Glutamine and Ornithine Conjugation: Conjugation processes with glutamine and ornithine occur in man and bird, respectively.

Inhibition of Biotransformation Mechanisms. The microsomal enzyme systems can be inhibited by several compounds in concentrations which by themselves have little pharmacologic activity. The most common example of such inhibitor is SKF 525A (diethyl aminoethanol ester of diphenyl propyl acetic acid). Thus, the inhibition of the microsomal enzyme system would result into the inhibition of the metabolism of toxic foreign compound, the presence of which may cause increased toxicity. On the other hand, when the product of metabolic transformation is of greater toxicity than that of the parent compound, the inhibition of the microsomal enzymes by SKF525A would be expected to protect the animal from toxicity resulting from metabolism of the parent compound.

Induction of Biotransformation Mechanisms. The total quantity of microsomal drug metabolizing enzymes can be increased in humans and in animals by prior administration of large variety of chemical substances. Such substances include the anesthetics, such as nitrous oxide; the sedatives, such as barbiturates; the analgesics, such as phenylbutazone; and the insecticides, such as chlordane. Induction of increased enzyme activity usually involves repeated exposure to the inducing agent. It is usually temporary and lasts for two to four weeks following the administration of the inducing chemical. Since metabolic transformation has been shown to result in the formation of more or less toxic products as compared to the parent compound, enzyme induction may be protective to the animal (when detoxication is involved) or detrimental to the animal (when toxication is involved).

POISON EXCRETION

Excretion (elimination). The final step in the disposition of a xenobiotic is excretion, whereby the xenobiotic or its metabolites are removed from the body via a number of different routes.

Renal excretion is the most common means by which xenobiotics and the products of their biotransformation are eliminated from the body, but toxicants can also be excreted in the feces (biliary excretion or elimination of unabsorbed xenobiotic), saliva, sweat, cerebrospinal fluid, or even the milk, which is clinically relevant in xenobiotic-exposed bitches or queens nursing offspring. In instances of exposures to toxic vapors or volatile xenobiotics, exhalation can also be a major route of elimination from the body. Xenobiotics and their metabolites can be excreted by more than one route of elimination, and the total excretion is generally broken down into renal and nonrenal routes.

Toxicokinetic aspects of xenobiotic elimination. With regard to toxicokinetics, "elimination" of a xenobiotic generally incorporates both the processes of biotransformation and excretion. "Clearance," which is expressed for the whole body and individual organs in terms of the volume of blood that is cleared of the chemical per unit time, is an indicator of the body's ability to eliminate a given toxicant from the body by processes such as metabolism, excretion, and exhalation. The toxicokinetic aspects of xenobiotic elimination are clinically relevant to the management and diagnosis of veterinary toxicoses. These quantitative indices can be used to predict the duration of a toxicosis and the time period necessary for therapeutic intervention.

Toxicokinetic aspects of xenobiotic elimination can also be used to determine the time frame and biological samples that are best suited for diagnosing a specific toxicosis.

When developing toxicokinetic models, assumptions are often made with regard to whether a given xenobiotic best fits a "one-compartment" or a "multicompartment" model. A one- compartment model is the simplest toxicokinetic model and assumes that changes in xenobiotic concentrations in the blood or plasma are accurate reflections of what is occurring in the tissues. Assuming that a one-compartment model is appropriate for a particular xenobiotic, elimination of this compound is most likely via first-order kinetics, where the involved processes are most likely nonsaturable and the rate of elimination at any given time point is proportional to the amount of compound that remains in the body at that point in time. With first order kinetics in a one- compartment model, it is possible to calculate the elimination "half-life" of a xenobiotic using the volume of distribution and the clearance for a given xenobiotic. In this instance, half-life indicates the time required for the blood or plasma concentration of the xenobiotic to be reduced by one half, with approximately 97% of a xenobiotic being eliminated from the circulation in five half- lives.

The term "half-life" can also be used in terms of elimination of xenobiotic from body storage depots rather than from the blood or plasma. It is important to know the context in which this particular term is being used and the compartmental model involved to understand what process in the xenobiotic's disposition is actually being discussed.

There are some xenobiotics for which the processes involved in their elimination are saturable and the rate of elimination is independent of the amount of chemical remaining in the body at a given point of time. Under these circumstances, the pathways of elimination for a given xenobiotic can be described in terms of zero order kinetics. Only a finite amount of xenobiotic can be eliminated per unit time.

Following are the major ways through which poisons and their metabolites are excreted.

1. Faecal excretion. Ingestion of a relatively insoluble poison (e.g., lead arsenate) is followed by excretion of the major part in the faeces. Substances may also find their way into faeces via bile; metals stored in the liver are slowly excreted in this way.
2. Pulmonary excretion. Volatile poisons may be mainly excreted in the expired air, e.g., CS2 cyanide. In phosphorus poisoning the breath may smell of garlic odour and glow in the dark. Diagnosis of hemlock poisoning may be made from the characteristic "mouse-like" odour of coniine in the exhaled air (and also in urine). The lesions found in the lungs in paraffin poisoning are probably due to irritant effect caused by pulmonary excretion.
3. Urinary excretion. This is the most important pathway of the excretion of a poison. Irritant poisons cause damage to kidney. Urine is often a convenient material for diagnostic analysis. In veterinary field it is of great importance in detecting the pasture contamination with fluorides.
4. Milk and dermal excretion. Excretion can also take place through skin, e.g., arsenic, and in lactating animals in milk. Many of the insecticides are fat soluble and it has been shown that DDT, aldrin, and several other chlorinated hydrocarbons can be detected in minute amounts in cow's milk.

Interactions between xenobiotic toxicodynamics and disposition/toxicokinetics. In contrast to toxicokinetics, the toxicodynamics of a particular xenobiotic describe what that compound actually does to adversely affect an animal's health rather than how the animal handles the exogenous chemical.

However, a xenobiotic's toxicodynamics and toxicokinetics are not mutually exclusive. What a toxicant does physiologically, biochemically, and molecularly to a living organism following exposure is not only dependent on that xenobiotic's mechanism of action and its dose-response relationship but also on its disposition and/or toxicokinetics within an exposed animal.

The first step in the development of a toxicosis is the delivery of the "ultimate toxicant" to its site of action or "target." "Ultimate toxicant" refers to the parent xenobiotic, its metabolite, or even a generated reactive oxygen species that actually causes cellular damage. The term "target" is often used to describe a molecule that interacts with the ultimate toxicant, resulting in adversely affected biological processes within an organism. "Target(s)" can also be an inclusive term referring to the cell types, organs, or tissues most susceptible to the effects of a toxic xenobiotic. The distribution and biotransformation of a xenobiotic often limit the delivery of the ultimate toxicant to susceptible target cells, organs, or tissues. Distribution of xenobiotics to storage depots that are physically removed from potential target sites is one means by which the disposition of a toxicant can be protective and can limit the adverse effects of a particular xenobiotic on an animal. Presystemic elimination or the first-pass effect prevents toxic xenobiotics from ever reaching the general circulation and therefore many potential sites of action. Most biotransformations produce metabolites that are more water soluble and as a result more readily eliminated from the body.

In contrast to circumstances where the disposition of a xenobiotic decreases the risk of toxicosis, there are also instances where the distribution and biotransformation of a given toxicant actually increase the likelihood that an ultimate toxicant will be delivered to the site of action. A chemical's toxicity can be enhanced by specialized transport mechanisms and by physiochemical characteristics that facilitate the accumulation of ultimate toxicants within susceptible cells. The toxicity of a xenobiotic can also be facilitated by processes, such as enterohepatic recirculation, that increase its bioavailability. Xenobiotic biotransformations that result in lethal synthesis or bioactivation predispose animals to toxicoses and can, in some instances, actually occur within target cells. While some biotransformations result in metabolites that react more efficiently with target enzymes or receptors, it is more common for intoxication to result in chemical species, such as electrophiles, free radicals, nucleophiles, and redox-active compounds that are indiscriminately reactive with endogenous molecules.

General mechanisms of xenobiotic action. The basis for most toxicoses is cellular damage, and this damage is often most dramatic in cells with high rates of metabolism and replication. A toxic xenobiotic's "mode" or "mechanism of action" is the activity of that compound or its metabolites at the molecular or cellular level that results in adverse effects. Without specific mechanisms of action of toxicants to which small animals are commonly exposed, there are a number of general ways in which toxic xenobiotics adversely affect cellular structure and function.

Although a toxic xenobiotic can adversely affect cells by changing their biological microenvironment through alterations in pH or occupation of a particular receptor site, as mentioned previously, ultimate toxicants generally interact with target molecules or cells.

Some xenobiotics mimic the actions of normal nutrients and endogenous hormones or neurotransmitters. Specific receptors can be stimulated or blocked, and enzymes can be inactivated or inhibited.

Electrophiles, free radicals, nucleophiles, and redox-active compounds are often generated through biotransformations, and these chemical species can react indiscriminately with target macromolecules to exert their toxic effects. At the cellular level, chemicals can alter cellular maintenance, both internally and externally, by adversely affecting membrane integrity and the ability of cells to regulate their volume and/or their energy metabolism. Cellular injury and death often result from the impaired cellular synthesis of ATP, uncoupling of oxidative phosphorylation, and the inability of cells to regulate their intracellular calcium concentrations. The cellular production of vital proteins and the regulation of gene expression within cells can also be disrupted by toxicants. Ultimately, high enough exposures to toxic xenobiotics cause cellular dysfunction and injury and, sometimes, disrepair, and these adverse effects can be observed clinically as abnormalities in the structure and/or function of different organs and tissues.

3

TOXICODYNAMICS MODE OF ACTION OF TOXICANTS

In a broader sense, mode of action of poisons may be divided as physical and chemical disruption of the living process.

Physical Action: Because of their physical properties such as lipid solubility, some substances exert a non-specific inhibitory effect on enzyme systems by virtue of the fact that their physical nature is such as to bring about their accumulation in vital parts of cells where they depress cellular functions. Many compounds, including hypnotics and anaesthetics such as hydrocarbons, chlorinated hydrocarbons, alcohols, ethers, ketones exert their inhibitory effects in this way.

Chemical Action: Majority of poisons produce their effect as a result of their chemical interactions with cell components. It is the enzymes concerned in cell oxidation and oxidative phosphorylation which seem to be most vulnerable to the action of poisons.

Enzymes have active sites and the toxic compounds may occupy those active sites thus preventing normal substrates to combine. The enzyme inhibition caused by a toxicant may be irreversible (as in the case of certain organophosphates) or reversible (as in the case of carbamates). Some of the factors that may affect the enzyme inhibition are: chemical structure of the inhibitor, cell permeability, enzyme inhibitor concentrations, the presence of an antagonist of the inhibitor, lethal synthesis, presence of an antidote.

FACTORS AFFECTING ACTIONS OF TOXICANTS

Many factors inherent in the toxicant, the animal, or the environment can alter a toxicity value determined under defined experimental conditions. The toxicity of a compound will vary with the route of exposure. Usual routes of exposure are oral, dermal, inhalation, intravenous, intraperitoneal, and subcutaneous. In addition, the most potent routes of exposure are usually the intravenous, intrapulmonary, and intraperitoneal routes. In clinical veterinary toxicology, oral and dermal routes of exposure are the most common, and these routes generally delay the absorption and diffuse exposure over a longer period of time. A daily dosage of toxicant mixed in food and consumed over a 24-hour period may cause much less effect than that same dosage given as a bolus at one specific time. However, retention in the gastrointestinal tract, including enterohepatic cycling, and dermal or hair retention of poisons can significantly prolong the exposure or exposures. Another factor that can accentuate the toxic effects of a compound is concurrent organ damage as a result of other causes. This is most important for diseases that alter liver or kidney function, leaving the animal with insufficient resources to metabolize and excrete toxicants.

Species and breed differences exert important influences on toxicity. The familiar example of cats and their intolerance to phenolic compounds results directly from their lack of glucuronyl transferase, which is necessary to produce glucuronides for the excretion of phenolic metabolites. A common example is acetaminophen, which is quite toxic to cats partly as a result of ineffective excretion of the toxic metabolite. In addition, the amino acid and sulfhydryl content of feline hemoglobin and a relative lack of methemoglobin reductase in erythrocytes makes it more susceptible to oxidant damage. As a result, the cat is more likely to be poisoned by agents that induce methemoglobinemia. Occasional differences within a species can increase the probability of toxicosis. The anthelmintic ivermectin provides an example of breed susceptibility differences, with collies and individuals in other herding breeds being more susceptible.

Many environmental and physiological factors can influence the toxicity of compounds, and one should remember that such factors, or others possibly unknown, can substantially influence an individual's response to toxicants. Entire publications are devoted to drug and chemical interactions, and the reader is encouraged to be aware of toxicological interactions that are illustrated throughout this text.

Numerous factors influence the action of poisonous substances. In addition to those already described (e.g., route of absorption, biotransformation), they include (1) dosage, (2) the physical and chemical nature of the poison, (3) the source of the poison, (4) repeated exposure to the poison, (5) species, (6) size, age and sex, (7) general state of health of the animal.

Dosage. Harmful effect of a toxic compound is largely dependent on the amount of that compound absorbed into the body.

Physical and Chemical Nature. The physical state, e.g., whether solid, powder, or in solution will affect the dose of a poison. Coarsely crystalline arsenic trioxide is slowly absorbed and so has relatively low toxicity; finely powdered arsenic is highly toxic. Many substances are readily absorbed from oily than from aqueous solution, e.g., insecticides.

Chemical nature is important in regard to toxicity, yellow phosphorus is a most poisonous substance, its allotrope red phosphorus is inert when taken in body, it is in soluble and is excreted unchanged. Compounds containing trivalent arsenic are much more toxic than those containing the pentavalent form. Barium carbonate is intensely toxic than barium sulfate.

Sources of Poisons. Under certain circumstances poisoning from a.particular compound may be enhanced or reduced. Some plant poisons are destroyed by drying or storage and hay contaminated by them is harmless (e.g., buttercups). Presence of oils in the diet will enhance the absorption and so toxicity of poison e.g., phosphorus. Accumulation of copper in the liver may be mobilized by administering molybdenum and vice versa.

Repeated Exposure. It is logical that several doses of a poison will be effective than a single dose. The degree of harmfulness of repeated small doses also depends on whether the poison accumulates in body and whether its effects are cumulative (e.g., carcinogen). Carcinogens are the example of such chronic toxicity.

Species. There are extraordinary wide variations in response to a particular poison between species.

Size, Age and Sex. In general, the amount of a poison required to produce toxic symptoms is related to the weight of the animals. This relationship between weight and dose may vary between species. Very young and very old animals are usually more susceptible to poisons. There are few instances of sex difference in response to poisons in animals. For example, red squill has about twice the toxicity for female rats than for males.

General State of Heath. Debilitated animals are more susceptible to poisons and drugs because their general resistance and detoxication mechanisms are defective. For example, hepatic or renal disease may enormously increase the susceptibility to poisons.

Common causes of toxicoses Accidental poisoning occurs in animals and maybe divided roughly into:

(1) poisoning by naturally-occurring toxicants and

(2) poisoning by man-made chemicals.

But there is no clear cut dividing line between these categories. Naturally occurring toxicological hazards include poisonous minerals and poisonous plants; man-made hazards include industrial contaminants, pesticides, domestic materials, unsuitable food and water, use of drugs, etc.

Biological variation and toxicity data in veterinary practice. Biological variation is a significant factor in interpretation of clinical and diagnostic data used in toxicology. A single toxicity figure will not define the range of toxicity and effects in a given population.

Because LD50 or other values are usually defined in very similar animals (e.g., laboratory rats and laboratory beagles), the laboratory toxicity figure does not reflect the biological variation and differences in toxicity that may occur in a diverse group of breeds within the canine or any other species. For animals of veterinary importance there is usually insufficient information on the variability of effects from low or moderate exposures. Furthermore, individual environmental and husbandry conditions vary widely and can affect the severity of response in any particular group of animals for a specific toxicant and dosage. Therefore, thorough clinical and environmental investigation and good laboratory diagnostic procedures are essential to toxicological evaluation in a suspected exposure.

TOXICITY AND TOXICOMETRICS

Mammalian and other vertebrate toxicities are usually expressed as the amount of toxicant per unit of body weight required to produce toxicosis.

Dosage is the correct terminology for toxicity expressed as amount of toxicant per unit of body weight. The commonly accepted dosage units for veterinary medicine are milligrams per kilogram (mg/kg) body weight.

However, toxicity can also be expressed as moles or micromoles of agent per kilogram body weight. In some experimental studies, comparisons of large and small animals relate dosage to the body surface area, which is approximately equal to body weight. The use of body surface area dosages is advocated by some as a more accurate way to account for very different body sizes in veterinary medicine. For clinical toxicology generally is that as animals increase in weight, the body surface area increases proportionally less, and this may affect the rate of metabolism, excretion, and receptor interaction with toxicants.

For many toxicants, larger animals can be poisoned by relatively lower body weight dosages than can smaller mammals. However, other factors, such as species differences in metabolism or excretion or specific differences in receptor sites can alter this generalization. Dose is a term for the total amount of a drug or toxicant given to an individual organism. In veterinary medicine, the extreme ranges of body weight and surface area, even within some species, generally make the "dose" approach of little practical value.

A commonly used means to compare the toxicity of compounds with one another is the median lethal dosage, also known as the acute oral LD50 in a standard animal, such as the laboratory rat. The LD50 value is usually based on the effects of a single oral exposure with observation for several days after the chemical is administered to determine an end point for total deaths. The LD50 is a standardized toxicity test that depends on a quantal (i.e., all-or-none) response to a range of regularly increasing dosages. In some cases a multiple-dosage LD50 is used to show the acute effects (typically up to 7 days) produced by multiple dosages in the same animals. Increasing dosage levels are usually spaced at logarithmic or geometric intervals. When cumulative deaths are plotted on linear graph paper, the dose-response curve is sigmoid, and the most predictable value is usually around either side of the LD50.

The end point of an LD50 study is death, and the published LD50 value says nothing about the severity of clinical signs observed in the surviving animals or the nature of the clinical effects. Twenty or more animals may be used to arrive at a good estimate of the LD50, which limits the use of LD50 values in most animals of economic significance. In some species, such as birds and fish, the oral toxicity is often expressed on the basis of the concentration of the substance in the feed or water. The acute oral toxicity for birds is often expressed as the LC50, meaning the milligrams of compound per kilogram of feed. For fish, the LC50 refers to the concentration of toxicant in the water.

Other terms are used in the literature to define toxicity of compounds. The highest nontoxic dose (HNTD) is the largest dose that does not result in haematological, chemical, clinical, or pathological drug-induced alterations.

The toxic dose low (TDL) is the lowest dose to produce drug-induced alterations; twice this dose will not be lethal. The toxic dose high (TDH) is the dose that will produce drug-induced alterations; administering twice this dose will cause death. The lethal dose (LD) is the lowest dose that causes toxicant-induced deaths in any animal during the period of observation.

Various percentages can be attached to the LD value to indicate doses required to kill 1% (LD1), 50% (LD50), or 100% (LD100) of the test animals.

Another acronym occasionally used is MTD. It has been used to note the "maximum tolerated dose" in some situations or "minimal toxic dose." Thus one should read such abbreviations carefully and look for the specific term defined.

Acute toxicity is a term usually reserved to mean the effects of a single dose or multiple doses measured during a 24-hour period. If toxic effects become apparent over a period of several days or weeks, the terms subacute or chronic toxicity may be used. Subacute may refer to any effects seen between 1 week and 1 month, whereas chronic often refers to effects produced by prolonged exposure of 3 months or longer. These definitions obviously leave a large gap between 30 days and 90 days. The term subchronic is sometimes used to define this time period, although others avoid the problem in semantics by stating the time period involved. For example, a study could refer to a 14-day toxicity trial with the toxic dosage being 5 mg/kg.

Duration of exposure can greatly affect the toxicity. The single-dose LD50 of warfarin in dogs is approximately 50 mg/kg, whereas 5 mg/kg for 5 to 15 days may be lethal. In rats the single-dose LD50 of warfarin is 1.6 mg/kg, whereas the 90-day LD50 is only 0.077 mg/

kg. On the other hand, rapidly inactivated or excreted compounds may have almost the same 90-day LD50 as the single dose LD50. For example, the single-dose LD50 for caffeine in rats is 192 mg/kg and the 90-day LD50 is slightly lower at 150 mg/kg.

Conversely, animals may develop tolerance for a compound such that repeated exposure serves to increase the size of the dose required to produce lethality. The single-dose LD50 of potassium cyanide in rats is 10 mg/kg, whereas rats given potassium cyanide for 90 days are able to tolerate a dosage of 250 mg/kg without mortality. The ratio of the acute to chronic LD50 dosage is called the chronicity factor. Compounds that have strong cumulative effects have larger chronicity factors. In the foregoing examples the chronicity factors are as follows: warfarin, 20; caffeine, 1.3; and potassium cyanide, 0.04.

From a public health and diagnostic toxicology perspective, it is essential to know the exposure level that will not cause any adverse health effect. This level is usually referred to as the no observed adverse effect level (NOAEL).

It can also be thought of as the maximum nontoxic level. This is the amount that can be ingested without any deaths, illness, or pathophysiological alterations occurring in animals fed the toxicant for the stated period of time. Usually a NOAEL in laboratory animals is based on chronic exposures ranging from 90 days to 2 or more years, depending on the species. The no-effect level is the largest dosage that does not result in detrimental effects.

The concept of risk or hazard is important to clinical toxicology. Although toxicity defines the amount of a toxicant that produces specific effects at a known dosage, hazard or risk is the probability of poisoning under the conditions of expected exposure or usage. Compounds of high toxicity may still present low hazard or risk if animals are never exposed to the toxicant. For example, ethylene glycol antifreeze would be defined as low toxicity (2 to 5 mL/kg body weight), but because it is often readily available in homes, is voluntarily consumed by cats, and is difficult to reverse once clinical signs have developed, it is seen as a high-risk or highhazard toxicant.

Another way to define risk is to compare the ratio of the lowest toxic or lethal dosage (e.g., the LD1) with the highest effective dosage, which could be defined as the ED99. The ratio of LD1/ED99 is defined as the standard safety margin, and it is useful for comparing the relative risk of therapeutic drugs, insecticides, anthelmintics, and other agents applied to animals for their beneficial effects.

If all animals in an LD50 study were the same, then the LD50 would actually be a standard toxic dosage for all animals. However, at the same LD50 dosage, not exactly 50% of animals will die each time. This biological variation can be due to many factors and is the reason that veterinary clinicians must exercise judgment about the response of animals to a given toxicant.

Even more variability is expected because of the differences in species, age, body size, route of exposure, inherent differences in metabolism, and pregnancy and lactation effects. Remember also that the slope of the LD50 curve is important and Is not revealed from the LD50 value alone. An LD50 with a very steep dose-response slope indicates a toxicant or drug has a very narrow margin between no effects and maximal lethal effects. Although such compounds may be dangerous to use as therapeutics, they could be very effective pesticides because of lower probability of survival of target animals.

There are different types of toxicity:

1. Acute toxicity: Effect or response that chemical produces when given to an animal in a single dose. It is measured by the median lethal dose, or LD50. This is the dose that will kill 50% of a group of animals under stated conditions.
2. LC50: It is another way of exposing a species to a series of concentrations and recording mortality at those concentrations. Here, we measure the concentration at which 50% of the entire population is killed.
3. LD50: Here the animals are exposed to one dose and the time at which 50% of the population is killed is recorded.
4. Chronic toxicity: Type of toxicity that produces functional and/or anatomical changes in animals with repeated exposure or dosing. The response is measured for a smaller dose over a prolonged period of time, usually lasting 2 years for rodents such as rats.
5. Subacute or subchronic toxicity: Similar to chronic toxicity, but the duration lasts for 3 months or less.

Dose-response relationships. Frequency response: There are two of the more common methods of predicting responses to dosages by observation of large number of animals. If an adequate dose is given to a very large number of animals and fixed single effect (e.g., convulsions seizures) is measured, a variety of differences will be observed. The data obtained from such an experiment may be plotted in the form of a distribution or frequency-response curve that includes percent responding on its ordinate and dosage levels on its abscissa. Such a curve follows the laws represented by the normal Gaussian distribution pattern and permits the use of statistical procedure applicable to such curves. But, these curves are not particularly useful or determining LD50 values.

Cumulative response: It is generally more useful to plot cumulative response (such as death) against dosage. These curves are commonly known as dose-response curves. Such plot can be developed for only a single response, for example, death. From this plot, one can draw a horizontal line across from the 50% death mark to an intersection with the curve, then a vertical line to the dosage scale. The dosage indicated will be the LD50.

One of the criteria that maybe used to determine relative toxicities of the two compounds is LD50, or equal effect. It is common practice to use the term "potent" for a chemical if the lethal dose is small i.e. a few milligrams. Several guidelines have been developed as aids in classifying the relative toxicities of compounds.

Margin of Safety. The LD50 value is usually based on the effects of a single oral exposure with the rats observed for several days after the chemical administered. The slope of a dose-response curve is an indicator of the range of dosage levels that will elicit a desired effect. If the range is wider the slope will be flatter; if the dosage range is smaller the curve will be steep. The slope of the dose-response curve is therefore an index of the margin of safety of a compound. Thus, the margin of safety is the magnitude of the range of doses involved in progressing from a non- effective dose to a lethal dose. Any LD50 is a calculated value which represents the best estimation of the dose required to produce death in 50% of the animals and the error of the value can be estimated statistically since the curves follow the laws of normal Gaussian distribution.

LD50 is obtained graphically. Other values such as LD95 or LD5 may be obtained in the same manner. If LD84 (lethal dose for 84% of the animals) represents + 1 S.D. from the LD50, then the LD16 represents -1 S.D. from the LD50. The percent mortality may be converted to probits, which are numbers assigned to percentage so that 50% mortality equals a probit of 5, 50% mortality +2 S.D. equals a probit of 7 or, 3 etc. If we plot cumulative % response vs log dose, we get a sigmoid curve. If the cumulative % response units are converted to probits and plotted against log dose we get a straight line. Thus probits are useful to convert a sigmoid to a straight line plot.

Safety Testing. It involves assessment of the safety and toxicity of new drugs for human and veterinary use, new agricultural chemicals, new feed additives, and a wide variety of industrial chemicals. It is done prior to release of a new product. Testing costs millions of dollars per chemical. Toxicologic tests conducted to determine safety of a new product include:

1. General Tests-Acute, subacute and chronic toxicity
2. Special Tests-Specific effects on reproduction; fertility; potentiation; teratogenicity; carcinogenicity; mutagenicity; skin, eye, muscle, subcutaneous or intramammary irritation; hemolysis; and behavior.

In general, rats and dogs are the most commonly used animal species. Other animals used include mice, rabbits, guinea pigs, hamsters, and gerbils. Primates are frequently used. The compound should be chemically as pure as possible. If vehicle is used, it should be inert. The most common route of exposure for general testing is oral.

Types of poisoning

It has been customary to subdivide poisoning into acute, a sudden violent sindrome caused by a single large dose of poison and chronic, a persistent, lingering condition brought on by small repeated doses, with subacute poisoning somewhere in between. This subdivision, however, is not really tenable, as there are types of poisoning which would be difficult to fit into any of these categories. 'Chronic'copper poisoning in sheep only becomes manifesting an acute hemolytic crisis. Symptoms of bracken poisoning may not appear until months after the plant has been ingested. In addition to these types of poisoning, there are other unto-ward effects due to 'poisonous' substances.

(1) Allergy, an immunological response due to sensitization of the subject by a previous dose, although less common in animals than in man, is still quite well known.

(2) Carcinogenicity, in which the agent is responsible for the formation of neoplasia, bracken and the cycads are examples;

(3) Teratogenicity, in which material ingested by the mother at some definite stage in pregnancy produces abnormalities in the off-spring. The classical example in veterinary toxicology is the cyclopean malformation in lambs due to Veratrum californicum.

Common toxicological problem in domestic and farm animals:

Dogs and Cats: Pesticides, garbage, ethylene glycol, heavy metals, biotoxins (toads, snakes, ticks), phyto toxins, mycotoxins, drug reactions.

Poultry: Pesticides (very sensitive to insecticides), feed and water additives, fungi, bacterial toxins, gases and fumigants, heavy metals.

Zoo Animals: Largely malacious and quite variable situations, drug reactions, poisonous plants, accidental-organophosphates and warfarin baits.

Exotic Animals: Largely due to feed additives. Mink: Botulism, chronic lead, phenolic wood preservatives, stilbesterol, etc. Rabbits: milkweed, toxic plants, neck paralysis and in coordination common. Turtles: paint on shell produces lump-back deformities.

Cattle: Heavy metals, pesticides, dietary and environmental contaminants (e.g., urea, nitrate, cyanide, mycotoxins), poisonous plants; snake and insect bites, drug adverse reactions.

Sheep and Goats: Poisonous plants -photo sensitizers, cyanogenetic, selenium, oxalate, lupine, sneezeweed, laurels, white snake root, larkspur, etc; pesticides, anthelmintics, otherslead, nitrate, sulfur, fluoride.

Horses: Poisonous plants-oleander, bracken fern, castor bean, locoweed, lupine, selenium containing, groundsel, crotalaria, cyanide; pesticides, drug adverse reactions, snake and insect bites, other aflatoxins, heavy metals, toxic gases.

Swine: Salt, coal-tar (pitch)and petroleum products, nitrates, wood preservatives, heavy metals, organic arsenicals, fungal toxins, poisonous plants, gossypol, insecticides, botulism, edema disease (endotoxins), rodenticides.

4

TOXINS BY DAMAGING ORGAN SYSTEM

Substances Potentially Associated with Blindness. Ethylene glycol, ivermectin and ivermectin- like parasiticides (e.g., moxydectin, eprinomectin, selamectin, abamectin), lead, metaldehyde, paintballs, salt (e.g., homemade play dough, ice melts, water deprivation).

Substances Potentially Associated with Cardiac Abnormalities. Acetylcholinesterase-inhibiting compounds (e.g., organophosphate and carbamate insecticides), Aconitum (monkshood), amitraz, amphetamines, antidepressants, Apocynum (dogbane), Asclepias (milkweed, pleurisy root), Bufo sp. (toads), caffeine, calcipotriene or calcipotriol, cantharidin, cholecalciferol (and related products), Cassia (senna), cobalt, cocaine, Convallaria (lily-of-the- valley), cyanide, Digitalis (foxglove), digitoxin, digoxin, Eupatorium (white snakeroot), fluoroacetate and fluoroacetamide (Compounds 1080 and 1081), gossypol (Gossypium— cottonseed), ice melts (e.g., magnesium chloride, potassium chloride), ionophores (e.g., laidlomycin, lasalocid, lizard (Heloderma), monensin, narasin, salinomycin) Kalmia (laurel), minoxidil, mushrooms, Nerium (oleander), Persea (avocado), petroleum hydrocarbons (e.g., diesel fuel, gasoline, kerosene), Phoradendron (mistletoe), Pieris(Japanese pieris), pit viper venom, Rhododendron (rhododendron, azalea), Sassafras (sassafras), selenium, Taxus (yew), theobromine, tricyclic antidepressants, Urginea (red squill), Veratrum (false hellebore), Vicia (vetch), xylazine, Zigadenus (death camas).

Substances Potentially Associated with Gastrointestinal Abnormalities. Abrus (precatory bean), acetylcholinesterase-inhibiting organophosphate and carbamate pesticides, aluminum/zinc/magnesium phosphide, antimony, arsenic, Asclepias (milkweed, pleurisy root), cadmium, caffeine, cantharidin, cationic detergents, cholecalciferol (and related products), chromium, Colchicum (Autumn crocus), copper, corrosives, 2,4-D (2,4- dichlorophenoxyacetic acid), diethylene glycol, Digitalis (foxglove), diethylene glycol, digitoxin, digoxin, diquat, ethanol, ethylene glycol, Euphorbia family, fireworks, flares, fluoroacetate and fluoroacetamide (Compounds 1080 and 1081), gorilla glue, grapes, Hedera helix (English ivy), Helleborus (Christmas rose), Hyacinthus (hyacinth), Hydrangea (hydrangea), ice melts (e.g., urea, calcium carbonate, calcium magnesium acetate, sodium chloride, magnesium chloride, potassium chloride), Ilex (holly), iron, Kalmia (laurel), lead, Ligustrum (privet), mace, matches, mercury, mothballs, mushrooms, Narcissus (daffodil, jonquil), nicotine, nitrate, nonionic detergents, nonsteroidal antiinflammatory agents, nutmeg, oil of wintergreen, paintballs, paraquat, phenol, phenoxy herbicides (e.g., 2,4-dichlorophenoxyacetic acid), phosphorus, Phoradendron (mistletoe), Phytolacca (pokeweed), play dough (homemade), raisins, Rhododendron (rhododendron, azalea), Ricinus (castor bean — ricin), Robinia (black locust), Sassafras (sassafras), soaps, thallium, theobromine, trichothecene mycotoxins (e.g., deoxynivalenol — vomitoxin or DON), Viscum (mistletoe), zinc.

Substances Potentially Associated with Heinz Bodies and/or Hemolysis (see also Substances Potentially Associated with Methemoglobin Production). Acetaminophen, Allium (e.g., onions, garlic, chives), chlorate, copper, coral snake venom, Cruciferae family (e.g., kale, broccoli, cauliflower, rapeseed, mustard), drugs (e.g., acepromazine, benzocaine, bupivacaine, chloramphenicol, griseofulvin, lidocaine, prilocaine, tetracaine, trimethoprim sulfa), fireworks, flares, insect stings, matches, methylene blue, mothballs (naphthalene), mushrooms, nonionic detergents, overhydration, paracetamol, pit viper venoms, propylene glycol, skunk spray, spider venom (Loxosceles - brown recluse), zinc.

Substances Potentially Associated with Hemostasis Abnormalities. Anticoagulant rodenticides (e.g., brodifacoum, bromadiolone, chlorophacinone, coumafuryl, difenacoum, difethialone, diphacinone, phenindione, pindone, valone, warfarin), nonsteroidal anti-inflammatories, pit viper venom, sulfaquinoxaline, toxins associated with liver failure (e.g., aflatoxin, acetaminophen, copper, cyanobacteria, mushrooms, zinc), toxins associated with thrombocytopenia (e.g., antibiotics—cephalosporins, diuretics, heparin, phenol, quinidine, quinine).

Substances Potentially Associated with Hepatic Abnormalities. Acetaminophen, aflatoxin, carbon tetrachloride, carprofen, coal tar, copper, cyanobacteria (i.e., blue-green algae), diazepam (cat), germander, iron, lead, mebendazole, melarsomine, mothballs, mushrooms, nitrosamines, nonsteroidal antiinflammatory agents, paracetamol, pennyroyal oil, petroleum hydrocarbons (e.g., diesel fuel, gasoline, kerosene), phenacetin, phenobarbital, phenol, phenytoin, primidone, pyrrolizidine alkaloids, sulfonamides, phenol, phosphorus, quinine, Sassafras (sassafras), stanozolol, tannic acid, thiacetarsamide, toluene, trimethoprim sulfas, valium, vitamin A, zinc.

Substances Potentially Associated with Hyperthermia. Bromethalin, cocaine, dinitrophenol, disophenol, halothane, hops, pentachlorphenol, seizures (or muscle tremors).

Substances Potentially Associated with Methemoglobin Production (see Substances Potentially Associated with Heinz Bodies and/or Hemolysis). Acetaminophen, aniline dyes, benzocaine, chlorate, chloroquine, copper, dibucaine hydrochloride, gallic acid, lidocaine, naphthalene (mothballs), nitric and nitrous oxide, nitrite (nitrate), nitrobenzene, nitroglycerin, nitroprusside, phenacetin, phenazopyridine, phenol, prilocaine, primaquine, propitocaine, pyridium, pyrogallol, resorcinol, silver nitrate, sulfonamides, sulfone, tannic acid.

Substances Potentially Associated with Nervous System—Depression. Acetone, amitraz, barbiturates, benzodiazepine, cholecalciferol (and related products), citrus oils, diethylene glycol, ethanol, ethylene glycol, ice melts (e.g., potassium chloride, magnesium chloride), isopropanolol, ivermectin and ivermectin-like parasiticides (e.g. moxydectin, eprinomectin, selamectin, abamectin), lizard (Heloderma), marijuana (Cannabis), methanol, mushrooms, nicotine, opioids, phenothiazines, pine oil, piperazine, pit viper venom, propylene glycol, tranquilizers, turpentine, xylitol.

Substances Potentially Associated with Nervous System—Excitation. Acetylcholinesterase- inhibiting organophosphate and carbamate pesticides, 4-aminopyridine, amitraz, amphetamines, antidepressants, Asclepias (milkweed), Atropa (belladonna), atropine, bromethalin, Bufo sp. (toad) caffeine, camphor, chlorinated hydrocarbons (e.g., aldrin, chlordane, endosulfan, heptachlor), cholecalciferol, Cicuta (water hemlock), citrus oils, cocaine, cyanide, cyanobacteria (i.e., blue- green algae), Datura (jimsonweed), DEET, Dicentra (bleedingheart, Dutchman's breeches), dichloromethane, ethylene glycol, 5-fluorouracil, Hyoscyamus (henbane), ice melts (e.g., sodium chloride), Ipomea (morning glory), imidacloprid, ivermectin and ivermectinlike parasiticides (e.g., moxydectin, eprinomectin, selamectin, abamectin), khat, Latrodectus spider venom, lead, Lobelia (lobelia), LSD, mace, ma huang, melaleuca oil, mercury, metaldehyde, metronidazole, mushrooms, nicotine, nutmeg, opioids (cat), paintballs, pemoline, penitrem A, phencyclidine, play dough (homemade), potassium bromide, propylene glycol, pyrethrin, pyrethrum, pyrethroids (e.g., allethrin, tetramethrin, resmethrin, permethrin), roquefortine, rotenone, ricin (Ricinus—castor bean), salt (e.g., homemade bread dough, ice melts, water deprivation), Sassafras (sassafras), scopolamine, sodium fluoroacetate and fluoroacetamide (Compounds 1080 and 1081), Sophora (mescal bean), strychnine, Taxus (yew), theobromine (chocolate), theophylline, tricyclic antidepressants, valproic acid, xylitol, yohimbe, zinc/magnesium/aluminum phosphide.

Substances Potentially Interfering with Oxygen Transport and/or Hemoglobin Binding.

Carbon monoxide, cyanide (e.g., mining activity, plants), hydrogen sulfide.

Substances Potentially Associated with Renal Abnormalities. Acetaminophen, antibiotics (e.g., amphotericin B, bacitracin, gentamycin, neomycin, oxytetracycline, paromomycin, polymixin-B, sulfonamides), Aristolochia (birthwort), bismuth, boric acid, 2-butoxyethanol, cadmium, calcipotriene or calcipotriol, cantharidin, carbamate fungicides, carbon tetrachloride, cholecalciferol, chromium, citrinin, copper, diethylene glycol, diquat, ethylene glycol, grapes, lead, lilies (Hemerocallis, Lilium), mercury, mothballs, mushrooms, nonsteroidal antiinflammatory agents, ochratoxin, oxalic acid (e.g., Oxalis, Rheum, Rumex), paraquat, petroleum hydrocarbons (e.g., diesel fuel, gasoline, kerosene), phenol, raisins, toluene, uranium, vitamin D–containing plants (e.g., Cestrum diurnum, Solanum malacoxylon), zinc.

Substances Potentially Associated with Respiratory System. Acetylcholinesterase-inhibiting organophosphate and carbamate pesticides, á-napthyl thiourea, ammonia, chlorinated

hydrocarbons (e.g., aldrin, chlordane, endosulfan, heptachlor), formaldehyde, freon, hydrochloric acid, hydrofluoric acid, hydrogen sulfide, iodine, mercury, nitrogen oxide, opioids, overheated Teflon, paraquat, pennyroyal oil, petroleum hydrocarbons e.g., diesel fuel, gasoline, kerosene), pine oils, selenium, turpentine, zinc/magnesium/aluminum phosphide.

Substances Potentially Associated with Skeletal Muscle Abnormalities/Paralysis. Acetylcholinesterase-inhibiting organophosphates, arsenic, botulism, coral snake venom, ciquatera, cyanobacteria, curare, ionophores (e.g., laidlomycin, lasalocid, monensin, narasin, salinomycin), macadamia nuts, phenoxy herbicides (e.g., 2,4-dichlorophenoxyacetic acid), saxitoxin, spider (Latrodectus—black widow), succinylcholine.

Substances Potentially Associated with Excessive Salivation/Oral Irritation. Acetylcholinesterase-inhibiting compounds (e.g., organophosphate and carbamate insecticides), acids and alkalis (e.g., detergents, disinfectants, soaps), batteries, bleaches, Bufo (toads), cationic detergents, citrus oils, corrosives, cyanobacteria (i.e., blue-green algae), DEET, formaldehyde, glow jewelry (dibutyl phthalate), ivermectin and other macrolide antiparasitic agents (e.g., selamectin, moxidectin, doramectin, eprinomectin, abamectin, milbemycin), lizards, metaldehyde, mushrooms, nonionic detergents, insoluble oxalate-containing plants (e.g., Alocasia [alocasia], Arisaema [Jack-in-thepulpit], Calla [calla], Colocasia [elephant's ear], Dieffenbachia [dumbcane], Monstera [split leaf philodendron], Philodendron [philodendron]), phenol, pine oil, pyrethrins, pyrethroids, spider (Latrodectus—black widow), strychnine, superglue, tremorgenic mycotoxins (e.g., penitrem A, roquefortine), turpentine.

5

GENERAL PRINCIPLES FOR MAKING POISONING DIAGNOSIS

History taking is a vital skill. Combined with performing a thorough physical examination, obtaining an appropriate minimum database, and establishing a differential diagnosis, it allows the clinician to potentially arrive at a correct diagnosis. It is a technique that must be continually improved and perfected, both consciously and constantly.

History taking is especially important in cases of suspected animal poisoning. Taking a complete toxicological history refines and focuses the trajectory of the interview in an attempt to detect the involvement of any potential poison. Let us begin this discussion by reviewing the basic history- taking techniques.

If any part of veterinary medicine is an art, it is the act of securing from an owner the facts surrounding an animal's clinical signs. The clinician must be sympathetic, gentle, and patient in an effort to quickly establish the trust of the person. Such trust will facilitate spontaneous volunteering of important information by the owner. If a person feels intimidated by the veterinarian, he or she may not offer pertinent observations that are crucial to the case, and valuable time will be lost. If the history is to provide any type of working diagnosis, the veterinarian's interview must be meticulous, caring, and thorough in scope.

For a variety of reasons, owners may give histories that are inaccurate, highly unreliable, and sometimes purposely deceitful. Veterinarians must realize that many owners may feel guilty about the duration a condition has existed, how long it has been since the last veterinary visit, how long an animal is left alone each day, how the animal actually came across a poison and how long it took the owner to realize it, or the level of care with which toxic substances are stored or disposed of in the home.

Owners frequently say things they think the veterinarian wants to hear in an attempt to be seen as a more responsible pet owner. Owners often deliberately falsify a history (as in the case of an animal's ingestion of an illicit drug) because of fear of legal repercussions and potential grounds for prosecution. Furthermore, the veterinarian must recognize the fears, anxieties, and emotional distress of many people as they face a potentially devastating health problem in their companion animal. The veterinarian must be a calming influence if a reliable account of events is to be obtained.

If it is not possible to obtain an adequate history from the pet owner, it may be necessary to question other family members, neighbours, and friends. Finally, owners have different emotional make-ups, different educational backgrounds, different intellectual levels, and different economic realities. Language differences, physical disabilities, and other barriers may prevent the veterinarian from communicating effectively. If the owner's primary language is not English, there may be a person fluent in the owner's language in the veterinarian's practice. A local person may be available who can act as an interpreter for the hearing impaired.

Veterinarians must be inventive and flexible in their approach to listening to and communicating with their clients. Clinicians must consciously strive to eliminate any preconceptions that they may have about owners that will bias the history and affect their diagnostic ability. The task of the veterinarian is to translate the owner's account into a comprehensive medical history. Remember that clients have not been schooled to give an accurate history in a precise chronological order, and they may have failed to recognize important changes in vital signs or the onset of clinical signs that veterinarians are trained to identify. Just as the clinician must avoid having his or her own preconceptions, incorrect perceptions of owners that their animal has been poisoned must be identified because these can lead veterinarians to search for a toxicological cause of a problem that is in fact nontoxicological in origin. Veterinarians must never suggest that a client's animal has been poisoned unless there is adequate evidence to support such a conclusion.

Last, it is up to the clinician to organize the history in an orderly and logical manner and to establish the exact chronology of events leading up to the animal's clinical presentation. For some veterinarians, a standardized history form is an effective aid in obtaining a complete, thorough, and objective history.

The history and all initial data obtained should be recorded at the time of the original presentation. The animal's records are a medicolegal document that can be subpoenaed, and they should be treated accordingly. Suspected poisoning cases have a particularly high potential for legal action because of possible liability and criminal activity. The recorded history should be organized, legible, and complete. A good rule of thumb is to not take any records that are incomplete or disorganized or that you would be ashamed to have reviewed by your peers or officers of the court. The history must be organized concisely and logically and must include any and all introductory data. Such data include species, breed, age, sex, reproductive status, vaccine history, previous or current medical problems, current medications, diet, home environment, presence of other animals in the house and any potential appearance of clinical signs in them, recent boarding or kennel history, any recent impoundment and potential exposure to sick or unvaccinated animals, and any recent application of herbicides, pesticides, household cleaners, finishing products, paints or stains (or spills thereof), or use of automotive products or any solvents.

Previous or referring veterinarian's notes or any laboratory data outlining previous medical problems or any recent veterinary treatments should be identified, examined, and added to the record. Additional helpful information can be obtained by calling the previous or referring veterinarian. This technique provides an opportunity for the interviewer to obtain supplemental information relevant to the present problem or to underscore the significance of previously obtained data. Next, the chief complaint should be identified, its duration noted, and the physical examination initiated.

Taking a toxicological history differs a little from the standard clinical history in that it attempts to more specifically establish the time of onset of clinical signs and link them with exposure to a particular toxin. Classically, in suspected poisoning cases, the clinician is faced with one of three scenarios:

(1) the animal has been exposed to a known toxin; (2) the animal has been exposed to an unknown substance that may be a toxin, or (3) the animal displays signs of disease of an uncertain or undetermined cause for which toxins must be considered as part of the differential diagnosis. The toxicological history focuses on the animal. The following questions must be answered: Are there predisposing factors that make the animal more sensitive to exposure? Is the situation compatible with a toxic exposure? Is there a potential source of toxins? Have there been any recent chemical applications?

Despite the often unreliable and possibly unknown nature of the owner's account of events leading to a suspected poisoning, veterinarians must try to obtain as definitive a history as possible. Just as veterinarians must not be misled when owners are convinced that their animals have been poisoned when other causes are actually responsible, clinicians must also never forget that preexisting infections or metabolic, congenital, and neoplastic conditions can mimic the clinical signs of a poisoning or can predispose the animal to a toxicosis. The correlation of an accurate history with the physical examination is crucial. Veterinarians must know the vital signs for the species they care for and must be able to recognize the telltale clinical signs and the characteristic "fingerprints" of specific poisonings.

Despite any possible flaws in the owner's account of the history, the history represents a record of events before and during the onset of the illness. Veterinarians must obtain and organize this information in a logical and orderly manner. Specific criteria characteristic of a toxicological history include what poison or poisons are involved, when the exposure occurred, how much poison the animal was exposed to, and the route of the exposure (e.g., dermal, oral, inhalation, intravenous, subcutaneous, or intraperitoneal). An important further question is whether other animals at home could also have been exposed.

It is important to review the animal's entire environment. Is the patient an indoor-only cat? An outdoor-only dog? What are the animal's normal daily activities? Is it free to roam? How long is it gone each day? This line of questioning will provide helpful information, particularly since many clients may not recognize the potential toxicological hazards present in their house and yard unless specifically asked. Likewise, information about weather conditions and season and the activities, hobbies, and occupations of the owner may all be important and can provide important clues to the cause of an animal poisoning. It is of tremendous help if the owner can bring in the original container of the toxic substance. For most suspected poisonings, the exact quantity of toxin ingested is unknown. However, by examining the container in which the poison was stored and questioning the owner, the amount of toxin previously present in the container may be determined. Using this information, the largest amount the animal could have ingested can be estimated. This amount can be compared with the known lethal dosage for that size of animal. Not only amounts, but also active ingredients, potential antidotes, and related manufacturer information sometimes can be obtained from the package. The first line of defense in management of poisonings is the telephone.

For this reason, all telephone personnel at animal hospitals should be trained as much as possible in the most common small animal poisonings, their relative toxicities, how they are managed, and what to tell people about the treatment Some clinicians and veterinary hospitals use a prepared toxicology history form to help prompt themselves in asking the right questions and successfully directing the initial interview in cases of suspected poisoning. Such a standardized document includes the animal's age, weight, environment, and any present medications used. What toxin or toxins are suspected is of course a critical concern. How much poison may have been involved, route of exposure, and when the potential exposure episode occurred are all prominent topics on such a list of questions. If and when clinical signs (and their nature) were first noted is likewise tremendously important. The form also helps determine if other toxins may be implicated and if other animals in the home may also have been exposed.

This type of direct-line questioning may help the veterinarian quickly obtain valuable information and aid in more swiftly establishing a diagnosis. An example of such a toxicology history form is included in text below.

Sample Toxicology History Form:

- Animal's name, species, breed, sex, intact/neutered
- Age, weight
- Medications presently receiving
- Other pertinent medical history
- Suspected poison involved
- Maximum amount of toxin suspected (worst-case scenario)
- Was the original container found?
- Potential route of exposure suspected
- When did possible exposure occur?
- When clinical signs first were noted? Describe them?
- Could other poisons be involved?
- Could other animals have been exposed?
- Describe the animal's environment (where animal kept, how long left alone, hobbies of owner, anything that might lead to poisoning).

If these history-taking techniques are used and applied in toxicology cases, valuable information can be obtained that will allow the clinician to establish a working diagnosis and determine the most appropriate course of treatment. The toxicological history is combined with a thorough physical examination, an appropriate minimum database, and a sound list of differential diagnoses. The history is an integral part of the picture in formulating a correct diagnosis and a treatment plan. It is up to the veterinarian to organize it in a logical sequence and then, through correlation with the physical examination, clinical signs, and a complete list of differential diagnoses, identify the diagnosis and undertake an efficacious therapeutic regimen.

6

FIRST-AID TACTICS

According to the American Association of Poison Control Centers, 140 614 poison exposures were reported in animals in 1993-1994. These data do not include poisonings handled in veterinary hospitals and not reported, so the actual number of poisonings in animals in any one year is higher. The majority of reported poisonings (82%) occurred in dogs, whereas only 14% occurred in cats. More than 90% of canine and 75% of feline poisonings occurred by ingestion. Dermal exposure was the second most common route. Inhalation and ocular exposures comprised less than 2% of exposures in both species.1 Poisonings, real or suspected, are situations encountered often by emergency veterinarians. Treatment of poisonings can be very rewarding for the owner, pet, and veterinarian, but it is essential to have a plan in place for dealing with poisonings to ensure a successful outcome.

The aim of this chapter is to provide an overall approach to the acutely poisoned patient similar to the approach taken to any critically ill patient. When it is germane to the topic, antidotes will be mentioned. However, specific antidotal treatments for specific poisons are covered elsewhere in the text.

The first contact with the owner is often over the telephone. Species, size, age, and breed of animal should be determined. Historical informatikon collected should include the length of time elapsed since the exposure and, if possible, the type and amount of compound the animal was exposed to. It is impossible to determine over the telephone the condition of the animal, but it is important to try to establish the level of consciousness. If the time before arrival at the hospital will bring the time elapsed since exposure to more than an hour, it may be of benefit to have the owner induce emesis. Induction of emesis should not be recommended if the animal is mentally compromised, if the level of consciousness cannot be determined, if there is any respiratory distress, or if there is a preexisting condition affecting the patient's ability to protect its airway (i.e., laryngeal paralysis). Induction of emesis is also contraindicated in animals who have ingested caustic substances, such as acids or alkalis. Owners of animals that have ingested alkali or acid materials should be instructed to give the pet large amounts of water or milk to dilute the material and proceed as quickly as possible to the veterinary hospital for further care.

If it is determined that the animal is alert and not in respiratory distress, emesis may be indicated. Syrup of ipecac, table salt, and 3% hydrogen peroxide (H2O2) are most often used for inducing emesis at home. Salt and hydrogen peroxide appear to induce emesis by irritating the pharyngeal and gastric sensory neurons, which transmit impulses via the glossopharyngeal and vagus nerves, respectively, to the central nervous system. Excessive salt ingestion could potentiate hypernatremia and other metabolic derangements; therefore, its use as an emetic is strongly discouraged. Hydrogen peroxide has been associated with a few cases of fatal air emboli and mucosal erosions in humans. Syrup of ipecac is readily available over the counter. The recommended dose is 1 to 2 ml/lb PO in dogs and 3.3 ml/lb PO in cats. Ipecac directly stimulates the gastric mucosal receptors and indirectly stimulates the chemoreceptor trigger zone in the posterior medulla to induce emesis. Although ipecac syrup has a high margin of safety, many adverse side effects of its use have been reported, including prolonged vomiting and diarrhea, lethargy, fever, and irritability along with isolated cases of gastric rupture, intracranial hemorrhage, and diaphragmatic hernia in humans.

After a review of the available scientific literature related to emesis induction in poisonings, the American Academy of Clinical Toxicology and the European Association of Poison Centres and Clinical Toxicologists published a position statement regarding ipecac syrup. They found that clinical studies in people have not confirmed that ipecac syrup improves the outcome of poisoned patients, and its routine use should be abandoned. Its use may delay further gastrointestinal decontamination and administration of activated charcoal and oral antidotes caused by prolonged emesis. The effectiveness of the activated charcoal and antidotes may be reduced as well. If vomiting is induced, it should be in cases of witnessed toxin ingestion and should be performed within 60 minutes of ingestion, and the gastric contents should always be saved and brought to the hospital with the pet for visual and toxicological examination. Although it is often recommended to repeat the emetic in 15 minutes if no vomiting has occurred, it may be more helpful in the long run if the owners are on their way to the veterinary hospital by that time.

If the animal is having seizures or is trembling, care should be taken to avoid self-induced trauma while en route to the hospital. Owners should be advised that a pet with an altered mental state, or one that is anxious and uncomfortable, may bite or react unexpectedly, so owners should protect themselves from possible injury as well. The animal should be kept warm if it is recumbent and/or unconscious.

AT THE VETERINARY CLINIC

Management of an acute case of intoxication can be organized into five areas: (1) ABCs of the critically ill patient: airway, breathing, circulation; (2) gaining control of seizures or tremors; (3) assessment of metabolic and medical derangements and institution of a plan for their management; (4) gastrointestinal decontamination; and (5) supportive care. In this chapter we will discuss the first four areas; supportive care and further details on gastrointestinal decontamination are discussed in later chapters. To some extent, these first four areas should be dealt with at the same time; it is merely convenient for the sake of discussion to put them in order.

On arrival at the veterinary hospital, vital signs should be immediately assessed. The patency of the airway should be evaluated and reestablished with a cuffed endotracheal tube if necessary. Anticonvulsants should be administered as necessary (to be discussed in more detail later), and hydration and metabolic status should be assessed and addressed rapidly.

Poisoned animals may have severely altered homeostatic mechanisms and therefore may be hypothermic or hyperthermic. It is necessary to treat the patient and not the toxicosis. This means that although gastric decontamination is integral to the long-term outcome, there may be more immediate life-threatening concerns in any individual case.

The ABCs of critical patient care

Airway. Any animal that is unconscious, appears to have neuromuscular paralysis or paresis, or is in severe respiratory distress is a candidate for intubation. If there is no voluntary respiratory effort or if gastric lavage is planned, intubation is required. Anesthesia is required for intubation if the animal is conscious. These patients may or may not also require intermittent positive pressure ventilation. An appropriately sized cuffed endotracheal tube is placed, the cuff is inflated, and the tube is tied securely to the muzzle. When possible, a blood sample is collected for arterial blood gas determination; this helps in assessing the need for supplemental oxygen and assisted ventilation. Other methods available for monitoring ventilation include apnea alert monitors and end-tidal carbon dioxide ($CO2$) monitors.

Breathing. Hypoventilation indicates a need for assisted ventilation. Hypoventilation is identified by the presence of hypercapnia ($PaCO2 > 45$ mm Hg) with acidosis ($pH < 7.35$). If assisted ventilation is contraindicated or is unavailable, a degree of "permissive hypercapnia" may be well tolerated by the patient. Guidelines for permissive hypercapnia state that a $PaCO2$ of >50 mm Hg can be tolerated if the pH remains above 7.25 and cardiovascular function is adequate.

Hypoventilation is usually secondary to central nervous system and neuromuscular abnormalities. If pulmonary function in dogs and cats is normal, ventilation at normal minute volumes (100 to 200 mL/kg) with room air or a slightly increased FiO2 should return the blood gases to normal.

Permissive hypercapnia is contraindicated in patients with CNS disease or cerebral oedema because the resulting cerebral acidosis will cause cerebral vasodilation and can increase cerebral morbidity.

Hypoxemia (PaO2 < 65 mm Hg) should be treated with enriched oxygen of 40% or more as necessary to maintain the PaO2 at >65 mm Hg. If continued increases in the inspired oxygen concentration do not improve the PaO2, or if the effort required to maintain the PaO2 at that level is excessive and exhausting, assisted ventilation is required. If the toxin was not inhaled, a reason for the hypoxemia must be sought. Aspiration pneumonija or preexisting respiratory disease should be considered, and a work-up should be performed when the patient is stabilized. The management of patients on mechanical ventilation is beyond the scope of this chapter, and the reader is referred to the references at the end of the chapter.

Circulation (oxygen delivery). Adequate oxygen delivery (circulation) is dependent on the volume of blood in the vessels, the pumping function of the heart, the integrity of the blood vessels, and the oxygen content of the blood. An early assessment of the electrocardiogram (ECG) will aid in determining the function of the pump. Toxins such as oleander, foxglove, and other cardiotoxic plants, organophosphates, and overdoses of therapeutic drugs for cardiac disease directly affect the heart. Many hydrocarbons and industrial chemicals have arrhythmogenic properties as well. The vascular volume status of any intoxicated patient varies greatly among individuals. The presence of a deficit in interstitial or vascular volume depends on whether the animal has been able to eat and drink or has had vomiting or diarrhea, the type of toxin and whether it induces a diuresis, and the presence of preexisting medical conditions. As in any emergency situation, tachycardia, cold extremities, pale mucous membranes, and slow capillary refill time indicate a vascular volume deficit and perhaps hypovolemic shock. A patient with these signs should be resuscitated quickly with isotonic crystalloid solutions (60 to 90 ml/kg in dogs and 40 to 50 ml/kg in cats), colloid solutions (10 to 20 ml/kg in dogs and 5 to 10 ml/kg in cats), or a combination of these. If the poison was a vitamin K antagonist, the preferred fluid may be frozen plasma and packed red blood cells. The source of hemorrhage should be located and vitamin K1 therapy begun immediately. Interstitial deficits are identified by decreased skin turgor, dry or tacky mucous membranes, and perhaps mild azotemia. The serum sodium concentration may be high if the patient has had free water losses, or low if the patient has had losses from vomiting and/or diarrhea, but is continuing to drink water. The percentage of deficit should be estimated and the deficit volume calculated. For example, a 10-kg dog who is estimated to be 7% dehydrated has a fluid volume deficit of 10 kg × 0.07 or 0.700

kg (700 ml). The deficit can be replaced over 8 to 12 hours using an isotonic crystalloid solution. A continuing maintenance fluid should be administered at the same time to keep up with insensible losses. Ongoing fluid losses (caused by diarrhea, vomiting, or polyuria) should also be assessed and a replacement fluid added to the fluid plan. Nonsteroidal antiinflammatory agents (NSAIDs) are known to be nephrotoxic.9 If a patient has ingested toxic amounts of NSAIDs, fluid therapy planning should maximize renal perfusion. Blood urea nitrogen and creatinine levels and urinalysis should be monitored closely.

If fluid therapy does not restore adequate circulation, the heart may be unable to provide adequate cardiac output because of intrinsic damage to or disease of the myocardium. Inotropic drugs, such as dobutamine, may be justified in this setting if adequate monitoring is available. Dobutamine is delivered as a continuous IV infusion at 5 to 15 μg/kg/minute. Dopamine is an effective first line vasopressor at an initial IV dose of 3 μg/kg/minute. The dose can be titrated up to 15 μg/kg/minute. Higher doses cause dangerous tachycardia and vasoconstriction with no further improvement in cardiac output and blood pressure. Epinephrine can be used if dopamine is unsuccessful at improving blood pressure. Epinephrine is a potent alpha and beta agonist that increases cardiac output and can improve blood pressure. Higher doses increase vasoconstriction and heart rate without leading to further improvement in cardiac output. The IV dosage of epinephrine is 0.05 to 0.30 μg/kg/minute. Higher dosages of epinephrine should be avoided because they predispose the patient to ventricular fibrillation. Norepinephrine is indicated if the previously mentioned agents do not improve blood pressure. Norepinephrine is an extremely potent vasoconstrictor and may impair perfusion to the viscera and periphery. The infusion is begun at a dosage of 0.5 μg/kg/minute and can be titrated up to 1 μg/kg/minute.

Toxins may affect the blood vessels by altering the function of the smooth muscle in the vessel walls or by poisoning the endothelial cells directly. Vascular endothelium is exposed to any blood-borne toxin, and accumulation of toxin in endothelial cells can occur. Toxins can also induce the activation of inflammatory mediators produced by the endothelium, which can greatly affect vascular tone and incite the cascades responsible for the systemic inflammatory response. Blood pressure should be measured and maintained above a mean of 60 mm Hg. Volume replacement is the initial means of restoring blood pressure. When maximum safe volumes have been administered without satisfactory effect, vasopressor medications may have to be given.

If the blood pressure stays persistently elevated (>140 mm Hg mean arterial pressure) after adequate fluid restoration and there is evidence of increased peripheral vascular resistance, hydralazine or nitroprusside can be used to reduce the blood pressure. The use of these agents requires adequate myocardial contractility. Hydralazine is primarily an arteriolar vasodilator. It works by decreasing cytosolic calcium. Hydralazine can be administered enterally or parenterally at a dose of 0.5 to 3 mg/kg every 8 to 12 hours. In the acute care setting, a test dose can be given intravenously. Onset of action occurs in 10 to 30 minutes, and the effect lasts up to a few hours. Because of the decrease in systemic vascular resistance induced by this drug, heart rate can increase significantly and should be monitored closely. Nitroprusside is a venous and arteriolar vasodilator. It is given intravenously and its onset of action is immediate. The dosage of nitroprusside is 0.5 to 1 µg/kg/minute. This drug should not be used in patients with preexisting renal or hepatic disease. Administration of any vasoactive agent requires close monitoring of the ECG, blood pressure, and volume status of the patient. These drugs are all potentially arrhythmogenic and can dangerously elevate or reduce systemic blood pressure. Their use should not be considered if adequate monitoring is not available.

The oxygen-carrying capability of the blood is affected by several toxins.

The blood oxygen content is a product of oxygen bound to hemoglobin and oxygen dissolved in the blood. These can be maximized by instituting transfusions of red blood cells, when anemia is present, and by increasing the inspired oxygen concentration. Carbon monoxide (CO) and acetaminophen are two toxins that inhibit the ability of hemoglobin to transport oxygen. Initial management in poisoning with either of these toxins is to increase the inspired oxygen concentration and support ventilation if necessary. In CO poisoning, hyperbaric oxygen therapy may be required. Acetaminophen toxicosis is treated with acetylcysteine at an initial dose of 150 mg/kg PO or IV in dogs and cats followed by three to five additional treatments at 70 mg/kg every 4 hours. Blood transfusions may be indicated based on the clinical picture.

Parameters, such as heart rate, mucous membrane color, distal extremity temperature, and color and mentation, should be monitored along with clinical laboratory indicators to aid in adjustment of the fluid plan as the patient's condition changes. Urine output is an important indicator of vascular volume and blood pressure adequacy. Monitoring urine output accurately requires placement of a urethral catheter and a closed collection system. A urine output of less than 1 mL/kg/hour suggests a volume deficit, decreased blood pressure, or a change in renal function and should be evaluated and treated promptly.

Controlling seizures and tremors

Animals who are suffering seizures or are trembling are given anticonvulsants and sedatives to relieve the problem. Initially 0.5 mg/kg of diazepam is administered intravenously. If the neuromuscular activity is not controlled, or if the effect is short lived, the dose can be repeated. In general, if diazepam needs to be repeated more than twice for seizure control, constant rate infusion should be instituted or a different drug should be tried. Pentobarbital at 5 to 20 mg/kg IV, titrated to effect, will anesthetize the patient and reduce the visible seizure activity. If seizure breakthrough occurs again, a continuous infusion of pentobarbital can be given at 1 to 6 mg/kg/hour. If an anesthetic plane is reached and must be maintained, the patient must be intubated. As mentioned earlier, blood gases should be monitored for evidence of hypoventilation or hypoxemia.

If blood gases are not available, an end-tidal CO2 monitor is useful for monitoring the presence of hypercapnia. The end-tidal CO2 is well correlated with the PaCO2 and is therefore a marker of ventilation. If hypoventilation develops, the pentobarbital infusion should be discontinued. If the animal again requires anesthesia for seizure control, a lower infusion dose should be used. A patient that has been having seizures continually or in clusters is prone to develop cerebral edema and increased intracranial pressure. The brain has an obligate requirement for glucose as an energy source for aerobic metabolism. The increased metabolic requirements of the brain cells coupled with the reduced perfusion of the brain lead quickly to a depletion of energy for the cells and increased amounts of carbon dioxide and lactate.

Ionic channel alterations secondary to the seizures and energy depletion lead to intracellular hypercalcemia and hyperosmolality. These in turn lead to swelling of the neurons and may induce increased blood flow overall.10 Because the brain is encased within the rigid structure of the calvarium, there is little room for swelling, and the pressure within the brain increases quickly. By creating a brief increase in serum osmolality relative to the tissues, mannitol can draw brain water into the blood vessels. Mannitol infused IV at 0.5 to 1 g/kg over 20 minutes increases serum osmolality and can be very effective at reducing intracranial pressure.

A rapidly deteriorating neurological status, including the onset of anisocoria and depression of brainstem responses (e.g., physiological nystagmus, heart rate, and respiratory rate), indicates an increase in intracranial pressure. An animal poisoned by a tremorgenic toxin may need only diazepam, diazepam combined with muscle relaxants, or diazepam and some degree of barbiturate anesthesia to control the tremors and relax the muscles. Methocarbamol is a centrally acting muscle relaxant. It is a central nervous system depressant and can itself cause salivation, emesis, ataxia, and sedation, making it difficult to assess these monitored parameters. The manufacturer's recommended dose in dogs and cats is 44 to 220 mg/kg IV, generally not to exceed 330 mg/kg/day. The lower end of the dose range is usually effective. As with seizures, intubation and monitoring of blood gases or end-tidal CO2 is necessary after the patient has been anesthetized. During this period the body temperature can undergo wide fluctuations. Temperatures of >106° F secondary to seizures or trembling are not uncommon. The high temperature need not be treated per se because it will come down when the excessive muscle activity is stopped. Sedatives, anesthetics, and muscle relaxants may cause the body temperature to drop quickly and steeply, making the animal hypothermic. Monitoring for temperature fluctuations and heating or cooling as necessary is very important. Rectal temperature probes can be placed for continuous monitoring.

ASSESSMENT OF METABOLIC STATUS

The availability of inexpensive rapid tests for blood chemistry and blood gases has made it much easier to provide optimal care for critically ill and poisoned patients. The poisons themselves and the secondary consequences of their ingestion (e.g., vomiting, diarrhea, seizures, and tremors) can induce varied derangements in acid-base and electrolyte values as well as azotemia, and, in some cases, specific organ failure. Mild metabolic acidosis or alkalosis can often be corrected by instituting fluid therapy and treating the underlying disease. For instance, the metabolic acidosis present in a patient having seizures is in large part caused by increased lactic acid production.

Lactic acid is produced in large amounts in the muscle in response to the excessive muscle activity and the relative deficit of energy during the seizure. Lactic acidosis quickly resolves once the seizures have ceased. Mild changes in bicarbonate and pH usually resolve with fluid therapy and treatment of the underlying disease. Metabolic acidosis with a pH of <7.2 or a base excess of greater than 8 may require treatment with a sodium bicarbonate infusion. The formula commonly used to calculate the amount of sodium bicarbonate to be administered is:

mEq of HCO3" = body weight (kg) × base deficit × 0.3 (0.3 is the volume of distribution of the bicarbonate).

Thirty to 50 percent of the calculated dose is administered over 30 minutes or is added to a fluid infusion lasting several hours. The serum bicarbonate level is rechecked after administration of this dose, and further bicarbonate is given as necessary.

Respiratory acidosis is a result of hypoventilation and can occur in the acutely poisoned patient owing either to the toxin's affect on the central nervous system (CNS) and muscle function or, as discussed earlier, to the medications used to control seizure activity. Respiratory acidosis is identified when the PCO2 is greater than 45 mm Hg and the pH is less than 7.35. Adequate minute ventilation must be restored with intermittent positive pressure ventilation. In most cases, an inspired oxygen concentration of 21% at normal tidal and minute volumes is necessary to restore the PCO2 to the 35 to 45 mm Hg range. If a hypoventilating patient is hypoxemic, even after adequate ventilation has been restored, another respiratory disease must be considered and investigated.

Metabolic alkalosis in the presence of toxicoses is usually due to gastrointestinal losses of chloride. Treating the underlying disorder and controlling vomiting and diarrhea generally restores the normal acid-base balance. It is rarely necessary to treat metabolic alkalosis in the acutely poisoned patient. Respiratory alkalosis is a result of hyperventilation (PCO2 <35 mm Hg). It usually results from disorders that cause hypoxia or stimulate the respiratory center. It may also result from excessive mechanical ventilation. It is necessary to identify and treat the underlying disorder to correct the problem.

Other metabolic parameters. It is sometimes useful to calculate the osmolality in an acutely poisoned patient. If the osmolar gap (the difference between the calculated and the measured osmolality) exceeds 15 mOsm/kg, an unmeasured osmole should be sought. In the context of poisonings, that osmole is most commonly ethylene glycol. The formula for calculating osmolality is:

2(Na + K) + BUN/2.8 + glucose/18

The osmolar gap is calculated by subtracting the calculated from the measured osmolality. The chemistry panel that provides the measured osmolality will also provide the blood glucose, sodium, potassium, and blood urea nitrogen (BUN) data required to calculate the gap. The osmolality cannot be calculated without these data. A high osmolar gap in a patient with the appropriate history, signs, and supporting laboratory data helps in making a diagnosis of ethylene glycol poisoning.

Blood glucose, calcium, sodium, potassium, and chloride concentrations can all be altered by various poisons. In the case of specific toxins or overdoses, such as calcium channel blockers, digoxin, and vitamin D–based rodenticides, the alterations in these chemical measurements are caused directly by the toxin. In other cases, these abnormalities are a result of the body's response to the disease process. For instance, hypoglycemia may develop with prolonged seizures because of the increased energy requirements of the muscles and brain. Although hypoglycemia and hypercalcemia or hypocalcemia must be treated quickly and specifically, it is again the treatment of the underlying disorder, withdrawal of the agent, and restoration of adequate circulation that will provide long-term improvement.

Hypoglycemia can be treated with 0.25 to 0.5 mL/kg of 50% dextrose (125 to 250 mg/kg) administered slowly intravenously. It is detrimental to increase the blood glucose level above normal because the osmotically active glucose molecules cross the blood-brain barrier, creating an osmotic shift that may lead to cerebral edema. By giving smaller increments of glucose and repeating the glucose measurements, the blood glucose level can be brought back to normal slowly without overshooting. The addition of 2.5% to 5% dextrose to the maintenance fluids, based on repeated glucose measurements, may be enough to maintain blood glucose within the physiologic range.

Gastrointestinal decontamination

Gastrointestinal decontamination can begin if the patient arrives alert and stable or after the animal has been stabilized. Decontamination consists of two stages in urgent cases: The first is gastric emptying, and the second is adsorbance of the remaining toxin.

Gastric emptying in the alert patient with no respiratory compromise can be induced with drugs. Vomiting should not be induced when strong acids or alkalis have been ingested. Vomiting is reported to be more effective at emptying the stomach than gastric lavage and is safer in the controlled environment of the veterinary hospital. Reviews of the literature, however, suggest that neither vomiting nor lavage is more effective, and neither may even be necessary. If emesis is to be induced, apomorphine provides consistent results in veterinary patients. Apomorphine is a centrally acting emetic agent that is available in tablet form. The tablets are 6 mg each and are easily dissolved in water or saline. The recommended dose is 0.04 to 0.08 mg/kg. The pill can be dissolved in saline and administered into the conjunctival sac. Once emesis has occurred, the sac should be thoroughly flushed to prevent further absorption of the drug and reduce its irritant effect. Apomorphine induces emesis within 15 minutes. In cats xylazine is an effective emetic if apomorphine is not available. A dose of 0.44 mg/kg given intramuscularly (IM) is recommended. These drugs may have profound effects on the central nervous system. For this reason, close monitoring of the patient after induced emesis is imperative, especially when toxins that affect the CNS are suspected. Gastric contents should be saved and submitted for laboratory analysis if the toxin is unknown. Serum, whole blood, urine, and faeces are all body fluids that can be analyzed for toxins and should be saved.

Once the gastric contents have been removed, activated charcoal (AC) can be given. Activated charcoal has a very large surface area that adsorbs many compounds and allows poisons to move through the gastrointestinal tract without being absorbed so that they can be eliminated. It has a tendency to cause constipation and is therefore often prepared with cathartics to speed transit time and subsequent elimination. If the AC movement through the gut is delayed, adsorbed compounds can move off the charcoal and can be absorbed by the gut. Activated charcoal preparations without cathartics are available and may be preferable if repeated dosing is planned. Repeated dosing with the preparations containing cathartics can cause diarrhea and associated metabolic disorders. Some animals willingly swallow AC alone, and others may eat it mixed with baby food or other foods. If the patient refuses to eat it, a dosing or catheter-tipped syringe can be used to feed it to the compliant patient. Otherwise, AC administration may require the passing of a stomach tube as described subsequently. Activated charcoal is given at a dose of 2 to 8 g/kg initially, and in many cases the dose should be repeated. If the patient has altered mentation or respiratory compromise, gastric lavage is the preferred method of gastrointestinal decontamination.

As with emesis, gastric lavage is contraindicated in patients that have ingested caustic substances or hydrocarbons because of the high potential for further damage of the mucosa and for aspiration. The equipment required includes a tube of the appropriate length, a lavage pump, and a container for the removed material. The patient must be intubated with a cuffed endotracheal tube to protect the airway. This requires induction of an anesthetic state. There are many drug protocols for inducing anesthesia. Important considerations in the choice of drug are the likelihood of respiratory or cardiovascular compromise in the presence of the suspected toxin's effects on these organs, operator familiarity with the drug's effects, preexisting disease, whether the drug is reversible, and drug availability. The tube used for lavage should have as large a lumen as can safely fit in the patient's esophagus. This allows for maximal infusion of lavage fluid and removal of larger ingested particles. The measurement necessary for proper placement of the tube in the stomach is the length from the rostral aspect of the mouth to the last rib. White tape is placed at the measured length to indicate to the operator when the tube is in far enough.

Once the airway is secured, a mouth gag is placed, the end of the tube is lubricated, and the tube is advanced into the stomach. Once the tube is in the stomach, one can commonly hear gas bubbling through the tube and can smell the gastric contents. If placement of the tube is uncertain, direct examination of the caudal pharynx and larynx in the anesthetized animal is required to ensure proper placement and prevent administration of lavage fluid into the airways.

The ideal lavage fluid is warm water. It is instilled and drained repeatedly until the returning fluid is clear and particle free. When the fluid is clear, an adsorbent such as activated charcoal can be administered. The AC can be delivered directly to the stomach through the tube already in place; it is followed by a warm water flush sufficient to clear the volume in the tube. When removing the tube, it must be kinked upon itself to prevent any material remaining in the tube from falling out and possibly going down the airway. Even a cuffed endotracheal tube is not guaranteed to prevent aspiration.

In one particular instance, it has been recommended to delay charcoal administration to allow administration of an antidote. This is in the case of acetaminophen poisoning. The AC is thought to adsorb N-acetylcysteine (NAC), and therefore these drugs should be staggered. Although this is true in vitro16 and studies show a statistical significance, it is not clear that this finding has clinical significance. The studies were done in humans who were given oral NAC, and the finding may not be relevant in veterinary patients receiving parenteral NAC.

Repeated doses of AC are advocated in patients with many toxicoses for two reasons. First, many compounds undergo enterohepatic circulation and are therefore available for reabsorption in the gut. By giving AC every 4 to 6 hours, some reabsorption can be avoided. Second, many toxins and drugs move down their concentration gradient from the bloodstream back into the gut lumen where they are again absorbed. Cathartics should be used with AC as mentioned earlier, although their use has not been thoroughly studied. In a position statement on the use of cathartics in poisoned patients, the American Academy of Clinical Toxicology and the European Association of Poison Centres and Clinical Toxicologists reviewed the literature and could not endorse the routine use of cathartics because of the lack of scientific data. One study performed in pediatric patients found that sorbitol was the most effective cathartic of those evaluated. Sodium sulfate (Glauber's salt) and magnesium sulfate (Epsom salts) are often suggested for use as cathartics.

They are osmotic cathartics and therefore could cause the patient to become dehydrated. Dangerous hypermagnesemia has also been reported to develop with the use of magnesium sulfate. The ingestion of alkalis and acids is a special case in that emesis reexposes the mucosa to the toxin, which is extremely erosive and may worsen injury. Alkali or acid ingestions should be treated by diluting the poison with large amounts of water or milk. Administration of an acidic or basic substance to neutralize the poison is contraindicated. The issue of administering steroids to reduce the likelihood of stricture formation is controversial. Several studies in children and rabbits suggest that significant benefit can be derived from using antiinflammatory doses of prednizone (1 mg/kg PO) or dexamethasone (0.1 to 0.25 mg/kg IM, IV, PO) for 21 days in patients with second- or third-degree burns of the esophagus. Dexamethasone may be better at reducing stricture formation than prednisone and may reduce the severity and frequency of strictures best when given in the immediate postburn period.

Acute management of non-ingested toxins

- Topically applied toxins. Many poisons can be applied topically. Insecticides are the most obvious example, but paints and oils containing heavy metals, industrial products containing various hydrocarbons, and household products can all be applied topically either on purpose or by accident. Bathing the animal with a mild soap shampoo is the most important initial step.
- Continued absorption through the skin or ingestion by grooming will prolong morbidity if the pet is not bathed. If the pet is not alert or is in distress of any kind, bathing should be done at the veterinary hospital rather than by the owner at home to ensure the protection of the airway and prevent injury to the owner.
- Since there is a possibility of skin absorption and subsequent enterohepatic circulation, and ingestion by grooming, activated charcoal is indicated in cases of topical poisoning. For the same

reasons as mentioned earlier, repeated dosing is recommended. If the skin has been burned by a caustic substance, copious lavage with warm water or physiological saline is required before further steps are taken. This prevents continuing damage by the toxin. Corrosive damage to the skin is different from contact dermatitis in that the destruction of tissue occurs by direct necrosis in the former and is secondary to the inflammatory process in the latter. The necrotic tissue can also act as a reservoir of the insulting compound, resulting in ongoing damage. The compound may also be absorbed and may lead to systemic injury. Ocular contact with irritating, caustic, and toxic materials should be treated with copious flushing. Two or three litters of physiological saline solution is recommended for flushing, which may require topical anesthetics and/or systemic sedatives or tranquilizers depending on the patient's degree of discomfort and compliance. After flushing, therapy is directed toward healing any ulcerations or inflammation of the eye. Further therapies for topical toxins are directed toward the patient's overall condition as outlined previously and toward the specific toxin.

7

BASIC PRINCIPLES OF TREATMENT OF POISONING

The poisoned patient will almost always require broad-based monitoring and supportive care to guard against direct and collateral organ damage induced by poison or intoxication. Such patients may have respiratory, cardiovascular, neurological, gastrointestinal, renal, hepatic, or hematologic abnormalities and these organ systems are the focus of the monitoring and support endeavors.

There are, in addition, important nursing care issues relative to the management of recumbent and comatose animals.

RESPIRATORY AID

Breathing rate, rhythm, nature and effort. Breathing rate can vary widely in normal animals and, except for extreme variations, is of limited value as a respiratory monitor. A change in breathing rate, however, is a sensitive indicator of a change in the underlying status of the patient. Bradypnea may be a sign of respiratory centre depression secondary to intracranial or extracranial disease. The underlying cause should be identified and treated if possible.

Arrhythmic breathing patterns are usually indicative of a problem within the central pattern generator in the brainstem, and such a finding suggests a poor prognosis. A weak breathing effort may be caused by central or peripheral neurological impairment or neuromuscular dysfunction, depending upon the specific poison or toxin. Bradypneic, arrhythmic, and weak breathing patterns may be associated with hypoventilation and such patients should be intubated and ventilated. Tachypnea is usually a sign of an underlying problem, such as hypoxemia, hypercapnia, hyperthermia, hypotension, sepsis, pain, or opioid administration, and seldom needs to be treated specifically.

Audible sounds. Low-pitched, snoring, or high-pitched, squeaky inspiratory noises heard without the aide of a stethoscope suggest an upper airway obstruction. Such patients should receive oxygen therapy and may need general anesthesia to gain access to the airway. Large airway obstructions must be removed or bypassed (by endotracheal intubation). Midpitched, wheezing, asthmatic sounds heard during inspiration and expiration suggest a fixed, big airway obstruction or narrowed lower airways, as would occur in bronchospasm or with the accumulation of airway exudate (pneumonia). Bronchodilators should be used if bronchospasm is thought to be the problem; positive pressure ventilation may be required if there is airway fluid. Auscultation of crepitation or bubbling are indicative of airway fluids, as would occur in pulmonary edema (transudate), pneumonia (exudate), or hemorrhage. Such patients may require positive pressure ventilation if oxygen therapy alone does not remedy the hypoxemia. Aspiration is a common problem in obtunded patients that vomit or regurgitate. The aspirated material may cause airway obstruction and sets the stage for pneumonia. Aspirated gastric fluid with a pH of less than 2 is also associated with epithelial necrosis and inflammation. Diminished, muted, or absent lung or heart sounds may indicate a pleural space disorder, such as pneumothorax or hemothorax. A radiograph may be a useful diagnostic tool for intrathoracic abnormalities, but should only be attempted if the respiratory distress is not considered life threatening at the moment. A diagnostic thoracentesis may be indicated if a pleural space disorder is suspected and the patient is too unstable for radiography. Pneumonia should be treated with broad-spectrum antibiotics (empiric at first, and later supported by airway secretion culture and sensitivity testing), nebulization and coupage, postural drainage and early ambulation (to the extent possible). Cough suppressants should be avoided.

Ventilometry. Ventilation volume can be estimated by visual observation of the chest wall and can be measured by ventilometry. Normal tidal volume ranges between about 8 and 20 mL/kg. Normal total minute ventilation ranges between about 150 and 250 mL/kg/min for dogs. PaCO2 defines alveolar minute ventilation and the measured minute ventilation should be appropriate. A large minute ventilation in combination with a normal (or high PaCO2) is indicative of dead space ventilation, as might occur in hypovolemia and pulmonary thromboembolism.

Partial pressure of carbon dioxide (PCO2). The arterial PCO2 (PaCO2) is a measure of the ventilatory status of the patient and normally ranges between 35 and 45 mm Hg. A PaCO2 in excess of 60 mm Hg may be associated with excessive respiratory acidosis and is often considered to represent sufficient hypoventilation to warrant positive pressure ventilation. PaCO2 values below 20 mm Hg are associated with respiratory alkalosis and a decreased cerebral blood flow, which may impair cerebral oxygenation. Venous PCO2 (PvO2) is usually 3 to 6 mm Hg higher than arterial in stable states and can generally be used as an approximation of $PaCO_2$. PvO_2 is variably higher in transition states and during hypovolemic or anemic states. An increased arterial-venous PCO_2 gradient suggests decreased tissue perfusion. $PaCO_2$ may also be estimated bymeasuring the carbon dioxide in a sample of gas taken at the end of an exhalation. End-tidal PCO_2 is usually 2 to 4 mm Hg lower than $PaCO_2$ in dogs.

Hypoventilation. Hypoventilation is treated by endotracheal intubation and positive pressure ventilation. Whether hand-bagging or using a mechanical ventilator, the general guidelines for positive pressure ventilation are the same. Peak proximal airway pressure should be about 10 to 15 cm H_2O; tidal volume should be about 10 to 15 mL/kg; inspiratory time should be about 1 second (or just long enough to achieve a full tidal volume); ventilatory rate should be about 10 to 15 times per minute. Diseased lungs are stiffer (less compliant) than normal lungs, and are therefore much more difficult to ventilate. It is a common finding that the above guidelines are insufficient to adequately ventilate a patient with diffuse pulmonary parenchymal disease. To improve ventilation, the proximal airway pressure could be increased in a step fashion up to 60 cm H_2O (or to the limit of the ventilator) and the ventilatory cycle rate could be increased in a step fashion up to 60 breaths per minute. The tidal volume should be decreased in an animal with diffuse lung disease because such lungs have a reduced vital capacity; a normal tidal volume could overdistend the reduced number of remaining lung units. Protective lung strategies currently aim for very small tidal volumes (i.e., 4 to 6 mL/kg). If oxygenation must be improved: (1) the inspired oxygen could be increased up to 100% for short periods of time or up to 60% for prolonged periods of time; (2) positive end-expiratory pressure (PEEP) can be added or increased (up to 20 cm H_2O [PEEP increases transpulmonary pressure and functional residual capacity, and keeps small airways and alveoli open during the expiratory phase and improves ventilation and oxygenation. PEEP also minimizes the repetitive collapse and reopening of small airways, a process that contributes to ventilator-induced injury]); or (3) the inspiratory time or the inspiratory plateau could be increased. The inspiratory/ expiratory [I/E] ratio must allow sufficient time for exhalation of all of the last breath, otherwise air trapping will occur.

Cyanosis. A blue to gray hue to the mucous membranes is caused by the presence of unoxygenated hemoglobin in the capillaries. Cyanosis is most often due to hypoxemia and should always be considered to be a late sign of Severe hypoxemia. Cyanosis is not a reliable sign of hypoxemia in the anemic patient. Cyanosis may also be caused by sluggish capillary circulation. Methemoglobinemia can cause a brownish to bluish discoloration.

Partial pressure of oxygen (PO_2). The arterial PO_2 (PaO_2) measures the tension of oxygen dissolved in the plasma, irrespective of the hemoglobin concentration. The PaO_2 is a measure of the oxygenating ability of the lungs. The normal PaO_2 is considered to range between 80 and 110 mm Hg when an animal is breathing room air at sea level, and >500 mm Hg when breathing 100% oxygen. Hypoxemia could be caused by low inspired oxygen, hypoventilation while breathing 21% oxygen, and lung disease such as pulmonary edema or pneumonia. A PaO2 <60 mm Hg is a commonly selected trigger for oxygen therapy. Venous PO2 reflects tissue PO2 and bears no correlation to PaO2. Mixed or central venous PO2 ranges between 40 and 50 mm Hg.

Values below 30 mm Hg may be caused by anything that decreases the delivery of oxygen to the tissues (e.g., hypoxemia, anemia, low cardiac output, vasoconstriction); values above 60 mm Hg suggest reduced tissue uptake of oxygen (e.g., septic shock, metabolic poisons, shunting). Venous blood for such evaluations must be taken from a central vein, such as the jugular, anterior vena cava, or pulmonary artery; peripheral venous PO2 values are highly variable and difficult to interpret.

Hemoglobin saturation with oxygen. When red to infrared light is transmitted through a blood sample, the various hemoglobins present in the blood sample will absorb a certain proportion of it as oxyhemoglobin, methemoglobin, carboxyhemoglobin, and reduced hemoglobin. A bench top co-oximeter measures and displays values for the first three. The displayed oxyhemoglobin is functional (i.e., it is expressed as a percentage of the amount of hemoglobin available for oxygen binding [total hemoglobin minus methemoglobin and carboxyhemoglobin]), as opposed to fractional oxyhemoglobin, which is expressed as a percentage of total hemoglobin irrespective of methemoglobin or carboxyhemoglobin. Normal methemoglobin and carboxyhemoglobin are normally less than 1% each and so usually functional and fractional oxyhemoglobin are quite similar. To the extent that either methemoglobin or carboxyhemoglobin is present in large concentrations, fractional oxyhemoglobin will be variably lower than functional. Hemoglobin saturation (SO2) measures the percent saturation of the hemoglobin and is related to PO2 by a sigmoid curve. In general a PO2 of 100 mm Hg is equivalent to an SO2 of about 98%; a PO2 of 80 mm Hg is equivalent to an SO2 of about 95%; a PO2 of 60 mm Hg is equivalent to an SO2 of about 90%; and a PO2 of 40 mm Hg is equivalent to an SO2 of about 75%. The clinical informatikon derived from the measurement of arterial SO2 (SaO2) is similar to that obtained from a PaO2 measurement in that they both are a measure of the ability of the lung to deliver oxygen to the blood. Pulse oximeters attach to a patient externally (tongue, lips, tail, toenail). For most clinical purposes, most pulse oximeters are sufficiently accurate approximations of oxyhemoglobin saturation; accuracy should be verified by an in vitro standard if possible. The accuracy of a pulse oximeter is greatest within the range of 80% and 95%, and is determined by the accuracy of the empirical formulas that are programmed into the instrument.

Tissue, venous and capillary blood, nonpulsatile arterial blood, and skin pigment also absorb light. A pulse oximeter must differentiate this background absorption from that of pulsatile arterial blood. It does this by measuring light absorption during a pulse and subtracting from that the light absorption occurring between the pulses. If the pulse oximeter cannot detect a pulse, it will not make a measurement. One of the common reasons for poor instrument performance is peripheral vasoconstriction. When a measurement is obtained, it may either be accurate or inaccurate. When inaccurate, it is usually inaccurately low. When a low measurement is obtained, particularly when it seems incongruous for the patient's condition at the time, retry the measurement in several different locations; and then either take the average or the highest reading. Methemoglobin and carboxyhemoglobin absorb light and impact the measurement made by a two-wavelength pulse oximeter designed to measure only oxyhemoglobin. Due to the biphasic absorption of methemoglobin at both the 660 and 940 nm wavelengths, abnormal accumulations tend to push the oximeter reading toward 85% (underestimating measurements when SaO2 is above 85% and overestimating it when below 85%). Carboxyhemoglobin absorbs light like oxyhemoglobin at 660 nm but hardly at all at 940 nm. When present in increased concentrations, it would increase the aparent oxyhemoglobin measurement.

Oxygen therapy. Oxygen therapy may be beneficial when the predominant cause of the hypoxemia is ventilation-perfusion mismatching or diffusion impairment. Oxygen therapy may not be substantially beneficial if the predominant cause of the hypoxemia is small airway and alveolar collapse. High inspired oxygen concentrations can be attained with a facemask. If the animal does not tolerate the facemask, the oxygen outlet should be held as close to the animal's nose as possible. Oxygen cages and hoods are commercially available or can be homemade; oxygen tents and infant incubators are available from used medical equipment suppliers. Enclosed oxygen environments should not be used during the initial stabilization stages, but are useful afterward when the patient needs to be maintained on oxygen. In any enclosed environment, it is important to measure and control the oxygen concentration, to eliminate the carbon dioxide produced by the animal, and to control the temperature. High humidity is acceptable as long as the temperature is controlled at a comfortable level. A convenient way to increase the inspired oxygen concentration is via insufflation. A soft, flexible catheter can be inserted into the nasal cavity, about to the level of the medial canthus of the eye. The catheter is sutured as it passes through the

lateral alar notch and again at points along the side of the face or the top of the head to keep the catheter out of the patient's view. An oxygen flow rate of 50 to 100 mL/kg is initially recommended; flow rates should be subsequently adjusted to the needs of the patient. Medical oxygen is anhydrous and should be humidified by bubbling it through warm water. The effectiveness of the oxygen therapy, however it is applied, should be evaluated shortly after beginning therapy. If therapy is not judged to be effective, positive pressure ventilation may be indicated.

CARDIOVASCULAR AID

Heart rate and rhythm. Heart rate and stroke volume are important to cardiac output. Slower heart rates are normally associated with larger end-diastolic ventricular volumes and larger stroke volumes. Up to a point, larger stroke volumes preserve cardiac output. Heart rate is too slow when it is associated with low cardiac output, hypotension, or poor tissue perfusion. This may occur when the heart rate falls below about 60 beats/minute in the dog; 90 in the cat. Common causes for bradycardia are excessive vagal tone secondary to visceral inflammation, distention, or traction, hypothermia, hyperkalemia, atrioventricular conduction block, end-stage metabolic failure, hypoxia, acetylcholinesterase inhibitors, organophosphate and carbamate poisonings, and digitalis overdose. Sinus tachycardia is primarily a sign of an underlying problem (e.g., hyperthermia, hypoxemia, pain, parasympatholytics such as atropine, sympathomimetics, supraventricular or ventricular ectopic pacemaker activity). In people, because of coronary artery narrowing, sinus tachycardia can increase myocardial oxygen consumption beyond oxygen delivery capabilities. In animals, where coronary artery disease is rare, tachycardia only becomes a problem for the patient when there is not enough time for diastolic filling, which results in a decrease in cardiac output. Specific treatment of sinus tachycardia may be indicated when the heart rate exceeds the low 200s for dogs or the high 200s for cats. Ventricular arrhythmias may be caused by sympathomimetics, hypoxia, hypercapnia, myocarditis, electrolyte disturbances (potassium and magnesium), and arrhythmogenic factors released from various debilitated abdominal organs, intracranial disease, or digitalis intoxication. Ventricular arrhythmias become a problem for the patient when they interfere with cardiac output, arterial blood pressure, and tissue perfusion, or when they threaten to convert to ventricular fibrillation when (1) the minute-rate equivalent exceeds the high 100s for dogs and the low 200s for cats; (2) the complexes are multiform; or (3) the ectopic beat overrides the T wave of the preceding depolarization. Total elimination of ventricular arrhythmia is not necessarily the goal of therapy since large dosages of antiarrhythmic drugs have deleterious cardiovascular and neurological effects. A simple decrease in the rate or severity of the arrhythmia may be a suitable endpoint to the titration of the antiarrhythmic drugs.

Ventricular arrhythmias can be caused by several mechanisms that are not readily apparent from the ECG appearance of the arrhythmia: (1) abnormal automaticity characterized by rapid, spontaneous, phase 4 depolarization; (2) reentry of depolarization wave fronts due to unidirectional conduction blocks; (3) early after-depolarizations caused by diminished repolarizing potassium currents resulting in prolongation of action potentials; and (4) delayed after-depolarizations caused by abnormal oscillations of cytosolic calcium concentrations after myocardial or Purkinje cell repolarization. A given antiarrhythmic may be effective in one mechanism and be ineffective, or even worsen, in another. Antiarrhythmic therapy is always a clinical trial. Lidocaine is usually a first choice antiarrhythmic because it selectively affects abnormal cells without affecting automaticity or conduction in normal cells.

Vasomotor tone and tissue perfusion. Peripheral and visceral perfusion is primarily regulated by vasomotor tone. Vasodilation improves peripheral perfusion while vasoconstriction impairs it.

Vasodilation is a potent cause of hypotension while vasoconstriction increases blood pressure. Vasomotor tone is assessed by mucous membrane color (pale = vasoconstriction; red = vasodilation), capillary refill time (<1 second = vasoconstriction; >2 seconds = vasodilation), toe-

web/core temperature gradient (>4° C = vasoconstriction; <2° C = vasodilation). Vasoconstriction may be caused by hypovolemia, heart failure, hypothermia, or vasoconstrictor sympathomimetics. Vasodilation may be caused by the systemic inflammatory response, hyperthermia, or the administration of vasodilator drugs. Vaso-corrective therapy may be necessary if treatment directed at the underlying cause has failed.

Pulse quality. Assessment of pulse quality by digital palpation involves an evaluation of both the height and width of the pulse pressure waveform compared with normal. Tall, wide pulse pressure waveforms are seen in sepsis; tall, narrow waveforms occur with a patent ductus and during CPR. Small, narrow pulse pressure waveforms are seen with small stroke volumes and vasoconstriction. Small stroke volumes occur with hypovolemia, poor heart function, tachycardia, and ventricular arrhythmias. The pulse pressure waveform is largely a reflection of stroke volume and vessel size. It is not a measure of arterial blood pressure per se, although in a general way vessels with low pressure are easier to collapse and vice versa. The weak, thready pulse that occurs with hypovolemia is caused by small stroke volumes and vessel constriction; such patients may be normotensive. Peripheral pulse quality (such as the dorsal metatarsal in the dog) decreases and disappears earlier than do larger, more central arteries (such as the femoral) with progressive hypovolemia. The relative pulse quality of more peripheral versus more central arteries may provide a rough index to the magnitude of the hypovolemia.

Central venous pressure. Central venous pressure (CVP) is the luminal pressure of the intrathoracic vena cava. Peripheral venous pressure (PVP) is slightly higher than CVP and may provide some useful information but is subject to unpredictable extraneous influences. CVP (or PVP) is the relationship between venous blood volume and venous blood volume capacity.

Venous blood volume is determined by venous return and cardiac output. For CVP, verification of a well-placed, unobstructed catheter can be ascertained by observing small fluctuations in the fluid meniscus within the manometer, synchronous with the heartbeat, and larger excursions synchronous with ventilation. Large fluctuations synchronous with each heartbeat may indicate that the end of the catheter is positioned within the right ventricle. The normal CVP is 0 to 10 cm H2O; PVP would be on average 2 to 3 cm H2O higher. Venous pressure is a measure of the relationship between blood volume and blood volume capacity and could be measured to help determine the end point for large fluid volume resuscitation. Below-range values suggest hypovolemia and that a rapid bolus of fluids should be administered. Above-range values indicate relative hypervolemia and that fluid therapy should be stopped. Venous pressure is also a measure of the relative ability of the heart to pump the venous return and should be measured whenever heart failure is a concern. Venous pressure measurements are used to determine whether there is "room" for additional fluid therapy in the management of hypotension.

Arterial blood pressure. Arterial blood pressure represents the relationship between blood volume and blood volume capacity. Cardiac output and systemic vascular resistance determine arterial blood volume. Arterial blood pressure is a primary determinant of cerebral and coronary perfusion. Systolic blood pressure is primarily determined by stroke volume and arterial compliance. Systemic vascular resistance and heart rate primarily determine diastolic blood pressure. Mean blood pressure is the average pressure: one half of the area of the pulse pressure waveform. If the pulse pressure waveform were a perfect triangle, mean pressure would be 1/3 of the difference between diastolic and systolic pressure. To the extent that the pulse pressure contour is not a perfect triangle—a tall, narrow pulse pressure waveform, for example—the mean pressure will be closer to diastolic. The mean arterial blood pressure is physiologically the most important since it represents the mean driving pressure for organ perfusion. Many clinical instruments, however, measure only systolic blood pressure. The relationship between systolic and mean arterial blood pressure is variable, depending upon the shape of the pulse pressure waveform; systolic blood pressure should always be assessed with this in mind. Arterial blood pressure can be measured indirectly by sphygmomanometry or directly via an arterial catheter attached to a transducer system. Sphygmomanometry involves the application of an occlusion cuff over an artery in a cylindrical appendage. The width of the occlusion cuff should be about 40% of the circumference of the leg to which it is applied. The occlusion cuff should be placed snugly around the leg. If it is applied too tightly, the pressure measurements will be erroneously low since the cuff itself, acting as a tourniquet, will partially occlude the underlying artery. If the cuff is too loose, the pressure measurements will be erroneously high since excessive cuff pressure will be required to occlude the underlying artery. Inflation of the cuff applies pressure to the underlying tissue and will totally occlude blood flow when the cuff pressure exceeds systolic blood pressure. As the cuff pressure is gradually decreased, blood will begin to flow intermittently. When the cuff pressure falls below systolic pressure: (1) systolic blood pressure can be estimated as the manometer pressure at which needle oscillations begin to occur during cuff deflation (caused by the pulse wave hitting the cuff); (2) systolic blood pressure can be estimated also as the manometer pressure at which one can digitally palpate a pulse distal to the cuff; (3) systolic blood pressure can be estimated as the manometer pressure at which the first blood flow sounds are heard via a Doppler ultrasound crystal placed over an artery distal to the occluding cuff; and (4) oscillometry analyzes the fluctuation of pressure in the cuff as it is slowly deflated and provides a digital display of systolic, diastolic, mean blood pressure, and heart rate. Most of these instruments can be set to recycle at discrete time intervals. Small vessel size and motion can interfere with measurements. All external techniques are least accurate when vessels are small, when the blood pressure is low, and when the vessels are constricted. Direct measurements of arterial blood pressure are more accurate and continuous compared with indirect methods, but require the introduction of a catheter into an artery by percutaneous or cut-down procedure. The dorsal metatarsal, femoral, and ear arteries are commonly used in dogs and cats.

The subcutaneous tissues around dorsal metatarsal and ear arteries are tight and hematoma formation at the time of catheter removal is rarely a problem. Once the catheter is placed, it is connected to a monitoring device. The catheter must be flushed with heparinized saline at frequent intervals (hourly) or continuously to prevent blood clot occlusion. The measuring device could be a long fluid administration set suspended from the ceiling. Fluid is instilled into the tubing via a three-way stopcock to a very high level and then allowed to gravitate into the artery until the hydrostatic pressure of the column of water is equalized with the mean arterial blood pressure of the patient. Since blood pressure oscillates, the system should be closed between measurements to prevent blood from entering into, and clotting, the end of the catheter. The measuring device could also be an aneroid manometer. Water or blood must not be allowed to enter the manometer. Sterile saline is injected into the tubing toward the manometer via a three- way stopcock until the compressed air increases the registered pressure to a level above that of mean blood pressure. The pressurized manometer system is then allowed to equilibrate with the mean blood pressure of the patient. The arterial catheter can also be attached to a commercial transducer and recording system. The extension tubing between the catheter and the transducer should not be excessively long and should be constructed of nonexpansible plastic to prevent damped signals. The transducer should be "zeroed" periodically and calibrated with a mercury manometer to verify accurate blood pressure measurements. The stopcock that is opened to room air for the zeroing process must be at the level of the heart. With modern patient monitors, the transducer can be placed anywhere with reference to the patient (the monitor will compensate internally with an "off-set pressure" for any vertical differences between the patient and the transducer). If the relative vertical position between the patient and the transducer changes, the transducer must be rezeroed.

Normal systolic, diastolic, and mean blood pressure are approximately 100 to 140, 60 to 100, and 80 to 120 mm Hg, respectively. In general, one should be concerned when the systolic blood pressure (ABPs) falls below 100 mm Hg or when the mean blood pressure (ABPm) falls below 80 mm Hg. In general, one should be very concerned when the ABPs falls below 80 mm Hg or the ABPm falls below 60 mm Hg. Hypotension may be caused by hypovolemia, poor cardiac output, or vasodilation.

Hypertension (high ABPm), when it occurs, is generally attributed to vasoconstriction. High ABPs, not associated with a high ABPm, is generally attributed to an inappropriate frequency response of the measuring system (for that patient and that time). Hypertension can cause increased hemorrhaging, retinal detachment, increased intracranial pressure, and high afterload to the heart and should be treated when ABPm exceeds 140 mm Hg. High ABPm may be due to a light level of anesthesia, hyperthermia, sympathomimetic drugs, hyperthyroidism (thyroxinecatecholamine synergy), renal failure (renin-angiotensin), pheochromocytoma (epinephrine), and an increased intracranial pressure. In the last case, the hypertension is most likely due to Cushing's response to maintain an adequate cerebral perfusion pressure and should not be treated.

Cardiac output. Cardiac output can be measured by a variety of techniques in clinical patients, but in veterinary medicine it is usually not measured. The concept of cardiac output must be on the forefront of one's monitoring and therapeutic considerations; it is after all the whole point of adequate cardiovascular function. Poor cardiac output is implied when preload parameters (CVP or PVP, pulmonary artery occlusion pressure, jugular vein distention, postcava distention on chest radiograph, and large enddiastolic diameter on cardiac ultrasound image) are high and the forward flow parameters (pulse quality, arterial blood pressure, signs of vasoconstriction, urine output, and physical and laboratory measures of tissue perfusion) are abnormal. Cardiac output is a flow parameter and can be low even when arterial blood pressure is normal. Cardiac output may be reduced by poor venous return and enddiastolic ventricular filling (e.g., hypovolemia, positive pressure ventilation, or inflow occlusion); by ventricular restrictive disease (e.g., hypertrophic or restrictive cardiomyopathy, pericardial tamponade, or pericardial fibrosis); by decreased contractility; by excessive bradycardia, tachycardia, or arrhythmias; by regurgitant atrioventricular valves; or by outflow tract obstruction. Poor cardiac output should be improved by correcting the underlying problem when possible. Preload should be optimized. When poor contractility is thought to be the problem, sympathomimetic therapy may be indicated.

Oxygen delivery. Oxygen delivery (DO2) is the product of cardiac output and blood oxygen content. Oxygen content is determined by hemoglobin concentration (most important) and PO2. Oxygen delivery must be adequate to meet metabolic requirements (oxygen consumption).

Excessive anemia, hypoxemia, bradycardia, arrhythmias, reduced stroke volume, poor heart contractility, valvular lesions, tamponade, and hypovolemia may cause inadequate oxygen delivery. Combinations of these abnormalities can compound the oxygen delivery deficit.

Excessive oxygen consumption can occur with hyperthermia and intense muscular activity, such as seizures. A disparity between oxygen delivery and oxygen requirement results in an increased oxygen extraction, low venous oxygen, metabolic (lactic) acidosis, and an increased arterial- mixed venous PCO2 gradient. Treatment of an oxygen delivery deficit should be directed at the underlying cause(s). If hypovolemia is thought to be the primary problem, fluids should be administered to reestablish an effective circulating volume. If contractility is thought to be the primary problem, dobutamine should be administered (dopamine if the patient is also hypotensive).

Fluid therapy. Poisoned animals are commonly dehydrated and hypovolemic, and such issues must be addressed early in the course of their management. Hypovolemia is defined as a low circulating blood volume. Dehydration is defined, for the purposes of this discussion, as a low extracellular volume caused by the loss of a crystalloid solution (sodium and water). A depletion of volume of all fluid compartments (intracellular and extracellular; total body dehydration) is caused by the loss of free water (water without electrolytes) and is identified in a patient by the measurement of sodium concentration (discussed under the heading of sodium). The clinical signs of hypovolemia and extracellular dehydration are different. Hypovolemia is identified by low preload (e.g., low CVP, PVP, or pulmonary artery occlusion pressure; collapsed jugular veins; radiographic appearance of a small postcava; and cardiac ultrasound appearance of a small end- diastolic diameter) and low forward flow (e.g., poor pulse quality, hypotension, vasoconstriction, oliguria in a patient that does not have renal disease, increased oxygen extraction, low venous oxygen, metabolic [lactic] acidosis, and an increased arterial-mixed venous PCO2 gradient).

Dehydration is identified by an acute loss of body weight, a decrease in skin elasticity, dry mucous membranes in a patient that is not open-mouth breathing and has not received an anticholinergic, and high urine specific gravity and low sodium concentration in a patient that does not have renal disease and has not recently received a bolus of colloids or a diuretic.

Hemoconcentration may occur, depending on the nature of the fluid loss. By definition, dehydrated patients have lost extracellular fluid and therefore all dehydrated patients are to some extent hypovolemic However, the magnitude of the dehydration and that of the hypovolemia do not necessarily correlate and so each should be evaluated independently.

Animals with evidence of subcutaneous edema may be associated with hypervolemia (e.g., heart failure, renal failure, and iatrogenic fluid overload) or hypovolemia (e.g., hypoproteinemia/hypocolloidemia and increased vascular permeability). In general, hypovolemia requires an initial, rapid (over a period of 10 to 60 minutes) fluid therapy plan, whereas dehydration requires a slow (over a period of 4 to 24 hours) fluid plan. Life-threatening hypovolemia should be addressed with large volume, relatively rapid fluid administration.

Isotonic crystalloids, such as lactated Ringer's, saline, or any other commercial intravenous solution with a normal-range sodium concentration, should be administered in approximately 20 mL/kg aliquots for dogs and 10 mL/kg aliquots for cats until the signs of hypovolemia are not severe. Complete normalization of cardiovascular signs is not necessarily the objective of this initial rapid fluid plan. This may require fluid doses in the range of 100 mL/kg in the dog and 60 mL/kg in the cat. The immediate objective is to achieve an acceptable circulating volume; fine tuning to normal should be accomplished more slowly. Life-threatening blood solute and electrolyte abnormalities (e.g., anemia, hypoproteinemia, sodium, potassium, and calcium) should also be addressed at this time (these are discussed later in this chapter). If the initial circulating volume was not judged to be a problem in the first place, the remaining fluid and electrolyte abnormalities can be dealt with. There are three categories to be considered when developing such a fluid therapy plan: (1) existing deficits or excesses; (2) replacement of the normal ongoing losses (commonly referred to as maintenance); and (3) replacement of the abnormal ongoing losses. First determine if the animal is dehydrated (decreased skin elasticity) or fluid overloaded (jelly-like subcutaneous tissues). If dehydrated, quantitate the magnitude of the deficit as either a specific number between 5% and 12% of the body weight or as mild (6% of body weight), moderate (9% of body weight), or severe (12% of body weight). This number is then multiplied by the animal's body weight to estimate the volume of fluids necessary to correct the deficit volume. If edematous, first determine if the animal is hypervolemic or hypovolemic. Edematous, hypovolemia patients may be better served with colloid therapy. Edematous, hypervolemic patients require fluid restriction and perhaps diuretic therapy. Usually an iso-osmolar, polyionic extracellular fluid (ECF) replacement crystalloid, such as lactated Ringer's or an equivalent solution with approximately normal extracellular concentrations of sodium, potassium, chloride, and a bicarbonate-like anion (e.g., bicarbonate, lactate, gluconate, or acetate) should be used to restore a deficit. These replacement fluids should be administered without alteration if they are to be administered rapidly or if it is known or suspected that there are no major electrolyte abnormalities. The volume of fluids required to replace the normal ongoing losses can be determined from predictive charts for the dog and cat. If a chart is not available, the maintenance volume can be assumed to be between 50 mL/kg per day for large dogs, 75 mL/kg per day for small dogs and cats, and 100 mL/kg for very small or young animals. The true nature of fluids used for replacement of normal ongoing losses (maintenance) are distinctly different from that of fluids used to replace extracellular volume deficits; the sodium concentration is only about 40 to 60 mEq/L and the potassium concentration is about 15 to 20 mEq/L. Administering a replacement solution, such as lactated Ringer's, to an animal for its maintenance requirements predisposes the animal to hypernatremia (which is usually not a problem because the kidneys can usually readily excrete the excess sodium) and hypokalemia (almost always a problem if the animal is not eating because the kidney is not very good at conserving potassium). In veterinary medicine, true maintenance fluids are seldom used. Isotonic, polyionic ECF replacement solutions, supplemented with potassium (20 mEq/L), are used and work well most of the time.

Maintenance solutions must not be used to replace extracellular volume deficits because they will cause hyponatremia and hyperkalemia when administered in large volumes. Abnormal ongoing losses, which occur via transudation into one of the major body cavities, into the tissues or via burn wounds contain similar electrolyte concentrations to that of the extracellular fluid compartment at the time and should be replaced with an unadulterated isotonic, polyionic ECF replacement solution. Losses caused by vomiting, diarrhea, or diuresis should be replaced with an isotonic, polyionic ECF replacement solution supplemented with potassium (10 mEq/L). When the fluid therapy plan is being constructed, it is usually not known how much fluid the animal will lose over the day. One could either leave this category blank for the time being or add fluid to the plan as losses occur. If the patient has a disease, which is known to be associated with unrelenting fluid losses, an estimated volume could be factored in at the time of the construction of the initial fluid plan and then adjusted upward or downward as the day progresses. The intravenous administration of fluids is often the preferred route since its effects are immediate and reliable.

The fluid prescription could also be administered subcutaneously in several divided daily dosages. The subcutaneous route is slower in onset than the intravenous route and is less efficacious in that some individuals will not absorb the fluids well or at all. The fluids could be administered orally or via stomach tube, in several divided daily doses, as long as the gastrointestinal tract is functional. Fluids could also be administered via the intramedullary route, if venous access is not possible because of the small size of the patient. A regular hypodermic needle works well in the very young animals, while a bone marrow biopsy needle works well in the older animal. The bony prominence at the proximal humerus and proximal tibia, and the trochanteric fossa of the femur are common sites for needle introduction. Fluids could also be administered intraperitoneally, although there is some danger of injury or perforation of an abdominal organ.

Temperature

Hypothermia. Core body temperatures down to 36 °C are not associated with detrimental effects to the patient and if protected from further heat loss, should rewarm spontaneously. Body temperatures of 32° C to 34° C also do not interfere with organ function, but some patients will require active rewarming. Body temperatures of 28° C to 30° C have marked CNS depressant effects and these patients will require active rewarming. Body temperatures of 25° C to 26° C are associated with unconsciousness, ventricular arrhythmias, and decreased tissue oxygen delivery out of proportion to decreases in oxygen requirements, resulting in anaerobic metabolism, lactic acidosis, and rewarming acidemia. Blood viscosity is about 200% of normal. Active rewarming can be achieved by circulating warm water or air blankets; infrared heat lamps (optimal distance 75 cm) or radiant heat warmers; hot water bottles placed under the drapes or blankets (avoid contact with skin if the water temperature exceeds 42° C); flushing the abdominal cavity or colon with warm, sterile, isotonic, polyionic fluids: or by extracorporeal techniques. Aggressive surface rewarming should be avoided in very cold patients because peripheral vasodilatation may induce excessive hypotension in the face of a cold-depressed heart. Ischemic peripheral tissues may have accumulated various metabolites, which may have deleterious cardiovascular effects when large quantities are washed into the central circulation. The rewarming rate should be limited to about 1° C per hour.

Hyperthermia. Fever is a reset thermostat and is caused by the release of endogenous pyrogens (interleukin-I) from monocytes in response to infection, tissue damage, or antigen-antibody reactions. Mild degrees of hyperthermia (less than 40° C) are not per se harmful to the patient and may represent an appropriate response to an underlying disease. Hyperthermia, without a reset thermostat, is pathological. It is usually caused by the excessive muscular activity associated with seizures. Cell damage starts to occur at body temperatures above 42° C when oxygen delivery can no longer keep pace with the racing metabolic activity and increased oxygen consumption. Severe hyperthermia causes multiple organ dysfunction and failure; renal, hepatic, gastrointestinal failure; myocardial and skeletal muscle damage; cerebral edema; disseminated intravascular coagulation; hypoxemia; metabolic acidosis; and hyperkalemia. Surface cooling techniques are most effective with room temperature fluids; it is the evaporation of the water from the skin surface that causes the cooling. Ice water causes vasoconstriction, which can impede heat loss from the core until skin temperature is <10° C, at which time vessel paralysis and vasodilation occur and core temperatures decrease precipitously. Convective heat loss can be enhanced with fans. Conductive heat loss can be enhanced with ice packs. Large volumes of cold crystalloid fluids, intravenously via the colon or stomach, or into a body cavity, are effective internal cooling techniques. Antipyretic drugs (antiprostaglandins, dipyrone, and aminopyrine) are generally effective for fever but are not effective for pathological hyperthermia.

Neurological aid

The neurological status of the patient should be characterized as part of the initial examination. Muscle weakness can be caused by an overdose of macrolide parasiticides (e.g., ivermectin) or aminoglycoside antibiotics, organophosphate and carbamate insecticides, lead, polyradiculoneuritis, tick paralysis, botulism, myasthenia gravis, cationic detergents (fabric softeners and sanitizers), or spinal cord disease. Muscle twitching, muscle tremors, hyperexcitability, and convulsions can be caused by amphetamines, herbal ephedra (ma huang or guarana root), methylxanthines (caffeine and theobromine), pseudoephedrine, LSD, phencyclidine, marijuana, cocaine, benzodiazepines (small dose), opioids (large dose), metaldehyde, mushrooms, moldy food stuffs (mycotoxins—penitrem A), blue-green algae, strychnine, 1080, bromethalin, ivermectin, organochlorine insecticides, vitamin D rodenticides, pyrethrins, permethrins, organophosphate and carbamate insecticides, nicotine, paraquat, lead, zinc phosphide, tricyclic antidepressants, 4-aminopyridine, 5-fluorouracil, cationic detergents (fabric softeners and sanitizers), hypocalcemia, and hypomagnesemia. Obtundation to coma can be caused by barbiturates, benzodiazepines (large dose), phenothiazines, opioids, ivermectin and other macrolide parasiticides, amitraz, marijuana, ethanol, methanol, ethylene glycol, lead, mushrooms, moldy food stuffs (mycotoxins - penitrem A), metabolic disease (e.g., hepatic encephalopathy, hypoglycemia, uremia, hypoxia, hypoosmolality or hyperosmolality (heat stroke), neoplasia, infectious or inflammatory disease, thromboemboli, or coagulopathies.

The symptomatic therapy of the weak muscle patient is first to protect them from self-induced injury if they are "flopping around" because they have enough muscle strength to be active, but not enough to move in a coordinated fashion. If they are mostly immobile, they should be managed as per the recumbent care protocol discussed later. If the muscle weakness is so severe that it interferes with ventilation, the patient will need to be intubated and ventilated until such time that adequate ventilatory muscle strength returns. Hyperexcitable patients must also be protected from self-induced injury. If the hyperexcitability is not too bad and the patient does not appear to be too uncomfortable, and if the poison or toxin is short acting, perhaps the best option would be no therapy and to either leave the animal alone or to physically hold it and "ride it out." Sedatives, however, are often required for these patients to facilitate their management. Almost any sedative or anesthetic except ketamine can be used; however, there are some important caveats. First, almost all sedatives or anesthetics have been associated with seizure activity in the human literature. Etomidate and propofol occasionally cause muscle twitching. Phenothiazines have a particular such reputation but are therapeutic at least for phencyclidine-induced hyperexcitability. Opioids cause CNS excitation (delirium) in modest dosages in cats and in high dosages in dogs, but the threshold for this side effect may be lowered in the face of hyperexcitable poisons and toxins. Benzodiazepines are not reliable sedatives in normal patients, causing hyperexcitability. They are more reliable in patients with preexisting CNS disease or sedation.

The benzodiazepines diminish the hyperexcitable effects of ivermectin and ivermectin-like intoxication, but are GABA-receptor agonists and might potentiate the CNS depressant effects of the ivermectin (and similar drugs), which is also a GABA-agonist. Unfortunately, barbiturates, propofol, and etomidate also have GABA-receptor agonist activity and so it is a difficult problem to escape. Inhalational anesthetics could be used to anesthetize such patients, but they must be intubated to minimize environmental pollution with the anesthetic. The goal of sedation is simply to take the "edge" off of the hyperexcitability, not necessarily to make it go away entirely or to anesthetize the patient. To this end and specifically because an advantageous effect without a disadvantageous effect of the anesthetic cannot be predicted, it is recommended to start with very small doses, titrated to effect. Comatose or anesthetized patients have lost any ability to take care of themselves and so these patients require the unconscious patient care protocol discussed at the end of this chapter.

Detection of deteriorating neurological status. The neurological status should be reevaluated at frequent, regular intervals. Deteriorating neurological function needs to be recognized early so that therapy can be implemented in a timely fashion.

1. Consciousness is maintained by the ascending reticular formation (in the midbrain and pons of the brainstem) and the cerebral cortex. Dysfunction of either area causes decreased mentation. Degrees of mental depression (when not asleep) may be classified as: (1) normal, alert; (2) mildly obtunded (depressed) but spontaneously aware; (3) moderately obtunded—does not spontaneously care about its environment but responds to loud or brusque external stimuli; responses may be inappropriate (disoriented, confused), or erratic (delirium); (4) severely obtunded and responds only to deep pain stimulation (stuporous); and (5) comatose—no conscious responses (reflexes are present). A loss of vision, menace, and obtundation while maintaining subcortical reflexes (e.g., pupillary light reflex [PLR], dazzle reflex, vestibulo-ocular nystagmus, corneal reflex, eyeball retraction, and palpebral reflex) is suggestive of cerebral disease.

2. The menace reflex is done by pretending to poke the eye with a finger; the animal should blink and may move the head away from the danger. The menace reflex tests the ability of the animal to see the danger (optic nerve), to interpret it (cortical), and to react to it (blink—facial nerve VII). Cerebellar lesions can also interfere with the menace reflex (unilateral cerebellar lesions cause ipsilateral loss of menace) via the corticotectopontocerebellar pathway in the rostral colliculi of

the midbrain without loss of vision. Corticocerebellar disease may result in the loss of the menace reflex while the subcortical corneal and dazzle reflexes remain.

3. Ocular position—both eyes should be looking ahead in the same direction at the same time. Strabismus is caused by lesions of the oculomotor nerve (ventrolateral), the abducens nerve (medial), or trochlear nerve. Strabismus may also occur in some positions with vestibular disease. Strabismus may also be caused by retrobulbar masses. Gaze deviation, when both eyes are looking in the same direction but off to the side, is attributed to severe cerebral injury.
4. Anisocoria is caused by an imbalance between the parasympathetic and sympathetic influences (oculomotor nerve) and by several intraocular diseases. Mydriasis is caused by adrenergic stimulation of the iris dilator muscle and simultaneous inhibition of cholinergic stimulation of the sphincter muscle. Miosis is caused by cholinergic stimulation of the sphincter muscle and inhibition of adrenergic stimulation of the dilator muscle. Direct examination of each eye must be made to rule out ocular causes, such as uveitis, which cause miosis, or glaucoma or retinal disease or atrophy, which cause mydriasis. Asymmetric or bilateral miotic pupils may represent either cerebral or brainstem disease. Mydriatic, nonlight responsive pupils represent irreversible brainstem disease. In general in order of increasing severity of lesion and decreasing prognosis is:

   (1) normal pupil size and PLR; (2) slow PLR; (3) anisocoria; (4) bilateral miosis, responsive to light; (5) pinpoint, unresponsive; and (6) bilateral mydriasis, unresponsive.
5. The pupils should dilate in the dark (or when the eyelid is closed) and should constrict in the light (or when a bright light is shined upon the retina [be sure to check both eyes—direct and indirect]). Observe: (1) latency of the response; (2) the speed of the response; and (3) the magnitude of the contraction. The intensity of the light stimulus is important; use the same bright light source each time. Diseases, such as iris atrophy, glaucoma, posterior synechia, high sympathetic tone or sympathomimetic therapy, or anticholinergic therapy will cause mydriasis, anisocoria, and diminish the pupillary light reflex. Retinal or preoptic chiasmal disease is suspected if pupillary dilation occurs when that retina is exposed to light at the same time that light is removed from the contralateral retina ("swinging flashlight test" or the "cover-uncover test").
6. The dazzle reflex is when one or both eyelids blink in response to the bright light shined upon the retina. It is a subcortical reflex.
7. The oculovestibular reflex (physiologic nystagmus; "doll's eyes") is normal. It occurs only when the head (or head and body) is being rotated; its absence indicates

vestibular nerve, brainstem, medial longitudinal fasciculus, or oculomotor/abducens nerve dysfunction. If the nystagmus continues after the head motion is stopped, vestibular disease should be suspected. Unilateral absence suggests ipsilateral oculomotor or abducens nerve lesion. Absence of the oculovestibular reflex in association with coma suggests brainstem injury.

8. Pathologic nystagmus (spontaneous; positional [occurs when the head position is static]) indicates inner ear, vestibular nerve, and brainstem or cerebellar dysfunction. Horizontal nystagmus when the head is in a normal position is commonly seen in peripheral vestibular dysfunction (fast phase is usually away from the side of the lesion). Vertical nystagmus when the head is in an abnormal position, such as lateral or dorsal recumbency, is more commonly seen in central vestibular, brainstem, or cerebellar dysfunction. Rotary nystagmus is not local.
9. Corneal blink reflex - an animal that blinks spontaneously but not in response to corneal stimulation has a sensory problem but not a motor problem. Head withdrawal requires conscious perception of the stimulus. A blink or head withdrawal before actually touching the cornea is the menace test and requires both vision and conscious perception. It would be better to stimulate the cornea with a puff of air from a syringe rather than direct digital stimulation of the cornea.
10. The palpebral reflex (blink), when the medial canthus is touched, is mediated by the trigeminal nerve (sensory) and the facial nerve (motor). Its absence suggests brainstem disease.
11. The nasal sensation is assessed by passing a cotton-tipped swab into the nose; it should evoke an avoidance response. Its absence suggests brainstem disease.
12. Swallowing and gagging are mediated by the glossopharyngeal and vagus nerves (both are both afferent and efferent). Their absence suggests brainstem disease.
13. Irregular breathing patterns (e.g., tachypnea, Cheyne-Stokes, apneustic, cluster-breathing, bradypnea, or apnea) suggest brainstem injury. Brainstem involvement carries a very poor prognosis. It is heralded by a constellation of signs: unconsciousness, bilaterally unresponsive miotic or mydriatic pupils, absent gag-swallow-laryngeal reflexes, strabismus, absent physiological nystagmus, spontaneous or positional nystagmus, irregular breathing rhythms/apnea, and decerebrate posturing (extensor rigidity of all four limbs and opisthotonus). Compression of the brain stem can result in abrupt changes in respiration, heart rate, and blood pressure, which are often the immediate cause of death.

Management of deteriorating neurological status. The intracranial contents consist of cellular structures (neurons and nerve tracts), interstitium, cerebral spinal fluid, and blood.

They are encased in a nonexpandable vault of bone. An increase in the volume of one must either be accompanied by a decrease in volume of another (usually CSF or blood volume) or an increase in intracranial pressure. The normal brain has some margin for safety in this process; early changes in volume are not associated with much of an increase in pressure. But in short order, the limit is reached such that additional, small, incremental increases in intracranial volume are associated with logarithmic increases in intracranial pressure. Increases in intracranial volume can accrue with cellular swelling (cytotoxic cellular edema); interstitial edema (vasogenic); an increase in cerebral spinal fluid (hydrocephalus); and an increase in blood volume (e.g., venous outflow obstruction, arterial vasodilation, or hypertension); hematoma formation; or neoplasia. The problem with elevated intracranial pressure is that it (1) interferes with neuronal function, (2) diminishes cerebral blood flow (CBF), and eventually, (3) causes subtentorial herniation of the cerebrum or foramen magnum herniation of the cerebellum, causing brainstem compression.

Cerebral perfusion pressure (CPP) is the difference between mean arterial blood pressure and intracranial pressure. For reasons of maintaining adequate CPP, it is commonly recommended to maintain the mean arterial blood pressure above 60 mm Hg in patients with normal ICP and more than 90 mm Hg in patients with suspected elevated ICP. Some patients with intracranial disease develop a spontaneous systemic hypertension and it is usually assumed to represent a compensatory response to intracranial hypertension and is not treated.

Support of respiration and circulation are the first priority in the patient with deteriorating neurological status. Good physiological management of the patient is of paramount importance. Avoid events and drugs that increase intracranial blood flow (e.g., aggressive fluid therapy, and drugs - á2-agonists and ketamine), cause cerebral vasodilation (e.g., hypercapnia, hypoxemia, hyperthermia, and drugs—inhalational anesthetics), or decrease outflow (e.g., jugular vein obstruction, aggressive positive pressure ventilation, and head-down positioning). The bloodbrain barrier of the cerebral capillaries prevents the movement of most particles, including sodium, from the vascular to the interstitial spaces.

Isotonic ECF replacement solutions are effective in supporting cerebral blood flow without increasing cerebral interstitial volume. The benefits of rapid, adequate fluid resuscitation in patients suffering from brain trauma have been demonstrated. While fluid overload is to be prevented, the practice of keeping head trauma patients "on the dry side" is not supported by current clinical and experimental evidence. Hypertonic saline and mannitol are both effective in decreasing intracranial pressure because they create osmotic gradients from the intracellular to the vascular space. The recommended dose of mannitol is 0.5 g/kg administered over 20 minutes.

Mannitol should be administered as intermittent 20-minute infusions and not as a continuous infusion. The rheological effects of mannitol and its beneficial effects upon CBF and CPP make it a useful part of the overall fluid therapy plan. The goal is to achieve a serum osmolality that exceeds calculated osmolality (2[Na+K] + BUN/2.8 + glucose/18) by about 20 mOsm/kg.

Hypertonic saline (4 to 6 mL/kg) is also effective at creating an osmotic gradient in the brain. Glucose containing solutions should be avoided unless they are specifically indicated to treat hypoglycemia. Hyperglycemia and the administration of glucose containing solutions (whether or not they produce hyperglycemia) have both been associated with worse neurological outcomes in brain-injured patients. Corticosteroids have only been found to be beneficial in vasogenic cerebral edema secondary to neoplasia. They have not been found to be useful in cytotoxic cerebral edema or traumatic brain injury. Deep barbiturate anesthesia reduces cerebral blood flow and metabolic oxygen consumption and may be protective in some instances of traumatic or hypoxic brain damage, but is not recommended for the poisoned patient. Moderate hyperventilation (to a $PaCO_2$ of about 25 mm Hg) results in vasoconstriction and reduces cerebral blood flow to areas of the brain with intact reactive vasoactivity. Extreme hypocarbia ($PaCO_2 < 20$ torr) causes cerebral vasoconstriction and compromises brain oxygenation. Damaged vessels lose the ability to respond to carbon dioxide, and blood flow to these regions may actually increase when the patient is hyperventilated secondary to a decrease in blood flow to the normal regions. Hyperventilation is not normally recommended but may represent a useful adjunct for the acute management of patients with deteriorating neurological function. Any beneficial effect of hyperventilation is, however, transient as the brain adjusts its internal pH after just a few (4) hours. Systemic or regional cooling has been experimentally demonstrated to reduce the magnitude of neuronal degeneration associated with traumatic or hypoxic brain injury. However, systemic cooling generates its own array of problems and is not recommended.

Hyperthermia, however, should be prevented. Reactive oxygen metabolite inhibitors (e.g., calcium channel blockers, allopurinol, super oxide dismutase and catalase, deferoxamine), scavengers (e.g., mannitol, dimethylsulfoxide), and membrane protectants (e.g., corticosteroids, 21-aminosteroids, vitamins E and C) have been demonstrated to have beneficial effects in the laboratory models of traumatic and hypoxic brain damage, but clinical efficacy remains to be determined. There is not enough known about the mechanisms of excitotoxic or inflammatory cascade-mediated neuronal damage to provide therapeutic recommendations.

Gastrointestinal help

Many poisons and toxins are associated with vomiting and diarrhea. Diarrhea may be caused by irritant or corrosive intoxicants, dietary indiscretion (moldy foodstuffs), arsenic, detergents, paraquat, heavy metal intoxication (e.g., iron, lead, zinc), mushroom poisoning, and by a host of other diseases not associated with poisoning or intoxication (e.g., infectious, food allergies, foreign bodies). Diarrhea is associated with an alkaline, crystalloid loss containing variable amounts of albumin and red blood cells. The gastrointestinal irritation or inflammation may be associated with abdominal pain and the translocation of endotoxin and bacteria into the systemic circulation. Symptomatic therapy includes activated charcoal, sucralfate, empirical antibiotics therapy with gram negative and anaerobic coverage, and appropriate fluid and electrolyte restoration.

Motility modifiers (opioids and anticholinergics) and antisecretory drugs are not generally indicated.

Vomiting. Vomiting is the active, forceful elimination of stomach contents caused by contraction of the abdominal muscles. The "vomiting center" in the reticular formation of the brainstem is stimulated by the chemoreceptor trigger zone, by ascending vagal or sympathetic influences, by the vestibular apparatus, or by cerebral cortex influences. Humoral substances, such as uremia and toxemias; digitalis, opioids, and other drugs; infections; and motion sickness primarily stimulate the chemoreceptor trigger zone. Mechanical stimulation/inflammation of the gastrointestinal tract, peritoneum, genitourinary tract, heart and lungs, and liver, activates the ascending vagal and sympathetic afferent nerves. Vomiting is associated with antiprostaglandins, ethylene glycol, mushrooms, organophosphate and carbamate insecticides, pyrethrins, permethrins, arsenic, iron, lead, zinc, phenols, detergents, disinfectants, bleach, zinc phosphide, chlorinated hydrocarbons, crayons, and organic solvents.

Vomiting sets the stage for the aspiration of material into the airways, specially in obtunded patients. This may be associated with physical obstruction of the airway, tracheobronchitis, and bronchopneumonia secondary to bacterial contamination of the respiratory tract, and, if the pH of the aspirant is less than about 2, an immediate contact necrosis of the airway epithelium occurs. If the vomiting is prolonged, it will cause dehydration and electrolyte abnormalities. Animals vomiting only gastric fluids tend to be hypernatremic, hypochloremic, hyperbicarbonatemic, and hypokalemic. Most dogs and cats, however, reflux considerable amounts of duodenal secretions into the stomach and lose a net alkaline solution.

These animals are generally hypernatremic, hyperchloremic, hypobicarbonatemic, and hypokalemic. These "duodenal vomiters" are by far the most common and can be verified by the observation of greenish or yellowish coloration in the vomitus or by measuring the pH of the vomitus. Animals that have been drinking water in association with the vomiting are commonly hyponatremia.

When the cause of the vomiting is not known and it is protracted, symptomatic therapy is indicated. Antiemetics should be administered on a trial basis and in sensible combinations until an effective therapy is established. Central and peripheral receptor activity responsible for mediating vomiting include D2-dopaminergic, á2-adrenergic,

5-HT3- serotonergic, M1- cholinergic, H1&2-histaminergic, and ENK-enkephalinergic. Many antiemetic drugs exhibit multiple receptor activity and therefore have multiple mechanisms (and sites) of potential effectiveness. An empty stomach and forward gut motility are important aspects of therapy Drugs that encourage gastric emptying include metoclopramide, erythromycin, domperidone, and bethanechol. Metoclopramide is a D2-dopaminergic and 5-HT3-serotonergic receptor antagonist that inhibits vomiting induced by gastrointestinal irritation and the chemoreceptor zone stimulation. The drug has some sedative effects and may cause extrapyramidal signs, disorientation, and excitement (especially in high doses). It perhaps should not be used in patients with seizure disorders. Metoclopramide also enhances gastric emptying by peripheral cholinergic sensitization and does so without stimulation of gastric, pancreatic, or biliary secretions. The drug increases lower esophageal sphincter tone (inhibiting gastric reflux) and relaxes the pyloric sphincter tone and increases duodenal and jejunal peristalsis (but not colon). It may cause abdominal pain and diarrhea. It should not be used in patients with intestinal obstruction. The dosage is 0.2 to 0.5 mg/kg q 6 to 8 hours, PO, SC, or IM, or 0.05 to 0.1 mg/kg/hr as an IV infusion. Prochlorperazine is a D2-dopaminergic, á1&2-adrenergic, M1-cholinergic, and H1&2-histaminergic receptor antagonist with tranquilizer, antiemetic, and vasodilatory effects. It may cause hypotension, sedation, and extrapyramidal signs (tremors and incoordination). It is generally considered to be a very effective antiemetic. The dose is 0.2 to 0.4 mg/kg q 8 hours, SC or IM. Diphenhydramine is an H1-receptor antagonist with sedative, anticholinergic, antitussive, and antiemetic effects. It may cause drowsiness, dry mouth, and excitation. The dosage is 2 to 4 mg/kg q 8 hours, PO or IM. Ondansetron is a 5-HT3-serotonergic antagonist that inhibits vomiting by actions on the chemoreceptor trigger zone and the vagal afferents. Side effects include sedation and head shaking. The dosage is 0.5 to 1 mg/kg PO or 0.5 to 1 mg/kg PO q 12 to 24 hours. Domperidone is a D2-dopaminergic antagonist that inhibits vomiting by actions on the gastrointestinal smooth muscle. It has both prokinetic and antiemetic activity (chemoreceptor trigger zone). It does not cross the bloodbrain barrier and is therefore devoid of other CNS effects (drowsiness, excitation, and extrapyramidal signs). It has no cholinergic activity and is not inhibited by atropine. The dosage is 0.1 to 0.3 mg/kg q 12 hours, IM. Scopolamine is an M1- cholinergic antagonist that inhibits vomiting by actions in the vestibular apparatus and chemoreceptor trigger zone. It also is an antagonist for M2&3-cholinergic antagonist and therefore inhibits gastric emptying and gut peristalsis. Side effects include dry mouth, gastrointestinal stasis, and drowsiness. The dosage is 0.03 mg/kg q 6 hours, SC or IM. Yohimbine is an á2-adrenergic antagonist, which inhibits vomiting by actions in the chemoreceptor trigger zone and emetic center. It may cause excitation and sedation. The dosage is 0.1 to 0.5 mg/kg q 12 hours, SC or IM. Sucralfate reacts with hydrochloric acid to form a complex that binds to proteinaceous exudates at ulcers and protects the site from further damage by pepsin, acid, or bile.

It may also stimulate prostaglandin E2 and I2 activity and therefore has a cytoprotective effect similar to misoprostol. It does not alter acid, trypsin, or amylase secretion. Sucralfate decreases the bioavailability and absorption of other drugs and may cause constipation. The dose is between 0.25 and 1 g for a small dog or cat to a large dog, respectively. Cimetidine, ranitidine, and famotidine block the H2 receptor on the basolateral membrane of the parietal cells of the stomach. They decrease acid production and do not alter gastric emptying time, lower esophageal or pyloric sphincter tone, nor pancreatic or biliary secretion. Cimetidine inhibits P450 microsomal enzyme function in the liver and may alter the metabolic rate of other commonly used drugs (e.g., â- blockers, calcium-channel blockers, diazepam, metronidazole, acetaminophen). Cimetidine binds H2 receptors on red cells and platelets and may be associated with anemia and thrombocytopenia. Cimetidine crosses the bloodbrain barrier and may be associated with mental confusion/depression. All these agents, by virtue of increasing gastric fluid pH, may allow repopulation of the stomach and mouth with potentially pathogenic organisms, which in turn predispose the animal to nosocomial pneumonia.

The dosage of cimetidine is 5 to 10 mg/kg PO, IV, IM q 6 to 8 hours; the dosage of ranitidine is 0.5 to 2 mg/kg PO, IV, IM q 8 to 12 hours; and the dosage of famotidine is 0.1 to 0.2 ng/kg PO, IV, SC, IM q 12 hours. Omeprazole is a gastric acid proton pump inhibitor. In an acid environment, it is activated to a sulfonamide, which binds irreversibly to the H+/K+ exchange ATPase enzyme on the secretory surfaces of the parietal cells. Recovery from the drug's effects depends upon the synthesis of new H+/K+ ATPase protein. The drug also inhibits the cytochrome P450 mixed function oxidase system in the liver and therefore inhibits metabolism of a variety of other drugs (sedatives and anesthetics). It may cause abdominal cramping, vomiting, or diarrhea. The dosage is 0.5 to 1 mg/kg PO q 24 hours.

Antacids have largely been supplanted by sucralfate and the H2-blockers because the latter work better and have fewer side effects. Antacids neutralizme gastric acid, inhibit the proteolytic activity of pepsin, and have a local astringent effect. They may be used as a supplement to the other drugs and are still recommended for the therapy of hyperphosphatemia in renal failure; they may cause hypophosphatemia in nonrenal failure patients. Magnesium products should not be used in renal failure Some products contain significant amounts of sodium and potassium and should be used with caution in patients with hypernatremia or hyperkalemia. Aluminum products may delay gastric emptying. Aluminum and calcium products may cause constipation while magnesium products may cause diarrhea. These drugs should also be given remote to other oral medicants. The dose of aluminum, calcium, or magnesium hydroxide is 1 to 10 mL, depending upon the size of the animal. Misoprostol directly inhibits gastric parietal cell acid secretion and is cytoprotective through the increased secretion of gastric mucus and bicarbonate. The drug also enhances mucosal defense mechanisms and healing of acid-induced injuries. It is specifically therapeutic for the GI complications of antiprostaglandin therapy, but does not interfere with the antiinflammatory/ analgesic effects of those drugs. Misoprostol enhances uterine contractions and specifically should not be used in a pregnant animal. It also enhances GI motility; cramps, diarrhea, or vomiting may be a problem. The dosage of misoprostol is 1 to 5 μg/kg PO q 8 hours.

Renal aid

Acute renal failure can occur as a direct result of some poisons and intoxications (e.g., ethylene glycol, vitamin D rodenticides, some mushrooms, some snake or spider envenomation, lead, and zinc) or as an indirect consequence of the systemic hemodynamic and inflammatory effects. Urine flow is an indirect measure of renal blood flow and renal blood flow is an indirect measure of visceral blood flow. Maintaining visceral blood flow and urine output is an important aspect of managing the poisoned or intoxicated patient. Renal blood flow is optimized by restoring and maintaining an effective circulating blood volume. Urine output can be assessed by serial palpation of the urinary bladder or by actual measurement following aseptic placement of a urinary catheter. Normal urine output should be about 1 to 2 ml/kg/hr. If oliguria or anuria persists after restoration of an effective circulating volume, a diuretic should be administered: furosemide (0.5 to 5 mg/kg bolus ± 0.1 to 0.5 mg/kg/hr) or mannitol (0.5 g/kg bolus ± 0.1 g/kg/hr).

Statistically, diuretic therapy does not prevent acute renal failure, but it does facilitate the medical management of the case. Oliguria or anuria leads to fluid and electrolyte retention (e.g., edema, hyperkalemia, and metabolic acidosis), and failure to excrete normal metabolic products (azotemia) and the poison or toxin. Many poisons and toxins are weak acids or bases.

Modification of urine pH, by increasing the ionized component of the product, can facilitate its elimination. Acidification of the urine can enhance the excretion of weak bases, such as amphetamine, phencyclidine, and strychnine. Ammonium chloride is administered at a dosage of 25 to 50 mg/kg q 6 hours in the dog and 20 mg/kg q 12 hours in the cat. Urine pH should be monitored to assure adequate urine acidification and plasma ammonia should be monitored to prevent hyperammonemia. Alkalinization of the urine can enhance the excretion of weak acids, such as salicylates, ethylene glycol, and barbiturates. Sodium bicarbonate is administered at a dosage of 1 to 2 mEq/kg q 4 hours. Urine and plasma pH should be monitored.

Hepatic aid

The liver plays an important role in the detoxification of exogenous and endogenous poisons and toxins. It is also a primary target of several poisons and toxins (e.g., acetaminophen, mothballs, and mushrooms). Hepatocellular damage and cholestasis can be ascertained by routine measures of hepatic enzymes and function tests (e.g., alanine aminotransferase, aspartate aminotransferase, alkaline phosphatase, and bilirubin). There are several important consequences of acute liver failure. Hypoglycemia, hypoproteinemia, and coagulopathy are discussed below.

Hyperammonemia may be associated with an encephalopathy. Acute renal failure may occur secondary to acute liver failure, but the mechanism is unknown.

HEMATOLOGICAL HELP

Hemoglobin. Anemia is associated with a number of poisons and toxins caused by intravascular hemolysis (e.g., bee stings, some snake or spider envenomation, anionic detergents, and zinc), extravascular hemolysis (e.g., mothballs and onions), gastrointestinal blood loss (e.g., antiprostaglandins, iron, arsenic, phenols, detergents, disinfectants, bleach, chlorinated hydrocarbons, and organic solvents) or the lack of red cell production (chronic lead poisoning). Anticoagulant rodenticide intoxications and disseminated intravascular coagulation can be associated with bleeding anywhere in the body. Historically, in humans, the trigger for a hemoglobin transfusion has been a hemoglobin concentration of 10 g/dL (a packed cell volume [PCV] of 30%). Recent studies in humans have suggested that a more relaxed trigger of 7 g/dL (PCV = 21%) is associated with at least as good, and perhaps better, morbidity and mortality statistics. In veterinary medicine, in animals with immune-mediated hemolytic anemia, it is well accepted to withhold blood transfusions until the hemoglobin concentration is below 5 g/dL (PCV = 15%). In human medicine, in Jehovah Witness patients, mortality rate does not increase significantly until the hemoglobin concentration is below 5 g/dL (PCV = 15%). There are many examples of human and veterinary patients surviving much greater levels of anemia. It may not actually be possible to define a minimum hemoglobin concentration, given the complexities of cardiac output and oxygen extraction compensatory mechanisms. An animal can tolerate greater degrees of anemia if it has the wherewithal to increase cardiac output. Metabolic markers of tissue oxygenation may help guide the need for hemoglobin transfusions. Blood may need to be administered in volumes of 10 to 30 mL/kg in the dog, depending upon the magnitude of anemia. Cats have a smaller blood volume (50 to 55 mL/kg) and bolus doses of all fluids should be approximately 50% of canine recommendations. The amount of blood to administer can also be calculated: (desired PCV − current PCV) × BWkg × 2 mL whole blood (or 1 mL packed red blood cells).

Oncotic pressure. Plasma oncotic pressure is an important vascular fluid retention force. When depleted, there is an increased risk of interstitial edema, but because of an offsetting decrease in perimicrovascular oncotic pressure, hypoproteinemia is not as edemagenic as might be expected. An increased capillary hydrostatic pressure and vascular permeability are in contrast potent causes of edema. Colloid osmotic pressure (COP) can be measured;

values in normal animals are 20 to 25 mm Hg. Values of 15 to 20 mm Hg are common in critically ill patients but are not thought to be of important concern. Values in the low teens should trigger therapy and values in the single digits should cause great concern. COP can be qualitatively approximated from an albumin measurement (albumin normally accounts for about 70% of the COP). Albumin values in normal dogs and cats are 2.9 to 4.2 and 1.9 to 3.9 g/dL, respectively. A 50% decrease in albumin is associated with about a 50% reduction in COP. The cheapest way to augment COP is the administration of an artificial colloid, such as dextran 70 or hetastarch, in bolus doses (if volume augmentation is also desirable) of 10 to 30 mL/kg or continuous infusions of 1 to 2 mL/kg/hr.

Plasma may be indicated if there are concurrent coagulation issues and whole blood may be indicated if there are concurrent hemoglobin issues.

Coagulation. Coagulopathies can be caused by coagulation or platelet problems. Coagulation is assessed by in vitro tests, such as partial thromboplastin time (PTT) (normal values are laboratory-dependent: approximately 9 to 18 seconds), prothrombin time (PT) (normal values are laboratory dependent: approximately 4 to 9 seconds), activated clotting time (ACT) (<120 seconds at 37° C), and whole blood clotting time (<4 minutes at 37° C; 8 minutes at room temperature).

The PIVKA test assesses for proteins induced by vitamin K antagonists (normal = 15 to 18 seconds). Elevated fibrin degradation products represent activation of the clotting/ fibrinolytic cascades; elevated d-dimer represents fibrinolysis. Decreased antithrombin (normal = 80% to 140% in the dog) may be indicative of a protein losing "-opathy" and a prothrombotic state, or may represent consumption and disseminated intravascular coagulation (DIC). Platelet numbers can be assessed with a platelet count or a platelet screen on a blood smear (normal = 12 to 25 per oil emersion field, in a good blood smear without platelet clumping; platelet count is estimated as 15,000 × number of platelets per oil emersion field). Platelet function can be assessed by examining for petechia or a buccal mucosal bleeding time (normal = <4 minutes). Thrombo- elastography provides an integrated assessment of clot formation that can be used to assess for hyper coagulopathy or hypocoagulopathy.

Coagulopathies may or may not need to be treated. If bleeding is minor, not into a vital organ, and blood can easily be replaced by transfusion, specific therapy may not be necessary. Specific treatment with fresh plasma is necessary if platelets are required; fresh frozen plasma is used if platelets are not required, but labile factors such as vonWillebrand's factor, factor 8, or antithrombin are required. For vitamin K antagonist poisoning, any plasma will suffice. The goal of plasma therapy is to stop the bleeding. Laboratory tests can be improved, but it is not the objective of therapy to push them to normal. Thrombocytopenia cannot be markedly improved with plasma therapy; this would be very expensive to try, and would probably not be possible as a result of hypervolemia.

Glucose. An adequate level of blood glucose is important for cerebral metabolism. Hypoglycemia might occur during hepatic failure or sepsis; hyperglycemia is also a common nonspecific hormonal response to stress. A blood glucose <60 mg/dL should be treated with a 2.5% to 5% glucose infusion. Severe hypoglycemia should be treated, in addition, with a bolus of glucose (0.1 to 0.25 g/kg). There is growing evidence that persistent moderate hyperglycemia (>200 mg/dL; >11 mM/L) in the intensive care setting is associated with a significantly poorer outcome. In this setting, it has been recommended to enforce glycemic control with insulin in quantities sufficient to maintain the blood glucose concentration below 150 to 200 mg/dL (8 to 11 mM/L).

Metabolic acid-base status. Lactic metabolic acidosis results from inadequate tissue oxygenation. The marker for metabolic acidosis is a decreased bicarbonate concentration (normal = 20 to 24 mEq/L in the dog; 18 to 22 mEq/L in the cat); a decrease in total carbon dioxide concentration (normally 1 to 2 mEq/L higher than bicarbonate); or an increase in the base deficit (normal = 0 to “4 mEq/L in the dog, “3 to “7 in the cat). Lactate is the marker for lactic acidosis (normal = <2 mM/L), which is usually presumed to represent inadequate tissue oxygenation, but can also be elevated as a result of catecholamine-stimulated Na-K-ATPase activity. Mild to moderate metabolic acidosis does not need to be treated specifically; correction of the underlying problem should suffice. Severe metabolic acidosis (a pH <7.20) may benefit from therapy with sodium bicarbonate: desired base deficit “ measured base deficit × BWkg × 0.3. These doses of bicarbonate should be administered over a period of at least 20 minutes.

Sodium. Sodium concentration is an expression of the relative numbers of sodium molecules to water molecules in the ECF (extracellular fluid). Sodium concentration is important to transcellular fluid flux. Abnormalities in sodium concentration (hyponatremia and hypernatremia) may occur in any combination with abnormalities in ECF sodium and water content (dehydration and edema). Abnormalities in sodium concentration can usually be attributed to changes in free water. Free water must be added to reduce a sodium concentration (hyponatremia = free water excess) and be taken away to increase it (hypernatremia = free water deficit). Free water may be gained by drinking water, may be administered in the form of 5% dextrose in water, or may occur secondary to high levels of antidiuretic hormone. Free water may be lost via evaporation (lungs and skin) or by losing fluids that are low in sodium compared with normal ECF (diarrhea, vomitus, and urine). Abrupt changes of sodium concentrations of more than about 15 to17 mEq/L (in either direction) should be avoided because they may be associated with untoward transcellular water shifts and unfavorable neurological consequences. The combination of hypernatremia, hypervolemia, and edema can be caused by excessive sodium intake or administration, or hyperadrenocorticism. The combination of hypernatremia and normovolemia can be caused by diabetes insipidus or a reset osmostat, or may be iatrogenic. The combination of hypernatremia, dehydration, and hypovolemia can be caused by water loss in excess of sodium (e.g., vomiting, diarrhea, osmotic diuresis, or insensible water losses), or insufficient free water intake.

Hypernatremia causes ECF hyperosmolality and intracellular dehydration. The cells first manifesting signs of dehydration are those of the central nervous system (e.g., depressed mentation, restlessness, irritability, muscle twitching/tremors, hyperreflexivity, muscle rigidity/spasticity, ataxia, myoclonus, tonic spasms, and coma). Tissue shrinkage can cause intracranial hemorrhage. Acute hypernatremia (<6 hours) should be treated by the rapid administration of free water (in the form of 5% dextrose in water). In time (12 to 24 hours) the intracellular compartment increases its intracellular osmoles to offset the effects of the extracellular sodium aberration and to restore intracellular water volume toward normal. Chronic hypernatremia (>24 hours) must be treated with caution so as to lower the sodium no faster than 1 mEq per hour so as to prevent water intoxication. The clinical signs of water intoxication are acute obtundation. Volume problems should initially be treated with a solution that has a sodium concentration that is close to that of the patient. The sodium concentration is then slowly decreased by administering 5% dextrose in water at the rate of about 3.7 mL of water/kg of body weight/h. The combination of hyponatremia, hypervolemia, and edema may be caused by heart failure or may be iatrogenic. The combination of hyponatremia and normovolemia may be caused by inappropriate antidiuretic hormone (ADH) secretion, primary polydipsia, or a reset osmostat. The combination of hyponatremia, hypovolemia, and dehydration may be caused by appropriate ADH secretion (hypovolemia and dehydration), natriuretic diuretics, hypoadrenocorticism, vomiting associated with water drinking, cerebral salt wasting syndrome, or whipworm infection. The sodium concentration can also be lowered by hyperglycemia. Hyponatremia causes intracellular edema. Hyponatremia has been associated with obtundation, anorexia, muscle weakness and wasting, and gastrointestinal signs. Common coexisting electrolyte problems include hypochloremia, hyperkalemia, and hyperphosphatemia. Mild hyponatremia requires no special consideration beyond therapy directed to the underlying disease process and volume restoration with any ECF replacement solution. Acute severe hyponatremia (<6 hours) should be treated by the rapid administration of saline. In time (12 to 24 hours) the intracellular compartment decreases its intracellular osmoles to offset the effects of the extracellular sodium aberration and to restore intracellular water volume toward normal. Chronic hyponatremia (>24 hours) must be treated with caution so as to raise the sodium level no faster than 0.5 mEq per hour to avoid myelinolysis. Volume problems should initially be treated with a solution with a sodium concentration that is close to that of the patient. The sodium concentration is then slowly increased by administering a hypertonic saline solution at a rate of about 0.3 mEq of sodium/kg/hr. The clinical signs of myelinolysis occur 2 to 7 days after correction of severe hyponatremia and include spastic quadriparesis, facial palsy, dysphagia, vocal dysfunction, mental confusion, and coma.

Potassium. Most of the potassium in the body is located in the intracellular fluid compartment (140 mEq/L), whereas very little of it is located in the extracellular fluid compartment (4 mEq/L). Repolarization of electrically excitable cells is largely attributed to the rapid efflux of potassium. Resting membrane potential is determined by the equilibrium between potassium moving out of the cell in response to the intracellular-to-extracellular potassium gradient and potassium moving back into the cell in response to the extracellular-to-intracellular electronegativity. Hyperkalemia is primarily caused by oliguric/anuric renal disease and hypoadrenocorticism, and may be iatrogenic, but may also be caused by rhabdomyolysis, metabolic (inorganic)/respiratory acidosis, or periodic familial hyperkalemia. It may also be falsely elevated if the blood sample is not analyzed for a period of time because of hemolysis or from platelet or white cell degradation.

Hyperkalemia causes membrane hypopolarization, which may result in extrasystoles/ fibrillation if the resting membrane potential is slightly more negative than threshold potential or asystole when the resting membrane potential is slightly less negative. Hyperkalemia also increases potassium permeability, which augments the repolarization phases of the electrocardiograph (tall, tented, narrow T wave) and diminishes the depolarization phases (small P waves, prolonged P-R intervals, bradycardia, and widened QRS complexes). Hyperkalemia may also be associated with peripheral muscle weakness, decreased contractility, and weak pulse quality. Finally, there is a blending of the QRS and T waves (a sinusoidal pattern), hypotension, and either ventricular asystole or fibrillation. Life-threatening hyperkalemia, defined by severe electrocardiographic disturbances, should be treated specifically. Calcium (0.2 mL of 10% calcium chloride or 0.6 mL of 10% calcium gluconate per kilogram of body weight), administered intravenously, by virtue of its effect on membrane threshold potential, antagonizes the effect of hyperkalemia and immediately returns the electrical performance toward normal. The effects of calcium are, however, short-lived, lasting only until the calcium is redistributed. Insulin and glucose (0.1 to 0.25 units of regular insulin/kg, administered as an intravenous bolus, and 0.5 to 1.5 g of glucose/kg, administered as an intravenous infusion over 2 hours) is the common treatment for hyperkalemia. Bicarbonate will also cause the intracellular redistribution of potassium if it is going to be administered for acidosis. Hypokalemia is primarily attributed to excessive abnormal losses (e.g., vomiting, diarrhea, and diuresis), dehydration (aldosterone-mediated renal losses), and lack of intake. It may also be caused by hypochloremia, hyperadrenocorticism, metabolic (inorganic) respiratory alkalosis, bicarbonatetherapy, â2-agonist administration, and familial periodic hypokalemia. Hypokalemia causes membrane hyperpolarization (electrical paralysis) and decreases potassium permeability (diminishes repolarization processes and enhances depolarization processes). Hypokalemia is associated with general muscle weakness (e.g., skeletal, gastrointestinal, and myocardial) and may be associated with ECG changes opposite to those of hyperkalemia (although the changes are not as characteristic as they are with hyperkalemia): flattened T wave, U waves (a positive deflection following the T wave), elevated P wave, increased R wave amplitude, and depressed S-T segment. Hypokalemia is also associated with CNS depression and an impaired ability of the renal nephrons to concentrate urine. A severely hypokalemic patient needs to be potassium loaded. As a general rule, potassium can be administered at rates up to 0.5 mEq/kg/hour. Moderately hypokalemic patients should be treated by supplementing the potassium concentration in the administered fluids to 20 to 50 mEq/L, depending upon the magnitude of the hypokalemia.

Calcium. Plasma calcium exists in three forms: ionized (55%), nonionized, chelated form (10%), and albumin-bound (35%). The ionized is physiologically the most important and the form that is regulated by the body. Normal total calcium concentrations in the dog and cat are about 9 to 11 mg/dL (2.2 to 2.8 mM/L); ionized calcium concentrations are about 1.1 to 1.4 mM/L (4.4 to 5.6 mg/dL). Hypocalcemia lowers (more negative) threshold transmembrane potentials and increases the excitability of the nervous system and muscles. This may be manifested by muscle tremors- fasciculations-twitching, muscle contractions, cramps and tetany, disorientation, restlessness, hypersensitivity to external stimuli, paresthesia and facial rubbing, panting and hyperthermia, prolapse of the third eyelid, arrhythmias, and hypotension. Hypocalcemia may be attributed to ethylene glycol poisoning, hypoparathyroidism, malabsorption syndromes, liver or renal disease, hyperphosphatemia, the administration of citrated blood products or EDTA, or necrotizing pancreatitis. Hypo(ionized)calcemia can be potentiated by sodium bicarbonate therapy, by in vitro heparin anticoagulation of the blood sample, and is commonly observed with hypothermia. There is no broad agreement as to when hypocalcemia should be treated, but as a general guideline, ionized concentrations below 0.75 mM/L should be treated. Calcium gluconate can be administered as a bolus (0.5 mL of the 10% solution [9.3 mg or 0.47 mEq per mL/kg] or as an infusion of 0.5 to 1.5 mL of the 10% solution/kg/h). Hypercalcemia impairs the function of most cells in the body by decreasing the threshold potential (less negative transmembrane potential) for excitable cells (making them less excitable and slows conduction), by increasing the contractile state of smooth and skeletal muscle, by increasing ATP use by cell membrane and endoplasmic reticulum membrane calcium pumps, and by interfering with ATP production associated with the mitochondrial accumulation of the calcium. This is manifested clinically by obtundation, poor diastolic heart function, increased arteriolar vasomotor tone, impaired nephron concentrating ability, lethargy and muscle weakness, arrhythmias, muscle twitching, and seizures. Chronic hypercalcemia is also associated with gastric hyperacidity and vomiting, and calciuresis, urolithiasis, and renal failure. Hypercalcemia can be caused by hyperparathyroidism, vitamin Drodenticides, lymphoid and anal sac apocrine gland malignancy, and chronic renal failure. The mainstay of hypercalcemia treatment is effective therapy of the underlying disease process.

Hypercalcemia should be symptomatically treated with volume support and saline diuresis. The latter can be augmented by furosemide; thiazides should be avoided. Sodium bicarbonate therapy will decrease the ionized calcium concentration. Corticosteroids may lower serum calcium if it is elevated because of neoplasia, hypoadrenocorticism, vitamin D rodenticides, or granulomatous disease. Corticosteroids can decrease intestinal calcium resorption, increase renal excretion of calcium, and decrease bone demineralization. Life-threatening hypercalcemia could be treated with chelating agents, such as sodium or potassium phosphate (0.25 to 0.5 mM/kg IV over 4 hours), EDTA (50 mg/kg/h IV to effect), sodium citrate, calcitonin, pamidronate disodium, or calcium-channel blockers. Peritoneal or hemodialysis could also be used to remove calcium from the body.

Magnesium. Magnesium is a cofactor for many intracellular enzyme systems and metabolic pathways, including formation and degradation of ATP, DNA synthesis and transcription, and nucleic acid polymerization. Magnesium exists in the plasma in three forms: ionized (55%), nonionized, chelated (15%), and bound to albumin (30%). The ionized fraction is the biologically active form. Normal total magnesium concentrations in the dog and cat are about 1.7 to 2.5 mg/dL (0.7 to 1 mM/L); ionized magnesium concentrations are about 0.8 to 1.3 mg/dL (0.3 to 0.55 mM/L). Hypomagnesemia lowers the threshold potential (more negative) for excitable cells, which increases their excitability, enhances the release of calcium from stores in the endoplasmic reticulum, facilitates the release of neurotransmitters, inhibits the Na-K-ATPase membrane pump, enhances the leakage of potassium from the cell (eventually raises resting membrane potential toward or beyond threshold potential), and is generally associated with widespread impaired cellular function. Hypomagnesemia is manifested by neural and neuromuscular excitability: hyperexcitability and noise hypersensitivity, muscle twitching, fasciculations, spasms, tetany, and eventually coma and muscle paralysis. Hypomagnesemia may be associated with refractory hypokalemia, hypophosphatemia, hyponatremia, and hypocalcemia. It may also be associated with ventricular arrhythmias; ECG changes may be similar to those of hyperkalemia.

Hypomagnesemia may be caused by malnutrition and malabsorption, diuresis (e.g., loop diuretics, osmotics, thiazides, and saline), aminoglycoside and penicillin antibiotics, cisplatin, cyclosporine, diabetic ketoacidosis, citrate-anticoagulated blood, and lactation. Hypomagnesemia should be treated if it is symptomatic, if the serum magnesium concentrations are less than 1 mg/dL (0.4 mM/L) total, or less than 0.2 mM/L (0.45 mg/

dL), ionized. An initial dose of magnesium sulfate (0.2 mEq/kg) can be slowly administered to see if clinical signs improve. Magnesium sulfate can be administered at a daily dosage of 0.25 to 1 mEq/kg/day (3 to 12 mg/kg/day) administered as a continuous rate infusion. There are many oral magnesium supplements available for longer term supplementation. Hypermagnesemia is associated with diminished neuromuscular transmission: skeletal muscle weakness and paralysis, respiratory depression, vasodilation, and hypotension. It may also be associated with heart block and coma. Hypermagnesemia can be iatrogenic or caused by the administration of magnesium-containing antacid or cathartics. Hypermagnesemia may be treated with aggressive fluid and diuretic therapy, and can be antagonized with 0.5 to 1 mL of 10% calcium gluconate/kg/h (0.45 to 0.9 mg calcium/kg/h).

GENERAL PRINCIPLES IN THE CARE OF CRITICALLY ILL PATIENTS

Infection control. All critically ill patients are susceptible to nosocomial infections. They are nonspecifically stressed by their disease process; they may have diseases, such as viral infections, diabetes mellitus, hyperadrenocorticism, or uremia, which lower their resistance to infection; they have high levels of immunosuppressive endogenous or perhaps exogenous corticosteroids; they are often subjected to invasive surgical diagnostic or therapeutic procedures and indwelling cannulas; they often suffer varying degrees of malnutrition; they may be receiving antibiotics, which predispose them to resistant bacterial or fungal overgrowth; they are surrounded by other infected patients and to environments contaminated with virulent organisms; and they are minimally mobile. The emphasis of nursing care protocols should be in maintaining asepsis. The patient and the immediate vicinity should be clean and antiseptic. All surgical, diagnostic, or therapeutic procedures should be completed under strict aseptic conditions using properly sterilized equipment. All indwelling cannulas should receive regular insertion site care. All fluids administered to the patient should be sterile; all fluids drained from the patient should be collected in sterile containers, which are completely closed to the atmosphere. All administration and collection aparatus should be changed at regular intervals. All mechanical therapeutic equipment, to which the patient is attached, should be properly sterilized and changed at regular intervals.

Immobile patients should be repositioned regularly (every few hours) and convalescing patients should be encouraged to ambulate early to minimize the accumulation of respiratory secretions in lower lung regions that predisposes to pneumonia. Personnel should wash their hands between patients. Soiled clothing should be changed immediately. Disposable gloves should be worn when handling patients with known infections. Patients and their excretions must be isolated from all other patients and from uncontaminated parts of the same patient. Soiled bedding and bandages should be placed in designated receptacles and not on the floor or counter. Floors, counters, and kennels should be regularly scrubbed with soap, water, and appropriate antiseptic solutions.

Insertion and maintenance of indwelling vascular catheters. The incidence of catheter-related infection increases with duration. A catheter-related infection should be suspected if an otherwise unexplainable fever or leukocytosis develops or if there is any evidence of an infection at the skin puncture site. The ideal treatment for a catheter-related infection is to remove it, culture the tip of the catheter, culture the blood, and aseptically replace a new catheter some place else. When there is no other location to place the new catheter, an over-the-wire catheter exchange is the compromise. Catheters should be aseptically placed and then taped (on appendages) or sutured (on flat surfaces) close to the skin puncture site to prevent it from sliding in and out of the vessel. Antibiotic/antifungal ointments (chlorhexidine and povidone iodine) at the skin puncture site can be used, but are not universally recommended. The site should be wrapped occlusively.

Indwelling catheters should be redressed and inspected every 24 to 48 hours. All soiled bandage material should be discarded. The puncture site should be cleaned with antiseptic solutions and the occlusive wrap reapplied. Infusion fluids and administration tubing must be sterile.

Connections should not be disconnected unless absolutely necessary and then must be done aseptically. All injection caps should be cleaned well with an antiseptic solution before needle insertion. The fluid bottles and all administration tubing should be changed every 24 to 48 hours. Tubing should be changed after blood or colloid infusion. The primary catheter or infusion line should not be used for the collection of blood samples except in emergencies.

Insertion and maintenance of indwelling urinary catheters. An indwelling urinary catheter is indicated if the bladder cannot be easily expressed or if urine needs to be quantitated or measured. Urinary catheters are occasionally used to maintain the cleanliness of recumbent patients. Urinary catheters must be placed aseptically and attached to a closed drainage system; all joints should be firmly attached or taped to prevent accidental disconnection. The collection system should be positioned so that urine drains downhill. Draining the collection reservoir or needle puncture of the collection tube to obtain a urine sample for analysis must be accomplished aseptically. The collection tubing should be taped to the hind leg or abdomen to prevent accidental traction on the catheter and suture sites. Enough slack should be provided in the tubing to allow the patient a full range of hind-leg motion. Antibiotic flushes have not been shown to prevent urinary infections, only to delay them and to select for resistant infections. The bag should be drained only every 8 to 12 hours unless more discrete output measurements are desired. The prepuce or vestibule should be flushed with a dilute chlorhexidine or povidone iodine solution three times daily to prevent migration of infectious agents into the bladder along the outside of the catheter. The urinary catheter should be removed when it is no longer necessary.

Patient comfort. The animal should be well padded, especially when it does not move around much; pressure points predispose the animal to decubiti. The animal should always be clean and dry; urine, feces, and other secretions are irritating to the skin and predispose to dermatitis. Make sure that the bladder is not distended with urine and that analgesics are administered if the animal is exhibiting signs of pain. The emotional needs of the patient also need attending. Critically ill patients must be allowed time to sleep. Make sure that some of the caregiver interactions are pleasant and not associated with poking, prodding, and injecting. Owner visitations are almost always helpful to the animal's sense of well being; only an occasional animal gets very upset because of separation anxiety when the owner leaves. The physiological discomforts, such as nausea and vomiting, must also be treated.

The recumbent patient

Positioning. Immobility and positional stasis for prolonged periods of time (1) predispose to tissue necrosis and decubiti over bony protuberances, (2) are associated with the accumulation of secretions and atelectasis in the lower regions of the lung, (3) are associated with contracture and stiffening of muscles and ligaments, and (4) may be associated with regional appendage edema because of poor lymphatic drainage. Make sure that the body and all appendages are padded and positioned "comfortably" and repositioned every 4 hours and perform passive range of motion exercises at 4-hour intervals. Lateral positions are easiest to maintain in animals and are usually associated with satisfactory lung function. In animals with lung disease, the prone position seems to provide the best and most consistent lung function.

Optimizing fluid therapy. The moistness and mobility of airway secretions depends a great deal upon adequate systemic hydration. With exudative secretions (pneumonia), patients should be maintained in the high range of normal hydration. Patients with transudative secretions (pulmonary edema) should be maintained in the low range of normal hydration to minimize fluid flux into the lungs. Animals that are being ventilated tend to retain sodium and water because of high aldosterone levels. Fluid ins and outs, body weight, and physical evidence of edema should be monitored closely.

The unconscious patient

Endotracheal intubation. Unconscious patients should be endotracheal intubated to (1) assure an open airway, (2) protect the airway should the animal vomit or regurgitate, and (3) provide a means of positive pressure ventilation should the need arise. Avoiding traction on the tube when the patient is moved or rotated prevents accidental tube dislodgement. Tubes should be secured by tying around the maxilla, mandible, or the back of head (for orotracheal tubes) or around the neck (for tracheostomy tubes). The tie around the tube needs to be tight (but not collapsing) and the tie around the face or neck needs to be snug (but not tight). Ties around the face should be moved every 4 hours to minimize pressure points and lip necrosis. The tying material could be roll gauze, but this soaks up secretions and is not easily cleaned; a length of intravenous fluid extension tubing works better. Endotracheal or tracheostomy tube cuffs are round, but the trachea is not.

To seal some portions of the circumference, other portions are likely to be subjected to excessive cuff pressure. Asymmetric tube cuffs and overinflated cuffs magnify the problem. Since the endotracheal tube may be in place for several days, prevention of cuff pressure–induced tracheal damage is of paramount importance. Use high volume, low pressure cuffs for long-term intubations. Inflate the cuff with just enough air to just barely stop the leak of air (auscult over the larynx) when positive pressure is applied to the airway or to just barely not stop the leak (a small leak is OK as long as the lungs can be inflated). The pilot balloon is only used to indicate that there is air in the cuff; it bears no correlation whatsoever with the amount of pressure being applied to the tracheal wall and must not be used for this purpose. It may be advantageous to periodically change the cuff pressure point. This can be accomplished with endotracheal tubes by deflating the cuff every 4 hours and moving the tube slightly inward or outward. The cuff is then carefully reinflated. Before cuff deflation, flush the mouth, pharynx, and the lumen of the trachea rostral to the cuff with saline. All fluid should be removed by suctioning before deflating the endotracheal tube cuff. Mucus and debris being elevated from the depths of the airways accumulate at the end of the tube. Intubated animals cannot cough and accumulated secretions can dry and obstruct the tracheal tube. The inner cannula of a tracheostomy tube can be easily removed and cleaned every 4 hours. Tracheal tubes without inner cannulas need to be suctioned and should be exchanged every 24 to 48 hours (depending upon the quantity and viscosity of the secretions). Tracheal suctioning should be done about every 4 hours irrespective of the tube type; somewhat more frequently if there are a lot of secretions; somewhat less frequently if there are scant secretions. There are Always some secretions; tracheal suctioning should never be ignored. Suction catheters should be soft and flexible, and should have more than one hole in their tip to prevent sucking an epithelial plug into a single hole, which would then be ripped away when the catheter is withdrawn. The suction catheter should have a proximal thumb hole so that suction can be applied in a controlled manner. The inside diameter of the catheter should be as large as possible to facilitate the removal of thick secretions. The outside diameter of the catheter should be no larger than 50% of the diameter of the endotracheal tube adapter. The air that is suctioned through the catheter must come from the room and must be able to flow freely down around the outside of the catheter to prevent excessive reduction in airway pressure and small airway and alveolar collapse. The suctioning procedure must be atraumatic and aseptic. The airway should be well humidified just before the suctioning. The animal should breath 100% oxygen for 5 minutes before suctioning to minimize the inevitable hypoxemia. Secretions can be mobilized into the central airways by chest coupage just before suctioning. Inject 0.2 mL/kg of saline into the tracheal tube and then manually hyperinflate the lungs several times. Gently insert the suction catheter into the trachea as far as it will advance. Suction is applied with a rotating and winding motion while the catheter is being withdrawn.

Suction should not be applied to the airways for more than about 5 seconds to minimize small airway and alveolar collapse. The suctioning procedure should cease immediately if discomfort, restlessness, or changes in cardiac or respiratory rhythm occur. The animal should be manually hyperinflated with oxygen after suctioning to alleviate the small airway and alveolar collapse. The procedure should be repeated several times if it is productive. Tracheal suctioning should always net some airway secretions. If not, it is because they are too dry. Better humidification and secretion liquefaction is necessary.

The presence of blood in the aspirant indicates an excessively traumatic procedure; try a more gentle aspiration technique with less pressure and lower flow rates, and perhaps a smaller suction catheter.

Airway humidification. Humidification is the provision of water vapor to help prevent the drying of airway secretions. Endotracheal intubation bypasses the upper nasal passages. The endotracheal tube is not nearly as efficient as the nasal passages at humidifying inspired gases. In-line humidifiers should be heated and sealed by a semipermeable membrane so that there is no water- air interface (warm water supports growth of infectious organisms, which are a source of nosocomial pneumonia). Unsealed humidifiers should be exchanged with sterile equipment every 24 hours. Cooling between the heated humidifier and the patient causes water to condense in the inspiratory tubing; it will need to be periodically drained. As an alternative to a commercial humidifier, sterile distilled water can be instilled into the inspiratory tubing and endotracheal tube about every 2 hours. There are commercial condenser/humidifiers that attach to the tracheal tube and function as an artificial nose. Medical oxygen is anhydrous, and when insufflated it must be humidified to prevent mucus and epithelial desiccation. This is usually accomplished with a bubble humidifier. Systemic hydration is of paramount importance in ensuring airway hydration; no airway humidification procedure will be very effective in a dehydrated patient. Nebulization is the provision of particulate water droplets to therapeutically moisten already thick secretions. This is usually accomplished with an ultrasonic nebulizer, which produces a dense mist of appropriately sized water droplets, which is then inhaled by the patient. The nebulization treatment is generally applied for about 20 minutes at about 4-hour intervals.

Mouth care. Unconscious patients do not eat, drink, or swallow. The mouth and pharynx accumulate secretions, which soon become infected and predispose the animal to nosocomial infections. The mouth and pharynx should be washed with sterile saline and suctioned at about 4 hour intervals. It is important to prevent torque or traction on the tracheal tube during this procedure. The pharynx and mouth should then be rinsed with a commercial chlorhexidine mouth wash solution. The tongue will dry out if it is allowed to flop out of the mouth cavity and it will develop pressure-induced ulcers if it is allowed to drape across teeth or if the pulse oximeter is left in one place for too long. The tongue should be cleaned along with the mouth. It should be left wholly within the mouth between cleanings. It can be wrapped in a saline and glycerinsoaked gauze sponge to keep it moist and to minimize sublingual edema.

Eye care. Corneal drying and ulceration is a common problem because of reduced lacrimal secretions and the absence of blinking. Artificial tears and bland ophthalmic ointments should be alternately placed into the conjunctival sac at about 2 hour intervals. The corneas should be fluorescein stained regularly to check for corneal ulcers, and if they develop, antibiotic ophthalmic ointment should be used.

Bladder care. Unconscious patients often do not urinate normally and when they do, the urine soils the skin. Human infant absorbent diapers, properly positioned, work very well for collecting the urine and preventing it from soaking the skin. The diapers can be weighed to quantitate the urine output. If the animal does not urinate, the bladder should be expressed; however, a urinary catheter is often placed to facilitate urine collection and sanitation, and to quantitate urine output. The catheter must be inserted and maintained in an aseptic manner as described above.

Colon care. Recumbent, sedated patients often do not defecate normally. The colon should be palpated daily to check for constipation. An enema of warm saline may be indicated. Long hair around the anus should be clipped to facilitate the cleaning of feces.

Nutrition. Nutritional support should be implemented as soon as it is apparent that the patient has not or will not eat for 3 days. There are no specific signs of acute malnutrition. Such animals in general just "do poorly." Enteral nutrition should be implemented if possible. Conscious animals can be coaxed to eat or may swallow the food if it is placed in the lip fold. Unconscious patients can be fed via nasogastric, esophagostomy, or gastrotomy tube if the stomach is motile. It is imperative that the stomach works properly if gastric feeding is to be continued. Gastric stasis resulting in the accumulation of previous feedings are a common problem and with continued feeding will result in gastric distention and regurgitation. Gastric residuals must be checked before each feeding and should not be allowed to exceed about 10 mL/kg. Gastric motility may require augmentation by metoclopramide, cisapride, or bethanechol. If the GI tract is not working, if there is a need for bowel rest, or if the animal suffers pancreatitis, intravenous nutrition should be implemented.

Treatment with antidotes

Specific antidotes exist for relatively few poisons. Antidotes react with the poison or its receptor or interfere with its metabolic or specific toxic pathway. For example, chelators react directly with metals, and atropine competes with organophosphates for muscarinic receptors. Fomepizole is an example of an enzyme inhibitor, which blocks the metabolic pathway that activates the poison.

Another way to block a metabolic pathway is to provide an alternative substrate for the enzymes involved; examples include giving ethanol for ethylene glycol toxicosis or acetates for sodium fluoroacetate (1080) poisoning. Pamidronate disodium is an example of an antidote that reverses the specific toxic effect of the poison, cholecalciferol. The use of an antidote does not lessen the importance of thorough detoxification or of supportive therapy. Many antidotes are potentially toxic and should be used with care and only when the diagnosis has been confirmed. Excluding venoms, which are covered later in this chapter, significant small animal poisons for which specific antidotes exist are listed in Table 1. These antidotes are discussed in alphabetical order below. Antidotes that work by preventing absorption are dealt with in a separate section. The suppliers named are suggestions only and may be relevant only to the United States. Some antidotes must be used "off-label" because they are licensed only for human use or only for large animal use. Certain older drugs, such as atropine and apomorphine, on which there is no patent, are sometimes unavailable because it is not commercially worthwhile for manufacturers to maintain continuous production of these drugs. Similarly, some antidotes are made by only one company and may be made in intermittent batches. In an emergency, human prescription drugs (e.g., Digibind, Desferal, Narcan) may be obtained from a hospital emergency room. Various compounding pharmacies across the country supply many of these antidote products.

Antidotes Description (Listed Alphabetically)

Abbreviations: IV, intravenous or intravenously; IM, intramuscular or intramuscularly; SC, subcutaneous or subcutaneously; PO, oral or orally. Dosages may vary with product throughout the literature; refer to a drug formulary or product insert and label for more detailed information if necessary. Except where otherwise indicated, dose rates are for dogs and cats.

Acetamide, 10% in 5% dextrose, at 7 to 10 mL/kg body weight over a 30-minute period, then continued at 5 mL/kg every 4 hours for 24 to 48 hours, for sodium fluoroacetate and sodium fluoroacetamide toxicoses.

Acetylcysteine, PO or IV, is given in a loading dose of 140 mg/kg followed by 70 mg/kg every 6 hours for up to seven treatments for acetaminophen toxicosis. Acetylcysteine is incompatible with oxidizing agents, including hydrogen peroxide, which is sometimes used as an emetic in poisonings. Acetylcysteine tastes bad, may cause vomiting, and may be adsorbed by activated charcoal. Suppliers include Bristol Myers Squibb, Mead Johnson, Cumberland Pharmaceuticals, Roxane, and others.

Atipamezole, at 50 µg/kg IM, has been shown to be an effective treatment of amitraz poisoning. Atropine sulfate, at 0.2 to 0.5 mg/kg, is given in cases of acetylcholinesterase-inhibiting organophosphate and carbamate poisonings. The calculated dose may be given

IV slowly to effect, or the practitioner may give one quarter of the dose IV, wait 15 minutes to observe effects, and then give the rest IM or SC. This agent should not be used in cyanotic cats. To minimize the toxic effects of atropine itself, the lowest dose should be used initially and increased only if the response is unsatisfactory. Heart rate and regularity should be monitored after IV administration and if high doses are used. Diphenhydramine, sometimes used to treat delayed nicotinic signs of organophosphate and carbamate toxicoses, may potentiate the effects of atropine, so concurrent use is not recommended. Atropine delays the actions of activated charcoal and cathartics by decreasing gastrointestinal motility. Suppliers include Vedco, Butler, Phoenix Pharmaceuticals, Western Veterinary Supply, and RX Veterinary Products.

Calcium disodium ethylenediaminetetraacetic acid (EDTA) is given at 110 mg/kg/day SC in four divided doses (27.5 mg/kg/dose) to a maximum of 2 g/day, diluted to 10 mg/mL in D5W, to chelate lead, zinc, inorganic mercury, and cadmium. IM injection is painful. The course of therapy should not exceed a maximum of 5 days because of the risk of nephrotoxicity. Allow 5 days before commencing a second course, if required. D-Penicillamine can be used between courses of calcium disodium EDTA. Make sure no lead is present in the gastrointestinal tract before using this antidote. Other side effects, which may be ameliorated by zinc supplements, include vomiting, diarrhea, and depression. In lead toxicosis, mobilization of lead may cause a transient increase in the severity of clinical signs. In human medicine dimercaprol is given beforehand to help alleviate acute neurological signs and to accelerate the excretion of lead. There are preparations of pharmaceutical grade calcium disodium EDTA for human use, or it may be obtained from chemical supply companies and compounded. Specify calcium disodium EDTA; do not attempt to use other salts of EDTA or the free acid. Calcium disodium EDTA is also known as calcium disodium edetate or calcium disodium versenate.

D-Penicillamine is given for acute lead, zinc, cadmium, or inorganic mercury poisoning and as a chelator of copper in inherited copper storage disorders in dogs. For dogs with lead poisoning, give up to 110 mg/kg PO on an empty stomach for 1 to 2 weeks. For cats, give 125 mg twice daily PO on an empty stomach for 5 days. For dogs with inherited copper storage disorders, give 10 to 15 mg/kg/day orally. D-Penicillamine may cause vomiting, in which case give smaller doses at more frequent intervals (same total daily dose) or reduce the dose to 33 to 55 mg/kg/day divided into three or four doses. For inherited hepatic copper storage disorder in dogs, D-penicillamine can also be given at 10 to 15 mg/kg PO twice daily on an empty stomach with concurrent ascorbic acid therapy (500 to 1000 mg PO per day). Suppliers include Merck & Co. and Wallace.

Deferoxamine (desferrioxamine, Desferal) is given for iron toxicosis. Emergency treatment consists of continuous IV infusion to 15 mg/kg/hr with close monitoring of cardiac function and blood pressure. If monitoring is not practical or if the animal is not in shock, a safer dose regimen is 40 mg/kg IM at 4- to 8-hour intervals to a total dose not exceeding 6 g/day. These dose rates have been established for dogs. There is a lack of information on appropriate doses for cats.

Treatment should be continued until serum iron levels fall to within the normal range. Oral ascorbic acid can be used concurrently to accelerate excretion. It is normal for the urine to turn reddish brown during deferoxamine therapy. The color is caused by excretion of the chelated iron. Deferoxamine has a number of adverse effects, the most clinically significant of which is pulmonary toxicity following continuous infusion. Deferoxamine is supplied by Novartis.

Digoxin immune Fab (Digibind) is used for toxicosis following ingestion of digoxin tablets and also in humans for ingestion of plants including Digitalis purpurea (foxglove), Convallaria majalis (lily of the valley), Kalanchoe blossfeldiana (Christmas kalanchoe) and Nerium oleander (oleander). Doses for animals have not been established, but as a guideline in humans the emergency dose is 400 mg of Digibind. Ideally, 1.7 mL of Digibind should be administered per mg of digoxin ingested. Ideally, Digibind should be administered IV over 30 minutes through a 0.22-μ filter, but it can be given as a bolus if cardiac arrest is imminent. The patient must be monitored for hypokalemia and anaphylaxis. A large dose of Digibind may cause a febrile reaction. To reconstitute Digibind, add 4 mL of distilled water to the vial and mix gently; then dilute to a convenient volume with normal saline. Do not save reconstituted Digibind. The patient should show improvement within 30 minutes and should recover within 4 hours. In veterinary cases, digoxin levels may not be readily available; it is sometimes recommended to administer 1 to 2 vials initially and then observe the effects. Digibind is manufactured by GlaxoSmithKline.

Dimercaprol (BAL [British Anti-Lewisite]) is given for acute arsenic and lead toxicosis and may also be used for copper chelation in dogs with inherited copper storage disease. In clinical arsenic toxicosis, a dose of 6 mg/kg IM is given every 8 hours for 3 to 5 days. For preclinical exposure to arsenic, 3 mg/kg IM is given every 8 hours. In clinical lead toxicosis, 2.5 to 5.0 mg/kg IM as a 10% solution is given every 4 hours for 2 days and then every 6 hours on the third day, followed by twice a day (b.i.d.) for 10 days. Because dimercaprol is nephrotoxic, renal function should be monitored during therapy, and the course should not be too long. Because dimercaprol is carried in a peanut oil vehicle, the injection is painful, and it should not be given SC or IV. Once arsenic is bound, the resulting compound is water soluble and is excreted in the urine.

Dimercaprol is manufactured by Becton Dickinson and by Hynson, Westcott and Dunning, Inc. Dimercaptosuccinic acid (DMSA or Succimer) is a suggested chelation therapy for dogs with lead or zinc toxicosis because it is less toxic than calcium disodium EDTA and can be given orally. Suggested dose regimens include 10 mg/kg PO given every

8 hours for 10 days or 10 mg/kg PO every 8 hours for 5 days followed by 10 mg/kg PO every 12 hours for 2 weeks. Recent experimental evidence suggests that DMSA may also be effective in removing organic mercury from the central nervous system (CNS) of mammals. DMSA is supplied as a human drug by Ovation Pharmaceuticals, Inc.

Diphenylthiocarbazone (dithizone) is used for thallium toxicosis. The dose is 70 mg/kg PO three times daily. Dithizone should be used within 24 hours of exposure to thallium. It is known to cause cataracts in dogs. If the animal is seen more than 24 hours after exposure, Prussian blue should be used rather than dithizone. Suppliers of dithizone include Sigma, Acros, and other chemical suppliers.

Ethanol (20%) is given for ethylene glycol (antifreeze) toxicosis. Guidelines are: dogs, 5.5 mL/kg IV every 4 hours for five treatments and then every 6 hours for four treatments; for cats, 5 mL/kg IV every 6 hours for five treatments and then every 8 hours for four treatments. In practice the dose is titrated to the desired effect, which is profound stupor (near coma) for up to 72 hours. Ethanol provides an alternative substrate for alcohol dehydrogenase and therefore inhibits the metabolism of ethylene glycol to its much more toxic metabolites. If metabolism is prevented, ethylene glycol will be excreted as the parent compound. Do not give a dog ethanol concurrently with or following 4-methylpyrazole (4-MP) because ethanol poisoning and fatal respiratory depression could result.

For sodium fluoroacetate (1080) and sodium fluoroacetamide (1081) toxicoses, 8 mL of 50% ethanol per kg PO is indicated in conjunction with 5% acetic acid (also 8 mL/kg PO). The purpose is to supply an alternative source of acetate to minimize conversion of fluoroacetate to fluorocitrate, which blocks the citric acid cycle by inhibiting aconitase. Clear spirits, such as gin, vodka, "white" rum (e.g., Bacardi) or Everclear, can be used as antidotes. To convert "proof " to percentage, divide by two. For example, a spirit labeled 100 proof is 50% ethanol. Alternatively, laboratory ethanol can be purchased from chemical companies, such as Aldrich. It may be listed as ethyl alcohol. Laboratory ethanol may be contaminated with benzene, so one should ensure that it is absolute ethanol (99.5+%) and deal only with reputable companies.

Glycopyrrolate may be used as an alternative to atropine in acetylcholinesterase-inhibiting organophosphate and carbamate poisonings. The dose rate for dogs is 0.05 μg/kg. A dose rate for cats has not been established. Glycopyrrolate takes somewhat longer than atropine to show an effect, but the effect lasts longer. It is supplied by Fort Dodge Animal Health and by various manufacturers of generic pharmaceuticals for human use.

4-Methylpyrazole (4-MP, fomepizole) is used for ethylene glycol (antifreeze) toxicosis in dogs. It is given IV in a 5% solution; the initial dose is 20 mg/kg, followed by 15 mg/kg at 12 and 24 hours and 5 mg/kg at 36 hours. Dosages used in cats are much higher. A specific inhibitor of alcohol dehydrogenase, 4-MP prevents the enzymatic metabolism of ethylene glycol to its more toxic metabolites. Do not use 4-MP concurrently with ethanol in dogs because fatal ethanol toxicosis may result. 4-MP can be purchased as the concentrated compound from chemical companies, such as Sigma-Aldrich; this compound should be refrigerated. A kit (Antizol-Vet) containing 4-MP and diluent, which is stored at room temperature, is marketed by Orphan Medical Inc., which also markets Antizol to the medical profession for use in methanol toxicosis. Monoacetin (acetin, glyceryl monoacetate) is used for sodium fluoroacetate (1080) and sodium fluoroacetamide (1081) toxicoses. It provides an alternative source of acetate to minimize conversion of fluoroacetate to fluorocitrate, which blocks the citric acid cycle. The dose is 0.55 g/kg IM given every hour to a total dose of 2 to 4 g/kg. Monoacetin is not widely available, but can be purchased from suppliers of fine chemicals, such as Acros. Because of the various names for this chemical, it may be most productive to search for it by CAS number, which is 26446-35-5.

Naloxone (Narcan) reverses a number of opioids including meperidine (pethidine), fentanyl, and oxymorphone. A dose of 0.4 mg naloxone HCl reverses 1.5 mg of oxymorphone hydrochloride, 15 mg of morphine sulfate, 100 mg of meperidine, or 0.4 mg of fentanyl. If the amount of narcotic is unknown, titrate naloxone to effect, using an initial dose of 0.04 mg/kg IV (for the most rapid response), SC, or IM. Naloxone has a wide margin of safety and virtually no agonist action.

However, small doses are recommended for severe narcotic overdose because of the risk of producing acute withdrawal syndrome. Patients should be monitored for relapse because the narcotic may have a longer duration of action than naloxone; repeat doses of naloxone can be given as needed. Naloxone is supplied by Endo Pharmaceuticals Inc.

Nicotinamide (niacinamide), given at a dose of 50 to 100 mg IM, replaces nicotinamide antagonized by pyriminil (Vacor), an old avicide that has been deregistered in the United States. There are various vitamin B complex injectables that include niacinamide available to veterinarians. These injections usually also contain thiamine, which can cause an anaphylactic reaction.

Pamidronate disodium (Aredia, Novartis) is one of a class of pharmaceuticals, bisphosphonates, developed to inhibit bone resorption in human beings. Pamidronate disodium has been identified relatively recently as a useful antidote for cholecalciferol and calcipotriene poisoning of dogs.

The treatment regimen is 1.3 to 2 mg/kg diluted in saline for IV infusion over 2 to 4 hours. A second infusion 4 to 7 days later may be required. Calcitonin, the antidote previously recommended for cholecalciferol poisoning, is of doubtful effectiveness, more likely to have undesirable side effects, and more tedious to administer than pamidronate sodium. It is not advisable to use both antidotes together, since limited experimental evidence suggests that the outcome is worse than if only one antidote is used.

2-Pralidoxime (pralidoxime chloride, 2-PAM) is useful to reactivate acetylcholinesterase in organophosphate toxicosis, but may not be effective in carbamate toxicosis. If it is uncertain whether the toxicant is an organophosphate or a carbamate, 2-PAM should be used unless it is likely that the toxicant is carbaryl, in which case there is evidence that 2-PAM may be harmful. The first dose of 2-PAM should be given as early as possible, since it is not effective once "aging" has occurred. Treatment may be repeated once or twice at 12-hour intervals. The dose rate is 20 mg/kg by IM, SC, or slow IV injection. 2-PAM is marketed under the trade name Protopam Chloride by Elkins Sinn.

Prussian blue (ferric ferrocyanide) is used in thallium toxicosis to bind thallium excreted in the bile and interrupt enterohepatic cycling. The treatment course in dogs is 3 g/day and given in three to six divided doses for 7 to 14 days. Studies have not been done in cats. Prussian blue can be obtained from chemicals suppliers, such as Acros Organics.

Pyridoxine (vitamin B6), given at 20 mg/kg IV, is used to replace pyridoxine antagonized by crimidine, an older rodenticide. There are numerous veterinary vitamin B complex preparations that contain pyridoxine, including IV preparations. Vitamin B complex injections usually also contain thiamine, which can cause an anaphylactic reaction.

Sodium sulfate is an effective antidote to acetaminophen poisoning and may be used instead of acetylcysteine. Sodium sulfate is given at a dose of 50 mg/kg IV every 4 hours as a 1.6% solution and is available from pharmacies or chemical companies.

Tolazine (tolazoline HCl), marketed as a reversal agent for xylazine, is also effective as an antidote for amitraz. The dose rate for both purposes is 4 mg/kg IV in dogs. Dose rates for cats have not been published. Side effects may include tachycardia and arrhythmia. Tolazine is supplied by Lloyd Laboratories.

Vitamin K1 (phytonadione) is used in anticoagulant rodenticide poisonings. Phytonadione is best administered PO, 2 to 5 mg/kg/day in divided doses with a fatty meal; it is given daily for 21 to 30 days. SC administration does not result in better or faster absorption than oral administration and may cause hemorrhage. Intravenous administration may cause an anaphylactic reaction and should only be used in emergencies and then by slow infusion over at least 15 minutes. Not all phytonadione preparations are suitable for IV use. It may not be necessary to give phytonadione for 21 to 30 days, particularly if gastrointestinal detoxification is prompt and thorough, or if warfarin is ingested. To test whether it is necessary, stop treatment after 10 days, wait 3 days, and then test clotting parameters. If they are normal, test again after another 3 days without phytonadione. If they are still normal, further phytonadione is not required. Testing clotting parameters 3 days after finishing a 21- or 30-day course of phytonadione is prudent if the toxicant is a second-generation anticoagulant. Oral and injectable forms of phytonadione are supplied by Merck and by veterinary pharmaceutical companies.

Yohimbine, marketed as a reversal agent for xylazine, is also effective as an antidote for amitraz. The dose rate for both purposes is 0.1 to 0.125 mg/kg IV in dogs. Dose rates for cats have not been published. Reported signs of overdose include tachycardia, agitation, and tremors. Yohimbine is marketed by Wildlife Pharmaceuticals.

Detoxicants

Detoxicants are antidotes that adsorb the poison, react with it before it reacts with tissue, act as barriers to absorption, or accelerate removal of the poison from the body. Detoxification is an essential part of treatment. The first step is to remove the animal from the source. Removing the poison from the animal is a high priority that should be undertaken as soon as emergency stabilization, such as control of seizures and preservation of cardiac and respiratory function, has been achieved.

External exposure. Poison on the skin or hair should be removed with copious warm water and soap or shampoo. For acid spills on the skin, a slurry of sodium bicarbonate (baking soda) can be used, and conversely, diluted vinegar will neutralizme base spills. However, the reaction is exothermic and may cause burning of tissue. Therefore neutralizers should be diluted and used with caution. Rinsing with copious amounts of water is safer and can be just as effective.

Emetics. Emesis is most likely to be effective within the first 4 hours after ingestion. Emesis should never be used in the presence of seizures, deep stupor, or coma and is contraindicated in cases involving caustic poisons, which may inflict further damage to the esophagus and mouth during emesis, or volatile poisons that could be aspirated. Emesis should not be relied upon to remove all poison from the stomach; removal of 40% to 50% of the stomach contents is regarded as a high recovery rate. Gastric lavage may be much more effective, especially if repeated large volumes of fluid are used. Enterogastric lavage (through-and-through enema) is likely to be even more effective and is the best option for poisons taken as pellets to prevent further absorption. For first-aid purposes, owners may wish to try 3% hydrogen peroxide, 2 to 5 mL/kg body weight or ipecac syrup, 1 to 2 mL/kg in dogs, 3 mL/kg in cats. For a more effective, controlled, and reversible response, the veterinary practitioner is better advised to use apomorphine or xylazine. Apomorphine is a suitable emetic for dogs, and can be used in cats. Dose rates are 0.04 mg/kg IV or IM, 0.08 mg/kg SC, or 0.25 mg dissolved in saline and given subconjunctivally. Apomorphine may cause CNS or respiratory depression, but this can be reversed by naloxone. Apomorphine for veterinary use may be obtained from compounding pharmacies that specialize in veterinary preparations.

Xylazine is used as an emetic for cats: 1.1 mg/kg IM. Once the cat has vomited, CNS or respiratory depression can be reversed with 0.1 mg yohimbine/kg IV. Xylazine is made by Bayer. There are also many generic versions.

Intestinal protectants

Milk and egg white. Caustic or corrosive poisons cause denaturation of the tissue proteins with which they come in contact, and the resulting lesion is a chemical burn. If such a poison has been swallowed, milk or egg white will form a protective coat on the intestinal mucosa while providing an alternative protein substrate (casein or ovalbumin) for denaturation. They are generally palatable to small carnivores, and both should be given ad libitum or as much as can be practically and safely administered. Dermal chemical burns can also be flushed with milk or egg white. Milk is a less rich source of protein than egg white, but in animals that do not routinely drink it, it may have additional therapeutic action as a cathartic. In eggs that are not fresh, the yolk may be too fragile to permit separation. To prevent rupturing the yolk, hold the egg over a bowl, crack it carefully, and tip the yolk gently between the half shells while the white falls into the bowl. In animals that have ingested soluble oxalates or oxalic acid, milk has the additional beneficial effect of providing calcium to react with unabsorbed oxalic acid. This promotes the precipitation of harmless calcium oxalate in the intestinal tract instead of allowing absorption of oxalic acid and the harmful precipitation of calcium oxalate crystals in the kidneys.

Kaolin, kaolin-pectin, and bismuth subsalicylate. These all act to coat the intestinal mucosa, thereby providing protection against irritant chemicals and limiting absorption. These preparations decrease peristalsis and may delay excretion of the chemical. There are numerous suppliers of veterinary preparations of kaolin-pectin and of bismuth subsalicylate.

Adsorbents and precipitants. Activated charcoal should be mixed to a slurry of 1 g/5 mL of water before being orally administered at a rate of 2 to 5 g of activated charcoal/kg body weight. Ready-prepared slurries should be used according to the manufacturers' directions. Activated charcoal is prepared by prolonged intense heating of charcoal under steam pressure. This results in the opening of thousands of pores in the charcoal, giving it a very large surface area for its volume. In 1830, before the French Academy of Medicine, a pharmacist named P.F. Touery swallowed 10 times the fatal dose of strychnine followed by 15 g of activated charcoal and survived unharmed. Although repetition of this demonstration is strongly discouraged, activated charcoal is the most important and widely used adsorbent in clinical toxicology. Activated charcoal should not be administered simultaneously with other oral antidotes because it may adsorb them, and it should be given at least 30 minutes before administering a cathartic. For poisons that undergo enterohepatic cycling, such as bromethalin, cholecalciferol, phencyclidine, and organochlorines, activated charcoal should be given repeatedly for 2 or 3 days.

- Administration every 3 hours is the ideal, and every 8 hours is a minimum. There are numerous suppliers of ready-to-use activated charcoal preparations.
- Bentonite (Fuller's earth), a porous clay, is the adsorbent for dipyridyl herbicides (paraquat, diquat), which have a strong affinity for clay. A slurry containing 300 g/L of water is administered by stomach tube. Bentonite powder is supplied by chemical companies, such as Sigma or Acros.

- Bentonite is used by actors to simulate aging skin and may be available from theatrical societies or their suppliers, but only the undyed powder should be used. Kitty litter, although usually made of bentonite, is not recommended because the coarse granules may cause intestinal obstruction and may contain deodorants or other additives. Bentonite is practically synonymous with montmorillonite. Attapulgite has very similar properties.
- Milk of magnesia (magnesium hydroxide) is a laxative and a weak base. Because it precipitates iron in the intestinal tract as insoluble iron hydroxide, it is the detoxicant of choice for iron toxicosis, for which activated charcoal is not effective. Milk of magnesia can also be used to neutralize ingested acids, although it should be used as a well-diluted solution because the reaction is exothermic and may cause burns. Suggested dose rates are 2 to 15 mL of milk of magnesia USP, which is 80 mg magnesium hydroxide per mL. Veterinary preparations commonly contain roughly 77 mg/mL. Some veterinary magnesium hydroxide boluses, if reconstituted according to the label, contain more than 10 times as much.
- Sulfate laxatives (see next section) precipitate lead in the gastrointestinal tract as insoluble lead sulfate.

Laxatives. Laxatives are often given 30 minutes after activated charcoal to accelerate the passing of the activated charcoal with the poison bound to it. If this is not done, the osmotic gradient is likely to promote release of the poison from the activated charcoal. Laxatives should not be given to animals with diarrhea. Osmotic cathartics include sorbitol and the saline cathartics sodium sulfate and magnesium sulfate (Epsom salts). Sorbitol is given as a 30% solution, 3 mL/kg PO. Sodium sulfate or magnesium sulfate is given as a 20% solution, 250 mg/kg. Osmotic cathartics draw water from the rest of the body into the gastrointestinal tract, so adequate fluids must be provided to prevent dehydration, sodium ion toxicosis from sodium sulfate, or CNS depression from excessive magnesium. Sorbitol and saline cathartics are readily available from pharmacies. Mineral oil is the preferred cathartic when the poison is oil soluble (e.g., white phosphorus [yellow phosphorus]). The dose rate is 1 to 2 mL/kg in adult dogs. Because of the risk of aspiration pneumonia, mineral oil must be administered with care, and vomiting should be controlled beforehand. Mineral oil is available from numerous veterinary suppliers.

SOME SYMPTOMATIC TREATMENTS IMPORTANT IN TOXICOLOGY

The following are not antidotes, but are among the drugs frequently used to treat serious effects in toxicoses. They are listed under the clinical signs or lesions for which they are used.

- Acidosis. Sodium bicarbonate. In the absence of precise acid-base measurement, 2 mEq/kg body weight is a useful initial dose. This drug is safest when administered by slow IV infusion as part of total fluid therapy.

- Anaphylaxis. Epinephrine (adrenaline) and antihistamines. The doses of epinephrine are: for dogs, 1 mL/10 kg body weight or 1:1000 solution IM or SC; for cats, 0.1 to 0.2 mL of 1:1000 solution IM or SC. For IV use, ¼ to 1/2 the above doses are used. As an intracardiac injection, the dose is 0.05 to 0.1 mL of 1:1000 solution. There are numerous veterinary suppliers of both epinephrine and antihistamines. The doses of antihistamines for dogs and cats are: 0.44 to 2.2 mg diphenhydramine HCl per kg IM; promethazine HCl 0.2 to 1 mg/kg IV, IM, SC, or PO.
- Cardiac Arrest. Epinephrine (adrenaline). Intracardiac dose is 0.05 to 0.5 mL of 1:1000 solution.
- Cerebral Edema. Mannitol, 2 g/kg IV, and dexamethasone, 2 mg/kg every 6 hours. Diuresis. Mannitol, 1 g/kg IV, or furosemide, 2 to 20 mg/kg IV in cases of acute renal failure. Fluids should be available when diuretics are used. In cases of pulmonary or cerebral edema,
- voluntary oral intake of fluids is the safest option. To maintain renal function and remove renally excreted poisons, parenteral fluids are likely to be more effective.
- Emesis (Control of). Aminopentamide hydrogen sulfate is given SC or IM according to package insert directions. Peristaltic activity and therefore passage of activated charcoal is reduced.
- Atropine also has an antiemetic action. Phenothiazine tranquilizers have antiemetic actions, but are seldom used in toxicology because they lower the seizure threshold. Metoclopramide is often used in dogs to control emesis.
- Gastric Irritation. Cimetidine and ranitidine are both histaminereceptor antagonists that control acid secretion in the stomach. The canine dose of cimetidine is 5 to 10 mg/kg PO every 8 hours. The ranitidine dose for dogs is 2 mg/kg PO or IV every 8 hours; for cats it is 3.5 mg/kg PO every 12 hours or 2.5 mg/kg IV every 12 hours. Cimetidine and ranitidine are supplied by GlaxoSmithKline. In toxicoses caused by nonsteroidal antiinflammatory drugs, misoprostol may be beneficial to inhibit gastric irritation and prevent ulceration.
- Gastric Ulceration. Sucralfate acts as a barrier over ulcers. Dose rates are 1 g (=1 tablet)/30 kg body weight three to four times daily for dogs and 1/4 to 1/2 tablet two to three times daily for cats. Sucralfate requires stomach acid to be effective, so it should be given at least 30 minutes before cimetidine, ranitidine, or antacids.
- Hypotension. Dopamine may be used if there is no response to fluid therapy or plasma expanders. The dose rate for small animals is 2 to 20 μg/kg/min by dilute IV injection or preferably by IV infusion in lactated Ringer's solution.

- Methemoglobinemia. Oxygen therapy or sodium sulfate or methylene blue (Basic blue, Urolene blue) or, in cats, ascorbic acid. Sodium sulfate, given at a dose of 50 mg/kg IV every 4 hours as a 1.6% solution in water for the methemoglobinemia associated with acetaminophen overdose in cats, is available from pharmacies or chemical companies. Methylene blue is given at a dose of 1.5 mg/kg IV as a 10% solution in saline. Repeated doses of methylene blue at therapeutic levels may cause hemolytic anemija in small animals. However, the dose may be safely repeated up to three times. Concurrent fluid therapy to prevent any hemoglobin precipitation in the kidneys is advisable. The complete blood count should be monitored for up to a week after treatment. Methylene blue is considered a possible carcinogen, but because of its value as a veterinary antidote, the Food and Drug Administration (FDA) Center for Veterinary Medicine has identified it as a substance that they would not normally object to being compounded for animal use. Ascorbic acid, 30 mg/kg PO four times daily, is safe for treating methemoglobinemia in cats, but its in vivo efficacy is limited, and its action is slow, so use of ascorbic acid alone is not recommended.
- Muscle Spasm. Methocarbamol or glyceryl guaiacolate. For mild muscle spasm in small animals, 44 mg/kg IV methocarbamol may be adequate. For tetanic muscle spasms, such as those seen in strychnine poisoning, an initial dose of 150 to 220 mg/kg IV is recommended followed by additional doses of 90 mg/kg as required to a total dose not exceeding 330 mg/kg/day. The recommended glyceryl guiaicolate dose is 110 mg/kg IV.
- Respiratory Depression. Doxapram 1 to 10 mg/kg IV. Doxapram may cause seizures if administered in overdose (70 to 75 times the therapeutic dose in healthy animals). Monitor blood pressure, reflexes, and respiratory rate and volume. Relapse may be rapid.
- Seizures. Diazepam, phenobarbital, or pentobarbital. An initial diazepam dose of 0.5 mg/kg IV may be repeated every 20 minutes up to three times. If diazepam is not sufficiently effective, use phenobarbital at 6 mg/kg IV or pentobarbital at 24 to 29 mg/kg IV. Barbiturate anesthetics may be increased to effect, but anesthesia should be as light as possible while controlling seizures. In strychnine poisoning, pentobarbital is the drug of first choice for seizure control.
- Urinary Ion Trapping. The theory of ion trapping is that reabsorption of basic poisons that are excreted renally may be prevented by acidification of the urine, and reabsorption of acidic poisons that are excreted renally may be prevented by alkalinization of the urine. The effectiveness of this practice has been questioned. The diuretic effect of the fluids in which the ions are given is probably the most beneficial aspect of this practice.
- Acidification. Ammonium chloride, 100 mg/kg PO for dogs and 20 mg/kg PO for cats, is used. Ammonium chloride should not be used in acidotic animals. Overuse may result in ammonia toxicosis.
- Alkalinization. Sodium bicarbonate is given in a dose of 0.5 to 2 mEq/kg IV every 4 hours.

8

PARTICULAR VETERINARY TOXICOLOGY

POISONING BY BARIUM COMPOUNDS

Barium is an alkaline earth metal, principally found as barite (barium sulfate) and witherite (barium carbonate) ores. Barium and barium compounds have a variety of uses including as getters in electronic tubes (barium alloys), rodenticide (barium carbonate), colorant in paints (barium carbonate and barium sulfate), and x-ray contrast medium (barium sulfate). Barium naturally occurs in food and groundwater. Barium concentrations in drinking water can typically average 30 μg/L, but can average as high as 302 μg/L. Low levels of barium are also found in ambient air; levels are typically less than 0.05 μg barium/m3.

There is little quantitative information regarding the extent of barium absorption following inhalation, oral, or dermal exposure. Available evidence indicates that barium is absorbed to some extent following inhalation, oral, and dermal exposure; however, in some cases, absorption is expected to be limited. For example, there is some evidence that gastrointestinal absorption of barium in humans is <5–30% of the administered dose. The general population can be exposed to barium via inhalation, oral, or dermal exposure; under most circumstances, oral exposure would be the predominant route of exposure.

An important factor affecting the development of adverse health effects in animals is the solubility of the barium compound to which the individual is exposed. Soluble barium compounds would generally be expected to be of greater health concern than insoluble barium compounds because of their greater potential for absorption. The various barium compounds have different solubilities in water and body fluids and therefore serve as variable sources of the Ba2+ ion. The Ba2+ ion and the soluble compounds of barium (notably chloride, nitrate, hydroxide) are toxic to animals.

Although barium carbonate is relatively insoluble in water, it is toxic to animals because it is soluble in the gastrointestinal tract. The insoluble compounds of barium (notably sulfate) are inefficient sources of Ba2+ ion and are therefore generally nontoxic to animals. The insoluble, nontoxic nature of barium sulfate has made it practical to use this particular barium compound in medical applications as a contrast media for x-ray examination of the gastrointestinal tract.

Barium provides an opaque contrasting medium when ingested or given by enema prior to x-ray examination. Under these routine medical situations, barium sulfate is generally safe. However, barium sulfate or other insoluble barium compounds may potentially be toxic when it is introduced into the gastrointestinal tract under conditions where there is colon cancer or perforations of the gastrointestinal tract and barium is able to enter the blood stream. There are a number of reports of serious health effects in individuals intentionally or accidentally exposed to barium carbonate or chloride. The predominant effect is hypokalemia, which can result in ventricular tachycardia, hypertension and/or hypotension, muscle weakness, and paralysis.

Barium is a competitive potassium channel antagonist that blocks the passive efflux of intracellular potassium, resulting in a shift of potassium from extracellular to intracellular compartments. The net result of this shift is a significant decrease in the potassium concentration in the blood plasma. Although the case reports did not provide information on doses, it is likely that the doses were high. In addition to the effects associated with hypokalemia, gastrointestinal effects such as vomiting, abdominal cramps, and watery diarrhea are typically reported shortly after ingestion. Similar effects have been reported in cases of individuals exposed to very high concentrations of airborne barium; the effects include electrocardiogram (ECG) abnormalities, muscle weakness and paralysis, hypokalemia, and abdominal cramps, nausea, and vomiting.

Several investigators have examined whether exposure to much lower doses of barium would adversely affect the cardiovascular system. A population-based study found significant increases in the risk of death from cardiovascular disease among residents 65 years of age and older living in communities with high levels of barium in the drinking water. However, these data cannot be used to establish a causal relationship because the study did not control for other cardiovascular risk factors or the use of water softeners, which would decrease barium levels and increase sodium levels. Two other studies did not find alterations in blood pressure and cardiac rhythm. In general, animal studies designed to assess cardiovascular function have not found significant alterations in blood pressure or ECG readings following low-dose oral exposure. One study did find significant increases in blood pressure in rats exposed to 0.80 mg barium/kg/day. However, the use of a low mineral diet with less than adequate levels of calcium may have influenced the study results.

The available animal data provide strong evidence that the most sensitive adverse effect of barium is renal toxicity. There are some reports of renal effects in case reports of individuals ingesting high doses of barium. Nephropathy has been observed in rats and mice following long-term oral exposure to barium. In both species, there is a steep dose-response curve for the incidence of nephropathy. For example, nephropathy was not observed in mice exposed to 205 mg barium/kg/day for an intermediate duration; at 450 mg barium/kg/day, 95% of the animals exhibited mild to moderate nephropathy. Data in mice also suggest that the severity and sensitivity to renal lesions is related to duration of exposure. As noted previously, a 205 mg barium/kg/day dose is a no effect level in mice exposed to barium chloride for 90 days; a 2-year exposure to 200 mg barium/kg/day resulted in moderate to marked nephropathy. The potential for barium to induce reproductive and developmental effects has not been well investigated.

Decreases in the number of sperm and sperm quality and a shortened estrous cycle and morphological alterations in the ovaries were observed in rats exposed to 2.2 mg barium/m3 and higher in air for an intermediate duration. Interpretation of these data is limited by the poor reporting of the study design and results, in particular, whether the incidence was significantly different from controls. In general, oral exposure studies have not found morphological alterations in reproductive tissues of rats or mice exposed to 180 or 450 mg barium/kg/day, respectively, as barium chloride in drinking water for an intermediate duration. Additionally, no significant alterations in reproductive performance was observed in rats or mice exposed to 200 mg barium/kg/day as barium chloride in drinking water. Decreased pup birth weight and a nonsignificant decrease in litter size have been observed in the offspring of rats exposed to 180/200 mg barium/kg/day as barium chloride in drinking water prior to mating. Several studies have examined the carcinogenic potential of barium following oral exposure and did not find significant increases in the tumor incidence. No studies have adequately assessed the carcinogenicity of barium following inhalation exposure. The EPA has concluded that barium is not classifiable as to human carcinogenicity, Group D. However, under EPA's revised guidelines for carcinogen risk assessment, barium is considered not likely to be carcinogenic to humans following oral exposure and its carcinogenic potential cannot be determined following inhalation exposure.

Toxicity of copper compounds

Acute or chronic copper poisoning is encountered in most parts of the world. Sheep are affected most often, although other species are also susceptible. In various breeds of dogs, especially Bedlington Terriers, an inherited sensitivity to copper toxicosis similar to Wilson's disease in humans has been identified. Acute poisoning is usually seen after accidental administration of excessive amounts of soluble copper salts, which may be present in anthelmintic drenches, mineral mixes, or improperly formulated rations. Many factors that alter copper metabolism influence chronic copper poisoning by enhancing the absorption or retention of copper. Low levels of molybdenum or sulfate in the diet are important examples. Primary chronic poisoning is seen most commonly in sheep when excessive amounts of copper are ingested over a prolonged period. The toxicosis remains subclinical until the copper that is stored in the liver is released in massive amounts. Blood copper concentrations increase suddenly, causing lipid peroxidation and intravascular hemolysis. The hemolytic crisis may be precipitated by many factors, including transportation, pregnancy, lactation, strenuous exercise, or a deteriorating plane of nutrition.

Phytogenous and hepatogenous factors influence secondary chronic copper poisoning. Phytogenous chronic poisoning is seen after ingestion of plants, such as subterranean clover (Trifolium subterraneum), that produce a mineral imbalance and result in excessive copper retention. The plants that are not hepatotoxic contain normal amounts of copper and low levels of molybdenum. The ingestion of plants such as Heliotropium europaeum or Senecio spp (Plants Poisonous to Animals) for several months may cause hepatogenous chronic copper poisoning.

These plants contain hepatotoxic alkaloids, which result in retention of excessive copper in the liver.

Acute poisoning may follow intakes of 20-100 mg of copper/kg in sheep and young calves and of 200-800 mg/kg in mature cattle. Chronic poisoning of sheep may occur with daily intakes of 3.5 mg of copper/kg when grazing pastures that contain 15-20 ppm (dry matter) of copper and low levels of molybdenum. Clinical disease may occur in sheep that ingest cattle rations, which normally contain higher levels of copper, or when their water is supplied via copper plumbing; cattle are more resistant to copper poisoning than sheep, and thus are not affected in these instances. Young calves or sheep injected with soluble forms of copper may develop acute clinical signs of toxicity. Copper is used as a feed additive for pigs at 125-250 ppm; levels >250 ppm are dangerous—although as for sheep, other factors may be protective, eg, high levels of protein, zinc, or iron. Chronic copper toxicosis is more apt to occur with low dietary intake of molybdenum and sulfur. Reduced formation of copper molybdate or copper sulfide complexes in tissues impairs the excretion of copper in urine or feces.

Clinical findings. Acute copper poisoning causes severe gastroenteritis characterized by abdominal pain, diarrhea, anorexia, dehydration, and shock. Hemolysis and hemoglobinuria may develop after 3 days if the animal survives the GI disturbances. The sudden onset of clinical signs in chronic copper poisoning is associated with the hemolytic crisis. Affected animals exhibit depression, weakness, recumbency, rumen stasis, anorexia, thirst, dyspnea, pale mucous membranes, hemoglobinuria, and jaundice. Several days or weeks before the hemolytic crisis, liver enzymes, including ALT and AST, are usually increased. During the hemolytic crisis, methemoglobinemia, hemoglobinemia, and decreases in PCV and blood glutathione are usually seen. In camelid species such as alpacas or llamas, no hemolytic crisis is observed, although extensive liver necrosis remains a predominant sign. Morbid animals often die within 1-2 days.

- Herd morbidity is often <5%, although usually >75% of affected animals die. Losses may continue for several months after the dietary problem has been rectified. Severe hepatic insufficiency is responsible for early deaths. Animals that survive the acute episode may die of subsequent renal failure.
- Lesions .Acute copper poisoning produces severe gastroenteritis with erosions and ulcerations in the abomasum of ruminants. Icterus develops in animals that survive >24 hr. Tissues discolored
- by icterus and methemoglobin are characteristic of chronic poisoning. Swollen, gunmetal-colored kidneys, port-wine-colored urine, and an enlarged spleen with dark brown-black parenchyma are manifestations of the hemolytic crisis. The liver is enlarged and friable. Histologically, there is centrilobular hepatic and renal tubular necrosis.

Diagnosis. Evidence of blue-green ingesta and increased fecal (8,000-10,000 ppm) and kidney (>15 ppm, wet wt) copper levels are considered significant in acute copper poisoning. In chronic poisoning, blood and liver copper concentrations are increased during the hemolytic period. Blood levels often rise to 5-20 μg/mL, as compared with normal levels of ~1 μg/mL. Liver concentrations >150 ppm (wet wt) are significant in sheep. The concentration of copper in the tissue must be determined to eliminate other causes of hemolytic disease.

Treatment and Control. Often, treatment is not successful. GI sedatives and symptomatic treatment for shock may be useful in acute toxicity. Penicillamine (50 mg/kg, PO, sid, for 6 days) or calcium versenate may be useful if administered in the early stages of disease. Experimentally, ammonium tetrathiomolybdate (15 mg/kg, IV, on alternate days) is effective for the treatment and prevention of copper poisoning. Daily administration of ammonium molybdate (100 mg) and sodium sulfate (1 g) reduces losses in affected lambs. Dietary supplementation with zinc acetate (250 ppm) may be useful to reduce the absorption of copper. Plant eradication or reducing access to plants that cause phytogenous or hepatogenous copper poisoning is desirable. Primary chronic or phytogenous poisoning may be prevented by top-dressing pastures with 1 oz of molybdenum per acre (70 g/hectare) in the form of molybdenized superphosphate or by molybdenum supplementation or restriction of copper intake.

Poisoning by lead compounds

In veterinary medicine, lead poisoning is most common in dogs and cattle. Lead poisoning in other species is limited by reduced accessibility, more selective eating habits, or lower susceptibility. In cattle, many cases are associated with seeding and harvesting activities when used oil and battery disposal from machinery is handled improperly. Other sources of lead include paint, linoleum, grease, lead weights, lead shot, and contaminated foliage growing near smelters or along roadsides. Lead poisoning is also encountered in urban environments, and renovation of old houses that have been painted with lead-based paint has been associated with lead poisoning in small animals and children.

Pathogenesis. Absorbed lead enters the blood and soft tissues and eventually redistributes to the bone. The degree of absorption and retention is influenced by dietary factors such as calcium or iron levels. In ruminants, particulate lead lodged in the reticulum slowly dissolves and releases significant quantities of lead. Lead has a profound effect on sulfhydryl-containing enzymes, the thiol content of erythrocytes, antioxidant defenses, and tissues rich in mitochondria, which is reflected in the clinical syndrome. In addition to the cerebellar hemorrhage and edema associated with capillary damage, lead is also irritating, immunosuppressive, gametotoxic, teratogenic, nephrotoxic, and toxic to the hematopoietic system.

Clinical Findings. Acute lead poisoning is more common in young animals. The prominent clinical signs are associated with the GI and nervous systems. In cattle, signs that appear within 24-48 hr of exposure include ataxia, blindness, and salivation, spastic twitching of eyelids, jaw champing, bruxism, muscle tremors, and convulsions.

Subacute lead poisoning, usually seen in sheep or older cattle, is characterized by anorexia, rumen stasis, colic, dullness, and transient constipation, frequently followed by diarrhea, blindness, head pressing, bruxism, hyperesthesia, and incoordination. Chronic lead poisoning, which is occasionally seen in cattle, may produce a syndrome that has many features in common with acute or subacute lead poisoning.

GI abnormalities, including anorexia, colic, emesis, and diarrhea or constipation, may be seen in dogs. Anxiety, hysterical barking, and jaw champing, salivation, blindness, ataxia, muscle spasms, opisthotonos and convulsions may develop. CNS depression rather than CNS excitation may be evident in some dogs. In horses, lead poisoning usually produces a chronic syndrome characterized by weight loss, depression, weakness, colic, diarrhea, laryngeal or pharyngeal paralysis (roaring), and dysphagia that frequently results in aspiration pneumonia.

- In avian species, anorexia, ataxia, loss of condition, wing and leg weakness, and anemia are the most notable signs.

- Lesions. Animals that die from acute lead poisoning may have few observable gross lesions. Oil or flakes of paint or battery may be evident in the GI tract. The caustic action of lead salts causes gastroenteritis. In the nervous system, edema, congestion of the cerebral cortex, and flattening of the cortical gyri are present. Histologically, endothelial swelling, laminar cortical necrosis, and edema of the white matter may be evident. Tubular necrosis and degeneration and intranuclear acid-fast inclusion bodies may be seen in the kidneys. Osteoporosis has been described in lambs. Placentitis and accumulation of lead in the fetus may result in abortion.
- Diagnosis. Lead levels in various tissues may be useful to evaluate excessive accumulation and to reflect the level or duration of exposure, severity, and prognosis and the success of treatment.
- Concentrations of lead in the blood at 0.35 ppm, liver at 10 ppm, or kidney cortex at 10 ppm are consistent with a diagnosis of lead poisoning in most species.

Hematologic abnormalities, which may be indicative but not confirmatory of lead poisoning, include anemia, anisocytosis, poikilocytosis, polychromasia, basophilic stippling, metarubricytosis, and hypochromia. Blood or urinary ä-aminolevulinic acid and free erythrocyte protoporphyrin levels are sensitive indicators of lead exposure but may not be reliable indicators of clinical disease. Radiologic examination may be useful to determine the magnitude of lead exposure.

Lead poisoning may be confused with other diseases that cause nervous or GI abnormalities. In cattle, such diseases may include polioencephalomalacia, nervous coccidiosis, tetanus, hypovitaminosis A, hypomagnesemic tetany, nervous acetonemia, arsenic or mercury poisoning, brain abscess or neoplasia, rabies, listeriosis, and Haemophilus infections. In dogs, rabies, distemper, and hepatitis may appear similar to lead poisoning.

Treatment. If tissue damage is extensive, particularly to the nervous system, treatment may not be successful. In livestock, calcium disodium edetate (Ca-EDTA) is given IV or SC (110 mg/kg/day) divided into 2 treatments daily for 3 days; this treatment should be repeated 2 days later. In dogs, a similar dose divided into 4 treatments/day is administered SC in 5% dextrose for 2-5 days. After a 1-wk rest period, an additional 5-day treatment may be required if clinical signs persist. No approved veterinary product containing Ca-EDTA is commercially available at present.

Thiamine (2-4 mg/kg/day SC) alleviates clinical manifestations and reduces tissue deposition of lead. Combined Ca-EDTA and thiamine treatment appears to produce the most beneficial response. d-Penicillamine can be administered PO to dogs (110 mg/kg/day) for 2 wk. However, undesirable side effects such as emesis and anorexia have been associated with this treatment. d- Penicillamine is not recommended for livestock. Succimer (meso

2,3-dimercaptosuccinic acid, DMSA) is a chelating agent that has proven to be effective in dogs (10 mg/kg, PO, tid for 10 days) and is also useful in birds. Fewer side effects have been associated with DMSA as compared with Ca-EDTA.

Cathartics such as magnesium sulfate (400 mg/kg, PO) or a rumenotomy may be useful to remove lead from the GI tract. Barbiturates or tranquilizers may be indicated to control convulsions.

Chelation therapy, in combination with antioxidant treatment, may limit oxidative damage associated with acute lead poisoning. Antioxidants such as n-acetylcysteine (50 mg/kg, PO, sid) have been used in combination with DMSA.

Mobilization of lead at parturition, excretion of lead into milk, and lengthy withdrawal times in food-producing animals raise considerable controversy regarding the rationale for treatment from both public health and animal management perspectives.

Arsenic poisoning

Arsenic poisoning in animals is caused by several different types of inorganic and organic arsenical compounds. Toxicity varies with factors such as oxidation state of the arsenic, solubility, species of animal involved, and duration of exposure. Therefore, the toxic effects produced by phenylarsonic feed additives and other inorganic and organic compounds must be distinguished.

Inorganic arsenicals. These include arsenic trioxide, arsenic pentoxide, sodium and potassium arsenate, sodium and potassium arsenite, and lead or calcium arsenate. Trivalent arsenicals, also known as arsenites, are more soluble and therefore more toxic than the pentavalents or arsenate compounds. The lethal oral dose of sodium arsenite in most species is from 1-25 mg/kg. Cats may be more sensitive. Arsenates (pentavalents) are 5-10 times less toxic than arsenites. Poisoning is now relatively infrequent due to decreased use of these compounds as pesticides, ant baits, and wood preservatives. Arsenites are used to some extent as dips for tick control. Lead arsenate is sometimes used as a taeniacide in sheep.

Toxicokinetics and Mechanism of Action. Soluble forms of arsenic compounds are well absorbed orally. Following absorption, most of the arsenic is bound to RBC; it distributes to several tissues, with the highest levels found in liver, kidneys, heart, and lungs. In subchronic or chronic exposures, arsenic accumulates in skin, nails, hooves, sweat glands, and hair. The majority of the absorbed arsenic is excreted in the urine as inorganic arsenic or in methylated form.

The mechanism of action of arsenic toxicosis varies with the type of arsenical compound. Generally, tissues that are rich in oxidative enzymes such as the GI tract, liver, kidneys, lungs, endothelium, and epidermis are considered more vulnerable to arsenic damage. Trivalent inorganic and aliphatic organic arsenic compounds exert their toxicity by interacting with sulfhydryl enzymes, resulting in disruption of cellular metabolism. Arsenate can uncouple oxidation and phosphorylation.

Clinical Findings. Poisoning is usually acute with major effects on the GI tract and cardiovascular system. Arsenic has a direct effect on the capillaries, causing damage to microvascular integrity, transudation of plasma, loss of blood, and hypovolemic shock. Profuse watery diarrhea, sometimes tinged with blood, is characteristic, as are severe colic, dehydration, weakness, depression, weak pulse, and cardiovascular collapse. The onset is rapid, and signs are usually seen within a few hours (or up to 24 hr). The course may run from hours to several weeks depending on the quantity ingested. In peracute poisoning, animals may simply be found dead.

Lesions. In peracute toxicosis, no significant lesions may be seen. Inflammation and reddening of GI mucosa (local or diffuse) may be seen followed by edema, rupture of blood vessels, and necrosis of epithelial and subepithelial tissue. Necrosis may progress to perforation of the gastric or intestinal wall. GI contents are often fluid, foul smelling, and blood tinged; they may contain shreds of epithelial tissue. There is diffuse inflammation of the liver, kidneys, and other visceral organs. The liver may have fatty degeneration and necrosis, and the kidneys have tubular damage. In cases of cutaneous exposure, the skin may exhibit necrosis and be dry or leathery.

Diagnosis. Chemical determination of arsenic in tissues (liver or kidney) or stomach contents provides confirmation. Liver and kidneys of normal animals rarely contain >1 ppm arsenic (wet wt); toxicity is associated with a concentration >3 ppm. The determination of arsenic in stomach contents is of value usually within the first 24-48 hr after ingestion. The concentration of arsenic in urine can be high for several days after ingestion. Drinking water containing >0.25% arsenic is considered potentially toxic, especially for large animals.

Treatment. In animals with recent exposure and no clinical signs, emesis should be induced (in capable species), followed by activated charcoal with a cathartic (efficacy of charcoal in arsenic toxicosis remains to be determined) and then oral administration of GI protectants (small animals, 1-2 hr after charcoal) such as kaolin-pectin, and fluid therapy as needed. In animals already showing clinical signs, aggressive fluid therapy, blood transfusion (if needed), and administration of dimercaprol (British antilewisite, 4-7 mg/kg, IM, tid for 2-3 days or until recovery). In large animals, thioctic acid (lipoic acid or á-lipoic acid) may be used alone (50 mg/kg, IM, tid, as a 20% solution) or in combination with dimercaprol (3 mg/kg, IM, every 4 hr for the first 2 days, qid for the third day, and bid for the next 10 days or until recovery). In large animals, the efficacy of dimercaprol alone is questionable. Sodium thiosulfate has also been used, PO, at 20-30 g in 300 mL of water in horses and cattle, one-fourth this dose in sheep and goats, and 0.5-3 g in small animals or as a 20% solution, IV, at 30-40 mg/kg, 2-3 times/day for 3-4 days or until recovery.

The water-soluble analogs of dimercaprol, 2,3-dimercaptopropane-1-sulfonate (DMPS) and dimercaptosuccinic acid (DMSA), are considered to be less toxic and more effective and could be given orally. d-Penicillamine has been reported to be an effective arsenic chelator in humans. It has a wide margin of safety and could be used in animals at 10-50 mg/kg, PO, 3-4 times/day for 3-4 days. Supportive therapy may be of even greater value, particularly when cardiovascular collapse is imminent, and should involve IV fluids to restore blood volume and correct dehydration. Kidney and liver function should be monitored during treatment.

Organic arsenicals. Phenylarsonic organic arsenicals are relatively less toxic than inorganic compounds or aliphatic and other aromatic organic compounds.

Aliphatic organic arsenicals include cacodylic acid and acetarsonic acid. These are generally used as stimulants in large animals, but their use is no longer common. Some aliphatic arsenicals such as monosodium methanearsonate (MSMA) and disodium methanearsonate (DSMA) are occasionally used as cotton defoliants or crabgrass killers. Persistence of MSMA or DSMA in the soil and their tendency to accumulate in plants creates a potential for arsenic poisoning, especially in grazing animals. Clinical signs, lesions, and treatment of aliphatic organic arsenicals are similar to those of inorganic arsenicals.

Aromatic organic arsenicals include trivalent phenylorganicals such as thiacetarsamide and arsphenamine for the treatment of adult heartworms in dogs and pentavalent compounds such as phenylarsonic acids and their salts. Thiacetarsamide and arsphenamine are no longer used commonly, especially since the recent introduction of melarsomine.

Phenylarsonic compounds are used as feed additives to improve production in swine and poultry rations and also to treat dysentery in pigs. The 3 major compounds in this class are arsanilic acid, roxarsone (4-hydroxy-3-nitrophenylarsonic acid), and nitarsone (4-nitro-phenylarsonic acid).

Etiology. Toxicosis results from an excess of arsenic-containing additives in pig or poultry diets. Severity and rapidity of onset are dose-dependent. Signs may be delayed for weeks after incorporation of 2-3 times the recommended (100 ppm) levels or may occur within days when the excess is >10 times the recommended levels. Chickens are tolerant of arsanilic acid; however, roxarsone can produce toxicosis in turkeys at only twice the recommended dose (50 ppm).

Roxarsone also has a higher toxicity in pigs as compared with other phenylarsonics.

Clinical Findings and Diagnosis. The earliest sign in pigs may be a reduction in weight gain, followed by incoordination, posterior paralysis, and eventually quadriplegia. Animals remain alert and maintain good appetite. Blindness is characteristic of arsanilic acid intoxication but not of other organic arsenicals. In ruminants, phenylarsonic toxicosis is similar to inorganic arsenic poisoning. There are usually no specific lesions present in phenylarsonic poisoning.

Demyelination and gliosis of peripheral nerves, the optic tract, and optic nerves are usually seen on histopathology. Analyses of feed for the presence of high levels of phenylarsonics confirm the diagnosis.

Phenylarsonic poisoning in pigs should be differentiated from salt poisoning, insecticide poisoning, and pseudorabies. In cattle, arsenic poisoning should be differentiated from other heavy metal (lead) poisoning, insecticide poisoning, and infectious diseases such as bovine viral diarrhea.

Treatment and Prognosis. There is no specific treatment, but the neurotoxic effects are usually reversible if the offending feed is withdrawn within 2-3 days of onset of ataxia. Once paralysis occurs, the nerve damage is irreversible. Blindness is also usually irreversible, but animals retain their appetite, and weight gain is good if competition for food is eliminated. Recovery may be doubtful when the exposure is long and the onset of intoxication slow.

Poisoning by fluorine compounds

Fluorides are widely distributed in the environment and originate naturally from rocks and soil or from industrial processes. Water supplies for human consumption have been adjusted to contain 1 ppm to prevent dental caries. Fluorine at 1-2 mg/kg in animal rations is considered adequate. The maximal tolerable level varies by species, eg, 40-50 ppm for cattle and horses, and 200 mg/kg for chickens. (The terms "fluorine" and "fluoride" are used interchangeably.)

Etiology. Toxic quantities of fluorides occur naturally, eg, certain rock phosphates and the superphosphates produced from them, partially defluorinated phosphates, and the phosphatic limestones. In certain areas, drinking water from deep wells may contain high levels of fluorides. Volcanic ash may be high in fluoride. Wastes from industrial processes, fertilizers, and mineral supplements are the most common causes of chronic fluorosis. The fluorine-containing gases and dusts from manufacturing of fertilizers, mineral supplements, metal ores (steel and aluminum), and certain enamelling processes may contaminate forage crops. Contamination of the surrounding area, particularly in the direction of the prevailing wind, may extend 5-6 miles.

Forage crops grown on high-fluorine soils have increased levels due to mechanical contamination with soil particles. Feed-grade phosphates must contain no more than 1 part of fluorine to 100 parts phosphorus. A 100-g tube of fluoride toothpaste may contain 75-500 mg of sodium fluoride, depending on the brand.

There is a general correlation between solubility of a fluoride and its toxicity. Of the common fluorides, sodium fluoride is the most toxic, and calcium fluoride the least toxic.

The fluorides of rock phosphates and most cryolites are of intermediate toxicity. Soluble fluorides originating from industrial fumes or dusts are more toxic than fluoride in rock phosphate.

Fluoride binds to Ca2+, Mg2+, and Mn2+, acting as a direct cellular poison (including bacterial cells, hence its use in dental hygiene). At high levels most fluorides are corrosive to tissue. In bone, fluoride binds calcium and replaces the hydroxyl groups in the mineral part of bone, which is mostly hydroxyapatite. In teeth developed during fluoride ingestion, the enamel is less soluble (protective) and more dense (brittle, if excessive). In addition, faulty mineralization of teeth and bones occurs when excessive fluoride interferes with intracellular calcium metabolism and damages ameloblasts and odontoblasts.

- Clinical Findings. Acute poisoning from inhalation of fluorine-containing gases or from ingestion of rodenticides or ascaricides containing fluoride is rare. Oral cleaning products present a danger to pets, especially dogs. The fatal dose of sodium fluoride is 5-10 mg/kg and toxic effects occur below 1 mg/kg. Fluoride (75-90% absorbed by 90 min) lowers serum calcium and magnesium.
- Clinically, gastroenteritis and cardiac (ventricular tachycardia and ECG abnormalities) and nervous signs may be followed within a few hours by collapse and death.
- The signs of fluorosis from chronic ingestion are the same regardless of the source of fluoride. Levels too low to produce skeletal signs can cause changes in the enamel of developing teeth, leading to chalkiness or mottling, staining, and rapid and irregular wear. When exposure occurs after dental development, the teeth remain normal even if severe skeletal fluorosis develops.
- Clinical signs, apart from mild tooth lesions, occur in many animals when bone fluoride reaches 4 000 ppm. Skeletal fluorosis results in accelerated bone resorption and remodeling with production of exostoses and sclerosis. Metabolically active bones (ribs, mandible, and long bones) and growing bones in the young are most affected. Affected animals are lame, and feed and water intake and weight gain are decreased. Severely diseased cattle may move around on their knees due to spurring and bridging of the joints in the late stages. When the skeleton becomes saturated (30-40 times normal bone content), "flooding" of the soft tissue occurs, which causes a rise in plasma fluorides and metabolic breakdown evidenced by a loss of appetite and listlessness.

Lesions. Acute ingestion of high levels of fluoride causes inflammation of the gut and degenerative changes in the lungs, liver, and kidneys. In chronic cases, mottling, staining, and excessive wearing occur in teeth that develop during the time of excessive fluoride ingestion. A more advanced stage of fluorosis is marked by skeletal abnormalities; the bones become chalky white, soft, thickened, and in the extreme, develop exostoses that may be palpated, especially along the long bones and on the mandible in animals exposed at any age.

Diagnosis. Urine fluoride levels are time dependent due to rapid elimination. In cases of known ingestion, serum calcium and magnesium levels are beneficial. Casual observation of affected animals may suggest chronic debilitating arthritis; osteoporosis; or deficiency of calcium, phosphorus, or vitamin D. Lameness in advanced cases may be wrongly attributed to an accident. Nonspecific staining seen in cattle teeth may be confused with incipient fluorosis. A developing fluoride toxicosis can be recognized by the following criteria (from most to least reliable): 1) chemical analyses to determine the amount of fluorine in the diet, urine, bones, and teeth; 2) tooth effects, in animals exposed at time of permanent teeth development; 3) lameness, as the result of fluoride accumulation in bone; and 4) systemic evidence as reflected by anorexia, inanition, and cachexia.

The normal levels of fluorine in livestock are considered to be <0.2 ppm in plasma, 1-8 in urine, 200-600 in bones, and 200-500 in teeth. Normal bovine urine contains <5 ppm fluorine; in borderline toxicity, urine contains 20-30 ppm, and in cattle with systemic signs, >35 ppm. In pigs, bones appear normal with 3,000-4,000 ppm fluorine, and levels of <4,500 ppm in compact bones from cattle are considered innocuous. In cattle, toxicosis is associated with levels of >5,500 ppm in compact bone and >7,000 ppm in cancellous bone; in sheep, levels are believed to be lower (2,000-3,000 ppm in compact bone and 4,000-6,000 ppm in cancellous bone).

Treatment and Control. Acutely exposed animals require calcium gluconate (IV) and oral magnesium hydroxide or milk to bind fluoride before absorption. In chronic exposure, control is difficult unless animals are removed from affected areas. It has been suggested that affected areas may be used for animals with a relatively short production life, eg, pigs, poultry, or finishing cattle and sheep. Feeding calcium carbonate, aluminum oxide, aluminum sulfate, magnesium metasilicate, or boron has either decreased absorption or increased excretion of fluoride, and thus could offer some control of chronic fluorosis under some conditions. However, no treatment has been shown to cure the chronic effects of fluorine toxicity.

Toxicity of selenium compounds

Selenium is an essential element that has a narrow margin of safety. Feed supplements containing 0.1-0.3 ppm selenium are added to the diet to prevent deficiency diseases such as white muscle disease in cattle and sheep, hepatosis dietetica in pigs, and exudative diathesis in chickens. The maximum tolerable level for selenium in most livestock feed is considered to be 2 ppm or as high as 5 ppm, although some believe that levels as high as 4-5 ppm can inhibit growth.

Selenium is a component of the glutathione peroxidase enzyme that acts as an antioxidant during release of energy. In excess, selenium has 2 general effects: the direct inhibition of cellular oxidation/reduction reactions, and the replacement of sulfur in the body. The inhibition of numerous cellular functions by high levels of selenium results in acute generalized cytotoxicity. The replacement of sulfur by chronic intake of selenium leads to altered structure and function of cellular components. Altered sulfur-containing amino acids (methionine, cystine) affects cell division and growth. Especially susceptible are the cells that form keratin (keratinocytes) and the sulfur-containing keratin molecule. Selenium therefore weakens the hooves and hair, which tend to fracture when subjected to mechanical stress.

Etiology. All animal species are susceptible to selenium toxicosis. However, poisoning is more common in forage-eating animals such as cattle, sheep, and horses that may graze selenium- containing plants. Plants may accumulate selenium when the element is found at high levels— generally in alkaline soil with little rainfall (<50 cm). Selenium accumulating plants have been categorized. Obligate indicator plants require large amounts of selenium for growth and contain high concentrations (often >1,000 ppm). Facultative indicator plants absorb and tolerate high levels of soil selenium accumulating up to 100 ppm under these conditions, but they do not require selenium. Nonaccumulator plants passively absorb low levels of selenium (1-25 ppm) from the soil. Poisoning may also occur in swine and poultry consuming grain raised on seleniferous soils or, more commonly, due to error in feed formulation. Selenium toxicosis after ingestion of selenium-containing shampoos or excess selenium tablets is rare in pets. Several factors are known to alter selenium toxicity; however, in general, a single acute oral dose of selenium in the range of 1-5 mg/kg is lethal in most animals. Parenteral selenium products are also quite toxic, especially to young animals, and have caused deaths in baby pigs, calves, and dogs at doses as low as 1.0 mg/kg.

Diagnosis. Severity of selenium toxicosis depends on the quantity ingested and duration of exposure. Poisoning in animals is characterized as acute, subchronic, or chronic. Diagnosis is based on clinical signs; necropsy findings; and laboratory confirmation of presence of high selenium levels in an animal's diet (feed, forage, grains), blood, or tissues (kidney, liver).

- Selenium levels in the diet >5 ppm may produce signs after prolonged exposure. Levels of 10-25 ppm could produce severe signs. In acute toxicosis, the blood selenium concentration may reach 25 ppm, and in chronic toxicosis, it may be 1-4 ppm. Kidney or liver may contain 4-25 ppm in both acute and chronic poisoning.
- Acute selenium poisoning due to consumption of plants with levels >50 ppm (dosages 3-20 mg/kg) is rare but has caused large losses in cattle, sheep, and pigs. Animals usually avoid these plants because of their offensive odor; however, when pasture is limited, accumulator plants may be the only food available. Young

animals are most susceptible to acute parenteral selenium toxicosis with dosages of 0.2-0.5 mg/kg. Clinical signs are different from those of chronic selenosis and are characterized by abnormal behavior, respiratory difficulty, gastrointestinal upset, and sudden death. Abnormal posture and depression, anorexia, unsteady gait, diarrhea, colic, increased pulse and respiration rates, frothy nasal discharge, moist rales, and cyanosis may be noted.

- Death usually follows within a few hours of consumption or injection. The major lesions are lung edema and congestion, and necrosis of multiple organs, including lung, liver, and kidney. Sheep usually do not show these signs, but instead become depressed and die suddenly.
- Blood selenium concentration in acute poisoning is much higher than in chronic poisoning. In acute cases, blood selenium may reach 25 ppm. Treatment consists of symptomatic and supportive care. Acetylcysteine to boost glutathione levels is beneficial.
- Subchronic selenium poisoning. Pigs fed a diet supplemented with selenium >20-50 ppm for >3 days develop a subchronic selenium toxicosis characterized by neurologic abnormalities. Animals are initially ataxic and uncoordinated followed by anterior paresis, then quadriplegia. Pigs continue to eat. The hooves show breaks and impaired growth similar to those seen in cattle; alopecia is observed. In sows, conception rate decreases and number of pigs born dead increases. Lesions of subchronic toxicosis include focal symmetric poliomyelomalacia, which is most prominent in the cervical and thoracic spinal cord. Death may result from complications of permanent paralysis. Hoof and hair damage is similar to but in most cases less severe than that observed in chronic selenium toxicosis. Treatment is similar to that for chronic toxicosis, but spinal lesions are usually permanent.

Chronic selenium poisoning usually develops when livestock consume seleniferous forages and grains containing 5-50 ppm of selenium for many weeks or months. Naturally occurring seleno- amino acids in plants are readily absorbed. Until recently, 2 types of chronic selenium poisoning were recognized—alkali disease and blind staggers. Blind staggers is no longer believed to be caused by selenium but by sulfate toxicity due to consumption of high-sulfate alkali water. Excess sulfate (>2% of diet) leads to polioencephalomalacia and the classical signs of blind staggers.

Animals consuming milk vetch (Astragalus bisulcatus) have demonstrated clinical signs similar to blind staggers. Although milk vetch contains high levels of selenium, evidence now indicates that the alkaloid swainsonine in milk vetch, responsible for locoism, produces the signs.

Clinical Findings. Alkali disease has been reported in cattle, sheep, and horses. Affected animals are dull, emaciated, and lack vitality. The most distinctive lesions are those involving the keratin of the hair and hooves. The animal has a rough hair coat and the long hairs of the mane and tail break off at the same level giving a "bob" tail and "roached" mane appearance. Abnormal growth and structure of horns and hooves results in circular ridges and cracking of the hoof wall at the coronary band. Extremely long, deformed hooves that turn upwards at the ends may be seen.

Subsequent lameness is compounded by degeneration of joint cartilage and bone. Reduced fertility and reproductive performance occurs especially in sheep. Reproductive performance may be depressed with a dietary level of selenium lower than that required to produce typical signs of alkali disease. Other lesions may include anemia, liver cirrhosis and ascites, and atrophy of the heart.

Birds also may be affected with chronic selenium toxicosis. Eggs with >2.5 ppm selenium from birds in high selenium areas have low hatchability, and the embryos are usually deformed.

Teratologic effects include underdeveloped feet and legs, malformed eyes, crooked beaks, and ropy feathers. This has been a problem with waterfowl in southern California, where selenium was leached by agricultural water and concentrated in lakes by runoff.

Blood levels of selenium in chronic cases are usually 1-4 ppm. Other changes in blood include decreased fibrinogen level and prothrombin activity; increased serum alkaline phosphatase, ALT, AST, and succinic dehydrogenase; and reduced glutathione. Hair may have >5 ppm selenium in chronic poisoning. A "garlicky" odor on the animal's breath may be noted. Treatment and Control. There is no specific treatment for selenium toxicosis. Eliminating the source and exposure and symptomatic and supportive care of the animal should be started as soon as possible. Addition of substances that antagonize or inhibit the toxic effects of selenium in the diet may help reduce the risk of selenium toxicosis. A high protein diet, linseed oil meal, sulfur, arsenic, silver, copper, cadmium, and mercury have reduced selenium toxicity in laboratory animals, but their use under field conditions is limited. Addition of arsenic salt at 0.00375% to enhance biliary excretion of selenium or use of a high-protein diet to bind free selenium may help reduce incidence of selenium poisoning in cattle.

Soil and forages should be tested regularly in high-selenium areas.

Poisoning by Mercurium Compounds

Mercury exists in a variety of organic and inorganic forms. The replacement of commercial mercurial compounds, including antiseptics (eg, mercurochrome), diuretics, and fungicides by other agents has decreased the likelihood of mercurial toxicosis; however, the possibility of exposure to environmental sources of organic methylmercury exists.

Inorganic Mercurials. These include the volatile elemental form of mercury (used in thermometers) and the salted forms (mercuric chloride [sublimate] and mercurous chloride [calomel]). Ingested inorganic mercury is poorly absorbed and low in toxicity. Large amounts of these mercurials are corrosive and may produce vomiting, diarrhea, and colic. Renal damage also occurs, with polydipsia and anuria in severe cases. In rare cases of chronic inorganic mercurial poisoning, the CNS effects resemble those of organic mercury poisoning. Mercury vapor from elemental mercury produces corrosive bronchitis and interstitial pneumonia and, if not fatal, may lead to neurologic signs as do organic forms. Emesis followed by initiation of chelation therapy (see below) is recommended after acute oral ingestion. Oral administration of sodium thiosulfate to bind mercury still in the gut may be beneficial.

Organic Mercury. Inorganic mercury is converted to the organic alkyl forms, methylmercury and ethylmercury, by microorganisms in the sediment of rivers, lakes, and seas. Marine life accumulate the most toxic form, methylmercury, and fish must be monitored for contamination.

There are reports of commercial cat food causing severe neurologic disturbances in cats fed an exclusive tuna diet for 7-11 mo.

The organic mercurials are absorbed via all routes and bioaccumulate in the brain and to some extent in the kidneys and muscle. Aryl mercurials (eg, phenylmercury fungicide) are slightly less toxic and less prone to bioaccumulation. Animals poisoned by organic mercury exhibit CNS stimulation and locomotor abnormalities after a lengthy latent period (weeks). Signs may include blindness, excitation, abnormal behavior and chewing, incoordination, and convulsions. Cats show hindleg rigidity, hypermetria, cerebellar ataxia, and tremors. Mercury is also a mutagen, teratogen, and a carcinogen, and is embryocidal. Differential diagnoses include conditions with tremors and ataxia as predominant signs, such as ingestion of other metals and insecticides and cerebellar lesions due to trauma or feline parvovirus.

Histologic lesions include degeneration of neurons and perivascular cuffing in the cerebrocortical gray matter, cerebellar atrophy of the granular layer, and damage to Purkinje cells. Laboratory diagnosis must differentiate between normal concentrations of mercury in tissue (especially whole blood, kidney, and brain) and feed (<1 ppm) and concentrations associated with poisoning.

Neurologic signs may be irreversible once they develop. Chelation therapy with dimercaprol (3 mg/kg body wt, IM, every 4 hr for the first 2 days, qid on the third day, and bid for the next 10 days or until recovery is complete) has been beneficial. When available, the water soluble, less toxic analog of dimercaprol, 2,3-dimercaptosuccinic acid, is the chelator of choice for organic mercury poisoning. Penicillamine (15-50 mg/kg, PO) may be used only after the gut is free of ingested mercury and renal function has been established.

Toxicity of molybdenum compounds

Molybdenum is an essential micronutrient that forms molybdenoenzymes, which are necessary for the health of all animals. In ruminants, the dietary intake of excessive molybdenum causes, in part, a secondary hypocuprosis. Toxicosis due to massive doses of molybdenum is rare. Domestic ruminants are much more susceptible to molybdenum toxicity than nonruminants. The resistance of other species is at least 10 times that of cattle and sheep.

Etiology. The metabolism of copper, molybdenum, and inorganic sulfate is a complex and incompletely understood interrelationship. It appears that the ruminal interaction of molybdates and sulfides gives rise to thiomolybdates (monoditri and tetrathiomolybdates). Copper reacts with thiomolybdates (primarily tri-and tetrathiomolybdates) in the rumen to form an insoluble complex that is poorly absorbed. On this basis, tetrathiomolybdate is used in treating and preventing copper toxicity in sheep. Some thiomolybdates are absorbed and decrease blood copper availability and also appear to directly inhibit copper-dependent enzymes. Therefore, the susceptibility of ruminants to molybdenum toxicity depends on a number of factors: 1) copper content of the diet and intake of the animal—tolerance to molybdenum toxicity decreases as the content and intake of copper decrease; 2) the inorganic sulfate content of the diet—high dietary sulfate with low copper exacerbates the condition, while low dietary sulfate causes high blood molybdenum levels due to decreased excretion; 3) chemical form of the molybdenum—water- soluble molybdenum in growing herbage is most toxic, while curing decreases toxicity; 4) presence of certain sulfur-containing amino acids; 5) species of animal—cattle are less tolerant than sheep; 6) age—young animals are more susceptible; 7) season of year—plants concentrate molybdenum beginning in spring (maximum level reached in fall); and 8) botanic composition of the pasture—legumes take up more of the element than other plant species.

Molybdenum toxicity associated with copper deficiency has been seen in areas with peat or muck soils, where plants grow in alkaline sloughs (eg, western USA), as a result of industrial contamination (mining and metal alloy production), where excess molybdenum-containing fertilizer has been applied, and where applications of lime appeared to increase plant molybdenum uptake.

In the diet of cattle, copper:molybdenum ratios of 6:1 are considered ideal; 2:1-3:1, borderline; and <2:1, toxic. Dietary molybdenum of >10 ppm can cause toxicity regardless of copper intake; as little as 1 ppm may be hazardous if copper content is <5 ppm (dry-weight basis). Mixing errors may occur; concentrations above 1,000 mg/kg (as sodium molybdate) cause growth retardation while concentrations of 2,000-4,000 mg/kg cause death within 40 days.

Clinical Findings and Diagnosis. Most of the clinical signs attributed to molybdenum toxicity arise from impaired copper metabolism and are the same as those produced by simple copper deficiency. Molybdenum toxicity in cattle is characterized by persistent, severe scouring with passage of liquid feces full of gas bubbles (peat scours or teart). Depigmentation, resulting in fading of the hair coat, is most noticeable in black animals and especially around the eyes, which gives a spectacled appearance. Other signs include unthriftiness, anemia, emaciation, joint pain (lameness), osteoporosis, and decreased fertility. Effects on reproduction, particularly in heifers, include delayed puberty, decreased weight at puberty, and reduced conception rates. It appears that fertility is uniquely vulnerable to the effects of molybdenum or thiomolybdates and alone responds indirectly to copper acting as an antidote. Some studies have suggested that relatively low levels of molybdenum may exert these direct effects on certain metabolic processes, particularly reproduction, independent of alterations in copper metabolism. Sheep, and young animals in particular, show stiffness of the back and legs with a reluctance to rise (called enzootic ataxia in Australia). Joint and skeletal lesions appear to be due to defects in development of connective tissue and growth plates. Clinical signs appear within 1-2 wk of grazing affected pasture.

- In molybdenum toxicity, low copper levels in blood and tissue and the occurrence of clinical signs of copper deficiency in cattle are poorly correlated. A provisional diagnosis can be made if the diarrhea stops within a few days of oral dosing with copper sulfate; the diagnosis is further supported if other causes of diarrhea and unthriftiness (including GI parasites) are ruled out.
- Diagnosis is confirmed by demonstrating abnormal concentrations of molybdenum and copper in blood or liver and by a high dietary intake of molybdenum relative to copper.
- The disease may be confused with many other enteritides and is commonly mistaken for internal parasitism, especially in young cattle. In pastured animals, it is not uncommon for the diseases to occur simultaneously.
- Effects in cattle and sheep poisoned with massive concentrations of molybdenum are unlike the chronic induced copper deficiency described above. Cattle lose appetite within 3 days and deaths begin to occur within 1 wk and continue for months after exposure ends. Animals appear lethargic, display hind limb ataxia that progresses to involve the front limbs, salivate profusely,
- and produce scant, mucoid feces. The molybdenum is toxic to hepatocytes and renal tubular epithelial cells, producing periacinar to massive hepatic necrosis and nephrosis.

Prevention and Treatment. Signs of severe acute toxicosis are reversed by providing copper sulfate in the diet. In areas where the molybdenum content of the forage is <5 ppm, the use of 1% copper sulfate ($CuSO_4 \cdot 5H_2O$) in salt has provided satisfactory control of molybdenosis. With higher levels of molybdenum, 2% copper sulfate has been successful; up to 5% has been used in a few regions where the molybdenum levels are very high. In areas where, for various reasons, cattle do not consume mineral supplements, the required copper may be supplied as a drench given weekly, as parenterally administered repository copper preparations, or as a top-dressing to the pasture. Copper glycinate injectable has been used successfully as an adjunct to therapy.

Toxicity of cadmium compounds

Cadmium accumulation in plants and animals is increasing from several sources of environmental exposure. The application of rock phosphate (which contains varying amounts of cadmium depending on the source) and sewage sludge fertilizers results in cadmium deposition in the soil of pastures. In addition to direct ingestion of soil containing cadmium, some forage plants extract cadmium from the soil. Although cadmium accumulation in the soft tissues of livestock has been demonstrated and there is ample experimental documentation of the toxicity of cadmium in animals, under natural conditions, documented cases of direct toxic or carcinogenic effects of cadmium in livestock have been uncommon.

Background. Pure cadmium is a soft, silver-white metal with an atomic number of 48 and a molecular weight of 112.41. It is a divalent transition metal with chemical properties that are similar to zinc and is usually found as a mineral in combination with other elements to form cadmium oxide, cadmium chloride, or cadmium sulfate. Numerous compounds are formed from cadmium and thus it is used in batteries, solders, semiconductors, solar cells, plastics stabilizers, and to plate iron and steel. All soil and rocks contain some cadmium. It can enter the environment from zinc smelting and refining, coal combustion, mine wastes, iron and steel production, and from the use of rock phosphate and sewage sludge as fertilizers. Cadmium accumulation in plants and animals is increasing from a variety of sources, being the most severe in the vicinity of zinc smelters. The use of cadmium-containing mineral supplements in feed (e.g. from calcium phosphate), the application on pastures and hay fields of phosphate fertilizers (which contain varying amounts of cadmium depending on the source) and sewage sludge results in cadmium deposition in the soil. Some plants readily extract cadmium from the soil thereby making it available for consumption. For example, cadmium concentrations in clover grown in soil fertilized with high cadmium rock phosphate were significantly higher than the concentrations in clover grown in soils treated with low cadmium phosphate fertilizer. A New Zealand National Survey of soils and plants, and random testing of kidneys from grazing animals revealed that there was an approximately two-fold increase in soil cadmium while over a 3-year period,

14–20% of cattle kidneys exceeded the New Zealand maximum residue level of 1 µg Cd/g. In a study where cattle were allowed to graze pastures treated with anaerobically digested sewage sludge for up to 8 years, cadmium was the only metal to accumulate consistently in increased amounts in the tissues of the cattle. It has been reported that cattle grazing on sewage sludge treated pastures consumed significantly more (up to 3 times) cadmium than cattle on control. In addition, a Swedish study has found a direct correlation between cadmium in feed and pig.

A recent study reporting the analysis of Wisconsin dairy feeds for heavy metals found that cadmium concentrations in complete dairy feed rations were the closest of the heavy metals to US maximum acceptable concentrations, suggesting that cadmium has the greatest potential to exceed those maximum standards if the amounts of However, several studies have failed to demonstrate any adverse clinical manifestations related to increased cadmium concentrations in the animals examined. In one study, although cattle on pasture fertilized with sewage sludge consumed increased amounts of cadmium and had increased fecal excretion and kidney accumulation of cadmium, there were no adverse health effects noted in these cattle. Similarly, when corn silage or corn that was grown on sewage sludge fertilized fields were fed to sheep or pigs, respectively, significant increases in kidney cadmium concentrations were measured, but no other adverse treatment-related effects were noted.

Although cadmium is of concern in the environment, and cattle grazing on cadmium-contaminated pastures have increased tissue concentrations of cadmium, two additional studines conclude that accumulation of cadmium in the liver and kidneys of cattle may be a moderately effective screen for the entry of cadmium into the human food chain, as long as liver and especially kidneys are not consumed. It has been reported that regardless of the concentrations of cadmium fed to livestock, the amount in meat, milk, and eggs is always lower than that in the diet that the animal was eating. Thus, foods derived from those products decrease human exposure.

This is fortunate as chronic cadmium poisoning has been documented in humans. In these cases, it has been associated with osteoporosis, renal lesions, tissue mineral imbalances, and death. In addition, the Department of Health and Human Services has determined that cadmium and cadmium compounds may be reasonably anticipated to be carcinogens.

Toxicokinetics. In animals, cadmium exposure is primarily through oral ingestion. Compared to other divalent cations such as zinc and iron, intestinal absorption of cadmium is relatively low, ranging from approximately 1% to 5% in most species, with up to as much as 16% in cattle, dependent on the dose. Interestingly, cadmium bound to metallothionein in foods of animal origin is absorbed less efficiently than cadmium salts and therefore, may be less available for uptake.

After absorption, cadmium is transported in the plasma bound to albumin and in lesser amounts to other serum proteins. It distributes throughout the body with the highest concentrations in the liver and kidneys, which account for approximately one-half of the total cadmium in the body. Muscle and bone do not accumulate high concentrations of cadmium. Blood cadmium concentrations are indicators of recent exposure while urine cadmium is a better indicator of the body burden.

Cadmium is not transported well into milk or eggs, or across the placental barrier. In pregnant and lactating livestock, the toxicokinetics of cadmium have been compared. In this study, the kinetics of cadmium were measured in lactating versus non-lactating ewes after a single intravenous or oral administration of cadmium chloride. The non-lactating ewes exhibited a low cadmium bioavailability (0.12–0.22%), a large steady-state volume of distribution (23.8–5.4 l/kg), and a low blood clearance (0.2–0.03 l/ kg/day) with a mean residence time of 113–28 days. The lactating ewes had a higher bioavailability (0.33–1.7%), and the mean residence time was close to that of the nonlactating ewes despite a greater blood clearance (0.46–0.013 l/kg/day) because the volume of distribution of cadmium in the body was larger. The cadmium clearance in milk remained low in the lactating ewes. In the body, cadmium is excreted very slowly, with daily losses of approximately 0.009% of the total via the urine and approximately 0.007% in the feces via the bile. Cadmium–protein complexes are excreted in the kidneys and then resorbed from the filtrate in the proximal tubules. This area of the renal cortex accumulates cadmium and is susceptible to damage and necrosis. Depending on the species, the biological half-life of cadmium can vary from months to years, which results in cadmium accumulating in animals as they age.

For example, several studies have documented age-related increases in cadmium in the kidneys of horses. In mammals and birds, cadmium accumulates in the liver and kidneys at concentrations of 0.1–2.0 and 1–10 mg/kg wet weight, respectively. It has been discovered that animals with long- life spans, such as horses, can accumulate large amounts of cadmium in their organs, particularly in their kidneys. In samples of renal cortex from old horses, concentrations of up to 200 mg/kg have been reported.

Mechanism of action. Experimentally, acute exposure to high doses of inorganic cadmium leads to its accumulation in many organs, eliciting liver and in some cases, testicular damage. Once inside the cell, free cadmium binds to protein sulfhydryl groups, disrupting the cellular redox cycle, depleting glutathione, and eliciting intracellular oxidant damage. In addition, its similarity to other divalent cations such as calcium interferes with their normal functioning. Cadmium ions can displace zinc and other divalent metals from their binding sites on metalloproteins. For example, in the testis, cadmium can interfere with zinc-proteins, leading to widespread apoptosis and necrosis. In the liver, acute cadmium toxicity results in widespread hepatocyte apoptosis, followed by varying degrees of necrosis depending on the dose. This is related, in part, to the effects of resident liver macrophages (Kupffer cells) to potentiate and increase the initial liver damage caused by cadmium alone. This has been demonstrated in several systems in which inhibition of Kupffer cells significantly decreases liver damage caused by a toxic dose of cadmium. Cadmium readily binds to, and induces the production of, metallothionein, a cysteine- rich, metal-binding protein. Binding to metallothionein does not have a major effect on the uptake of cadmium, but is, in part, responsible for retention of cadmium within cells and its long half-life (greater than 10 years in humans). Metallothionein does this by decreasing cadmium elimination, especially in bile. Within hepatocytes, metallothionein binds to cadmium, decreasing its hepatotoxicity. Experimentally, rats that have greater induction of metallothionein in the liver are somewhat protected from cadmium hepatotoxicity. However, in the kidneys the cadmium– metallothionein complex is nephrotoxic and it has been theorized that it may play a role in chronic poisoning in humans.

Toxicity. Increased exposure to cadmium in combination with zinc, lead, and/or other metals continues to occur in the vicinity of non-ferrous metal smelters and processing facilities. These exposures have resulted in toxicoses, although it can be difficult to separate the effects of cadmium from those of lead, zinc, and other metals. In one such case in the Netherlands, kidney cadmium concentrations were found to be twice those of cattle in control areas. However, although hemoglobin, blood iron concentrations, and ironbinding capacity were lower in the cadmium-exposed cattle compared to controls, no adverse clinical effects were observed. In an additional study in the Netherlands, bulls fed diets containing increased concentrations of cadmium, lead, mercury, and arsenic had increased concentrations of cadmium in the kidney and liver, but did not exhibit histological lesions related to the intake of heavy metals. However, more recently, deaths in horses exposed to cadmium, lead, and zinc from a non-ferrous metal processing plant in Eastern Europe were attributed to ingestion of these metals in their feed.

Analysis of tissues from a number of these horses revealed extremely high concentrations of cadmium (40–100 times normal) and 3–6 times the normal concentrations of lead. Toxicoses have also been reported in sheep and horses in the vicinity of nonferrous metal smelters in China.

Analysis of the tissues from these animals revealed lead and cadmium concentrations significantly higher than those of controls. A survey of cadmium concentrations in tissues from healthy swine, cattle, dogs, and horses in the midwestern United States was conducted in the mid-1970s. While the median cadmium concentration was low, at or below 0.6 ppm in the kidneys of cattle, swine, and dogs, the median concentration in the kidneys was 4 times greater in horses. One study has indicated that horses may be more at risk for cadmium toxicity than other species. In this Swedish study, the cadmium concentrations in the kidney cortices of 69 otherwise normal horses were measured and correlated to any histological lesions that were noted. In that study, renal cadmium concentrations ranged from 11 to 186 μg Cd/g wet wt., with an average of 60 μg Cd/g. This study found a correlation between increased chronic interstitial nephritis and increasing cadmium concentrations in the renal cortex. There was no obvious relationship between the age and the frequency of renal lesions. These same authors also found that cadmium concentrations in the kidney cortices were approximately 15 times greater than those in the liver of the same animals.

Age-dependent increases in kidney metallothionein and cadmium have also been reported in horses. One diagnostic investigation has reported lameness and swollen joints, i.e lesions of osteochondrosis, in addition to osteoporosis and nephrocalcinosis in horses near a zinc smelter in Pennsylvania. In the horses examined, kidney zinc and cadmium concentrations were elevated. It this case, it was postulated that the osteoporosis that was observed in one foal and the nephrocalcinosis seen in the foal and its dam were related to the elevated renal cadmium. When ponies were raised near a similar zinc smelter for periods of time up to 18.5 months, there were significant elevations in tissue zinc and cadmium concentrations. Increases in tissue cadmium concentrations were correlated with increasing age, although increases in tissue zinc concentrations were not. Generalized osteochondrosis was present in joints of the limbs and cervical vertebrae, and there was lymphoid hyperplasia. From this study, it was concluded that the development of osteochondrosis was associated with increased exposure to zinc and possibly cadmium. However, other lesions of cadmium toxicosis, such as renal damage or osteomalacia, were not present. In wildlife, white-tailed deer (Odocoileus virginianius) harvested within 20 km of zinc smelters in Pennsylvania had very high kidney concentrations of cadmium and zinc. These deer were also reported to have had jodint lesions similar to zinc-poisoned horses from the same area. In humans, occupational exposure to cadmium has been associated with renal dysfunction and osteomalacia with osteoporosis. One of the earliest effects of chronic cadmium exposure is renal tubular damage with proteinuria. Other chronic effects can include liver damage, emphysema (through inhalation), osteomalacia, neurological impairment, testicular, pancreatic, adrenal damage, and anemia. Tumorigenic effects have been reported in experimental animals.

Historically in the 1940s, high environmental exposure in one area of Japan from eating cadmium-contaminated rice resulted in itai-itai (ouch-ouch) disease. This was manifested by intense bone pains and pathological bone fractures, mainly in elderly women, with osteoporosis and renal dysfunction. In addition, studies in Europe and China have demonstrated that low to moderate exposure to cadmium from zinc smelters resulted in a decrease in bone density, and an increase in fractures in women. In one of these studies, cadmium concentrations in the blood and urine were taken as biomarkers of exposure. Experimental studines in animals have confirmed the adverse effects of cadmium on bones. Six mechanisms have been theorized to explain these effects: (1) interference with parathyroid hormone stimulation of vitamin D production in the kidney, (2) reduced renal vitamin D activation, (3) increased urinary excretion of calcium, (4) reduced intestinal calcium absorption, (5) interference with calcium deposition in bones, and (6) interference with bone collagen production.

Treatment. In animals, cadmium toxicosis is prevented by minimizing exposure in the environment and in feedstuffs.

TOXICITY OF ZINC COMPOUNDS

Zinc is an essential trace metal that plays an important role in many biologic processes. It is ubiquitous in nature and exists in many forms. The ingestion of some forms leads to creation of toxic zinc salts in the acidic gastric environment. Zinc toxicity has been documented in humans as well as in a wide range of large, small, exotic, and wild animals. It is seen commonly in pet dogs, possibly because of a higher degree of dietary indiscretion and greater levels of exposure to zinc- containing substances. Common sources of zinc include batteries, automotive parts, paints, zinc- oxide creams, herbal supplements, zippers, board-game pieces, screws and nuts on pet carriers, and the coating on galvanized metals such as pipes and cookware. One of the most well known sources of zinc that causes toxicity following ingestion is the USA Lincoln penny. Some pennies minted during 1983, and all pennies minted since, are 97.5% zinc by weight.

Pathogenesis. The low pH in the stomach causes the formation of soluble zinc salts. These are absorbed from the duodenum and rapidly distributed to the liver, kidneys, prostate, muscles, bones, and pancreas. Zinc salts have direct irritant and corrosive effects on tissue, interfere with the metabolism of other ions such as copper, calcium, and iron, and inhibit erythrocyte production and function. The mechanisms by which zinc exerts these toxic effects are not completely understood. The median lethal dose (LD50) of zinc salts in cases of acute toxicity has been reported to be ~100 mg/kg. Also, diets containing high levels of zinc (>2,000 ppm) have been reported to cause chronic zinc toxicosis in large animals.

Clinical Signs and Lesions. Clinical signs vary based on the duration and degree of exposure. Signs progress from anorexia and vomiting to more advanced symptoms such as diarrhea, lethargy, icterus, shock, intravascular hemolysis, hemoglobinuria, cardiac arrhythmias, and seizures. Large animals often show decreases in weight gain and milk production, and lameness has been reported in foals secondary to epiphyseal swelling.

Major histopathologic findings include hepatocellular centrolobular necrosis with hemosiderosis and vacuolar degeneration, renal tubular necrosis with hemoglobin casts, and pancreatic duct necrosis with fibrosis of the interlobular fat.

Diagnosis. Radiodense material is easily seen on radiographs of the GI tract in animals with zinc- containing foreign bodies. Changes in the CBC, chemistry profile, urinalysis, and coagulation profile reflect the degree of toxicity to various organ systems. The hemogram typically reveals a regenerative hemolytic anemia characterized by changes in erythrocyte morphology. The leukogram often shows a neutrophilic leukocytosis secondary to stress, pancreatitis, and a regenerative bone marrow. Serum chemistry changes that are seen secondary to hepatic damage include elevations in bilirubin, the transaminases, and alkaline phosphatase. As zinc accumulates in the pancreas, increases in amylase and lipase can be seen following pancreatitis and pancreatic necrosis. Glomerular damage and renal tubular epithelial necrosis result in elevations in BUN, creatinine, amylase, and urine protein. Hemoglobinuria can be differentiated from hematuria during urinalysis; the urine color will not clear after centrifugation in the presence of hemoglobinuria. Prolongation of prothrombin time and activated partial thromboplastin time can result from toxic effects on the synthesis or function of coagulation factors.

The hematologic and clinical findings in animals with zinc toxicosis are similar to the changes in animals with immune-mediated hemolytic anemia (IMHA). Misdiagnosis of a primary autoimmune disorder can lead to the inappropriate use of immunosuppressive drugs. Zinc toxicosis can cause the direct antiglobulin test (direct Coombs' test) to be positive in the absence of a primary autoimmune disorder. The direct Coombs' test is therefore not reliable when differentiating between zinc intoxication and IMHA.

Definitive diagnosis of zinc poisoning is achieved by measuring zinc levels in blood or other tissue. In dogs and cats, the normal serum zinc level is 0.7-2 μg/mL. Serum samples can be submitted in green-top heparinized tubes or in royal blue-top trace element tubes. Methods for quantifying zinc levels from saliva and hair have not been validated in domestic animals, and measuring zinc in urine is unreliable because elimination of zinc through the kidneys is variable. Differential diagnoses should include any infectious, toxic, immune-mediated, neoplastic, genetic, or other medical disorder characterized by clinical signs and laboratory test results similar to those seen in cases of zinc toxicity. These include IMHA, hypophosphatemia, splenic torsion, babesiosis, ehrlichiosis, heartworm disease, leptospirosis, hemobartonellosis, feline leukemia infection, hemangiosarcoma, lymphosarcoma, phosphofructokinase or pyruvate-kinase deficiency, and toxicity from acetaminophen, naphthalene, paradichlorobenzene, Allium, lead, or copper.

Treatment and Prevention. After stabilizing the animal with fluids, oxygen, and blood products as necessary, removal of the source of zinc as early as possible is paramount. This often requires surgery or endoscopy. Inducing emesis to remove chronic gastric zinc foreign bodies is typically not rewarding because zinc objects often adhere to the gastric mucosa.

Diuresis with a balanced crystalloid solution is indicated to promote renal excretion of zinc and prevent hemoglobinuric nephrosis.

There is debate regarding the necessity of chelation therapy in cases of zinc toxicosis. Animals can recover from zinc intoxication following only supportive care and removal of the source. However, chelation therapy enhances elimination of zinc and thus may accelerate recovery.

Calcium disodium ethylenediaminetetraacetate (Ca-EDTA) successfully chelates zinc when given at 100 mg/kg/day IV or SC for 3 days (diluted and divided into 4 doses), but may exacerbate zinc- induced nephrotoxicity. Although they have been used to treat animals with zinc toxicity, d- penicillamine and dimercaprol (British antilewisite) have not been specifically validated for this purpose. Reported doses are 110 mg/kg/day for 7-14 days for d-penicillamine, and 3-6 mg/kg tid for 3-5 days for dimercaprol. Chelation therapy with any of these agents should be monitored with serial serum zinc levels to help determine the appropriate duration of treatment.

If diagnosed early and treated aggressively, the outcome is often favorable for animals with zinc toxicosis. Eliminating sources of zinc from the environment is essential in preventing recurrence.

Poisoning by organic chlorine compounds

These compounds are poisonous to insects on contact, have good knock-down capacity, and kill quickly. They are absorbed through the chitinous cuticle of the insect, their residual effect is long lasting everywhere, including in the environment. Many chlorinated hydrocarbons are known or suspected to be carcinogenic. Due to tissue residues and chronic toxicity, use of these agents is drastically curtailed. Only lindane and methoxychlor are approved for use on or around livestock. Nevertheless, in a recent surveillance study, 51 % of the cattle (mainly originating from Colorado) had detectable residues of chlorinated hydrocarbon insecticides including heptachlor, heptachlor epoxide, lindane, and oxychlordane.

- Aldrin is a potent insecticide similar to dieldrin with the same order of toxicity. It is no longer registered in the USA but was used for termite control.
- Benzene hexachloride (BHC, hexachlorocyclohexane) was a useful insecticide for large animals and dogs· but is highly toxic to cats in the concentrations necessary for parasite control. Cattle in good condition have tolerated 0.2% lindane applications, but stressed, emaciated cattle have been poisoned from spraying or dipping in 0.075% lindane. Horses and pigs appear to tolerate 0.2- 0.5%, and sheep and goats ordinarily tolerate 0.5% applications. Emaciation and lactation increase the susceptibility of animals to poisoning by lindane; such animals should be

treated with extreme caution. Young calves are very susceptible to lindane and are poisoned by a single oral dose of 4.4 mg/ kg body wt. Mild signs appear in sheep given 22 mglkg, and death occurs at 100 mg/ kg.

- Adult cattle have tolerated 13 mg/kg without signs. BHC is stored in body fat and excreted in milk.
- Chlordane is no longer registered as an insecticide in the USA. Exposure occurs when livestock consume treated plants or when they come in direct contact through carelessness and accidents. Very young calves have been killed by doses of 44 mg/kg, and the minimum toxic dose for cattle is -88 mg/kg. Cattle fed chlordane at 25 ppm .of their diet for 56 days showed 19 ppm in their fat at the end of the feeding. Topical emulsions and suspensions have been used safely on dogs at concentrations up to 0.25%, provided freshly diluted materials were used; dry powders up to 5% have been safe. The no effect level in dogs in a 2-yr feeding study was 3 mg/kg. Pigeons and Leghorn cockerels and pullets suffered no effects after 1-2 mo exposure to vapors emanating from chlordane-treated surfaces.
- Dieldrin is not a registered pesticide in the USA. Residues limit its application, and it is one of the most toxic chlorinated hydrocarbon insecticides. Young dairy calves are poisoned by 8.8 mg/kg body wt, PO, but tolerate 4.4 mg/kg, while adult cattle tolerate 8.8 mg/kg and are poisoned by 22 mg/kg. Pigs tolerate 22 mg/kg and are poisoned by 44 mg/kg. Horses are poisoned by 22 mglkg. Because of its effectiveness against insect pests on crops and pasture and the low dosage per acre, dieldrin is not likely to poison livestock grazing the treated areas. Diets containing 25 ppm of dieldrin have been fed to cattle and sheep for 16 wk without harmful effects other than residues in fat, which are slow to disappear. Great care must be exercised in marketing animals that have grazed treated areas or consumed products from previously treated areas. There is a zero tolerance level for residues in edible tissues.

Heptachlor is not currently registered in the USA and is not recommended for use on livestock in the USA. Because it is very effective against certain plant-feeding insects, it is encountered from time to time in some geographic areas grazed by livestock. Young dairy calves tolerate dosages as high as 13 mg/kg body wt but are poisoned by 22 mg/kg. Sheep tolerate 22 mg/kg but are poisoned by 40 mg/kg. Diets containing 60 ppm of heptachlor have been fed to cattle for 16 wk without harmful effects other than residues in fat. Heptachlor is converted to heptachlor epoxide by animals and stored in body fat. For this reason, a specific analysis performed for heptachlor usually yields negative results, while that for epoxide is positive.

- Methoxychlor is one of the safest chlorinated hydrocarbon insecticides and one of the few with active registration in the USA. Young dairy calves tolerate 265 mg/kg body wt; 500 mg/kg is mildly toxic. While 1 g/kg produces rather severe poisoning in young calves, sheep are not affected. One dog was given 990 mg/kg daily for 30 days without showing signs. Six applications to cattle of a 0.5% spray at 3-wk intervals produces fat residues of 2.4 ppm; -0.4 ppm of methoxychlor is found in milk 1 day after spraying a cow with a O. 5% concentration.

Methoxychlor sprays are not approved for use on animals producing milk for human consumption. Cattle and sheep store essentially no methoxychlor when fed 25 ppm in the total diet for 112 days. If methoxychlor is used as recommended, the established tolerance in fat will not be exceeded. Commercial products are available for garden, orchard, and field crops and for horses and ponies. Numerous reports suggest that methoxychlor has negative reproductive effects in laboratory animal experiments, but this has not been seen in the field.

Toxaphene is no longer under active registration in the USA. It has been used with reasonable safety if recommendations were followed, but it can cause poisoning when applied or ingested in excessive quantities. Dogs and cats are particularly susceptible. Young calves have been poisoned by 1 % toxaphene sprays, while all other farm animals except poultry can withstand 1 % or more as sprays or dips. Chickens have been poisoned by dipping in 0.1% emulsions, and turkeys have been poisoned by spraying with 0.5% material. Toxaphene is primarily an acute toxicant and does not persist long in the tissues. Adult cattle have been mildly intoxicated by 4% sprays and severely affected by 8%. Adult cattle have been poisoned from being dipped in emulsions that contained only 0.5% toxaphene (an amount ordinarily safe) because the emulsions had - begun to break down, allowing the fine droplets to coalesce into larger droplets that readily adhere to the hair of cattle. The resultant dosage becomes equivalent to that obtained by spray treatments of much higher concentrations. Toxaphene is lethal to young calves at 8.8 mg/kg body wt but not at 4.4 mg/kg. The minimum toxic dose for cattle is -33 mg/kg, and for sheep between 22 and 33 mg/kg. Spraying Hereford cattle 12 times at 2-wk intervals with 0.5% toxaphene produced a maximum residue of 8 ppm in fat. Cattle fed 10 ppm of toxaphene in the diet for 30 days had no detectable toxaphene tissue residues, while steers fed 100 ppm for 112 days stored only 40 ppm in their fat.

Clinical Findings. The chlorinated hydrocarbon insecticides are general CNS stimulants. They produce a great variety of signs-the most obvious are neuromuscular tremors and convulsionsand there may be obvious behavioral changes common to other poisonings and CNS infections. Body temperature may be very high. Affected animals are generally first noted to be more alert or apprehensive. Muscle fasciculation occurs, becoming visible in the facial region and extending backward until the whole body is involved. Large doses of DDT, DOD, and methoxychlor cause progressive involvement leading to trembling or shivering, followed by convulsions and death.

With the other chlorinated hydrocarbons, the muscular twitchings are followed by convulsions, usually without the intermediate trembling. Convulsions may be continuous, clonic, or tonic lasting from a few seconds to several hours, or intermittent and leading to the animal becoming comatose. High fever may accompany convulsions, particularly in warm environments.

Behavioral changes such as abnormal postures (e.g., resting the sternum on the ground while remaining upright in the rear, keeping the head down between the forelegs, "head pressing" against a wall or fence, or continual chewing movements) may be seen. Occasionally, an affected animal becomes belligerent and attacks other animals, people, or moving objects. Vocalisation is common. Some animals are depressed, almost oblivious to their surroundings, and do not eat or drink; they may last longer than those showing more violent symptoms. Usually, there is a copious flow of thick saliva and urinary incontinence. In certain cases, the clinical signs alternate, with the animal first being extremely excited, then severely depressed. The severity of the signs seen at a given time is not a sure prognostic index .. Some animals have only a single convulsion and die, while others suffer innumerable convulsions but subsequently recover. Animals showing acute excitability often have a fever >41°C. The signs of poisoning by these insecticides are highly suggestive but not diagnostic; other poisons and encephalitis or meningitis must be considered.

Signs of acute intoxication by chlordane in birds are nervous chirping, excitability, collapse on hocks or side, and mucous exudates in the nasal passages. Signs of subacute and chronic intoxication are molting, dehydration and cyanosis of the comb, weight loss, and cessation of egg production.

Lesions. If death has occurred suddenly, there may be nothing more than cyanosis. More definite lesions occur as the duration of intoxication increases. Usually, there is congestion of va\tous organs (particularly the lungs, liver, and kidneys) and a blanched appearance of all organs if the body temperature was high before death. The heart generally stops in systole, and there may be many hemorrhages of varying size on the epicardium. The appearance of the heart and lungs may suggest a peracute pneumonia and, if the animal was affected for more than a few hours, there may be pulmonary edema. The trachea and bronchi may contain a blood-tinged froth. In many cases, the CSF volume is excessive, and the brain and spinal cord frequently are congested and edematous.

Diagnosis. Chemical analysis of brain, liver, kidney, fat, and stomach or rumen contents is necessary to confirm the poisoning. The suspected source, if identified, should also be analysed. Brain levels of the insecticide are the most usefuL Whole blood, serum, and urine from live animals may be analysed to evaluate exposure in the rest of the herd or flock. In food animal poisoning, if exposure is more than just the animals visibly affected, fat biopsies from survivors may be necessary to estimate the potential residue.

Treatment. There are no known specific antidotes. When exposure is by spraying, dipping, or dusting, a thorough bathing without irritating the skin (no brushes), using detergents and copious quantities of cool water is recommended. If exposure is by ingestion, gastric lavage and saline purgatives are indicated. The use of digestible oils .such as com oil is contraindicated; however, heavy-grade mineral oil plus a purgative hastens the removal of the chemical from the intestine. Activated charcoal appears to be useful in preventing absorption from the GI tract. When signs are excitatory, a sedative anticonvulsant such as a barbiturate or diazepam is indicated. Anything in the environment that stresses the animal-noise, handling etc-should be reduced or removed if possible. If the animal shows marked depression, anorexia, and dehydration, therapy should be directed toward rehydration and nourishment either IV or by stomach tube. Residues in exposed animals may be reduced by giving a slurry of activated charcoal or providing charcoal in feed.

Feeding phenobarbital, 5 g/day, may hasten residue removal. Examples of Organochlorines: DDT. Chemistry. Dichloro-diphenyl-trichloroethane (DDT); insoluble in water. Not to be used on livestock or premises where livestock are housed. Metabolism in animals. Poorly absorbed from the skin as powder or in water suspension. a. An oily vehicle increases absorption of DDT, b.

Poorly absorbed from the GI tract c. Fator fat solvents in the GItract increase absorption of DDT Distributes to all tissues, but selectively distributes to fat. Will store in fat over a period of time and reach levels higher than the level in the ration. Is cumulative. Partly biotrans formed by the liver. Excreted by kidney both free and in metabolites. Excreted by mammary gland in the butterfat of milk Toxicity. DDT has a relatively wide margin of safety under common uses because little is absorbed into the blood. Order of decreasing species-susceptibility to DDT toxicity: 1. Rodents 6. Monkey 2. Cat 7. Swine 3. Dog 8. Horse 4. Rabbit 9. Cow 5. Guinea pig 10. Sheep 11. Goat Young animals are especially susceptible. Fasted animals are especially susceptible. Toxic reactions: 1. Stimulation of CNS produce tremors, in coordinated walk, convulsions, coma, respiratory failure and eventually death from respiratory paralysis. 2. Sensory nerve is most sensitive; nerve ganglia are least sensitive. 3. Produces a change in the permeability of the nerve membrane. 4. Produces a prolongation of the falling phase of nerve action potential(i. e., negatyve after potential). 5. High body temperature. 6. Liver damage by high doses. Treatment of toxicity:1. Use a CNS depressant such as barbiturate anesthesia. Use with care. 2. Remove the source of exposure. To reduce a body load of DDT, which is particularly useful in lactating dairy cows, administer: 1. Phenobarbital in the feed. This speeds liver biotransformation of DDT via enzime induction. 2. Activated charcoal traps DDT and its metabolites in the in the GI tract for excretion with feces. Not greatly effective.

Analogs of DDT: Methoxychlor is one of the important analogs of DDT. It is less toxic, has little tendency to accumulate in body fat and there is little secretion into milk.

Cyclodienes. These are more complex chemically than DDT. They are very good insecticides. Many of them are synthesized by what is referred to as DielsAlder reaction. Chlordeneis not an insecticide by itself. It is important as a precursor of chlordane. Chlorination of chlordene to 68 to 69% gives chlordane which is a mixture of compounds. This mixture has three components; (1) betachlordane, (2) alfa-chlordane, and (3) heptachlor. Beta-chlordane is a cis isomer and is usually the active component. The other compounds of cyclodiene group include aldrin and dieldrin.

Toxaphene has a cyclic structure but it is different from cyclodienes in being a simple molecule. Chlordane: It is a dark brown viscous liquid and is soluble in organic solvents, insoluble in water. Its toxicity varies from preparation to preparation. The acute toxicity of chlordane is slightly less than DDT, the average single oral lethal dose being about 200-300 mg/kg for rats, rabbits, and dogs. Sheep are much more sensitive. The minimum oral toxic dose for young calves is about 25 mg/kg. The chronic toxicity of chlordane is considerably higher in animals like rabbits, sheep, goats and cattle.

Heptachlor: It is usually prepared in the form of waxy solid and is very stable and is a broad spectrum insecticide. The oral dose is of the order of 100 mg/kg. Its epoxide, heptachlor epoxide, is about ten times more toxic (orally) to dairy calves than the parent compound.

Aldrin: This is one of the most toxic of this group of insecticides. The single oral lethal dose is about 40 mg per kg in rats and slightly more in rabbits and dogs. For young calves the minimum toxic dose is about 5 mg per kg. It also has a high level of chronic toxicity. Isodrin is an isomer of. aldrin and is much more toxic, the single oral lethal dose being 12-17 mglkg.

Dieldrin: It is rather less acutely and chronically toxic than aldrin. The average single oral leth~l dose for all species is 50-90 mglkg. For one to two-week calves the single oral toxic dose is about 10-20 mglkg. It is used as a seed dressing and is reported that wildbirds eating the treated seeds are very susceptible to dieldrin. Endrin is the isomer of dieldrin, the single oral lethal dose is about 10-12 mg/kg. Endrin together with lindane, aldrin and dieldrin has been found to be effective as a systemic acaricide in cattle. Also, it is used for the control of cotton insect.

Toxaphene: It is also a mixture of compounds and is obtained by chlorinating camphene. Dogs seem to be particularly susceptible to the acute action of toxaphene. The oral lethal dose is about 20-40 mg/kg. In young calves, oral dose of about 5 mg/kg produced toxic symptoms. Chickens appear to be relatively resistant. Chronic toxicity is relatively low. Strobane is very similar to toxaphene and is used in control of soil inhabiting insects. Endosulfan and Telodrin are compounds related to cyclodienes. Mode of action. In general they have high degree of toxicity for most insects; topical toxicity to mammals is not very high. They are broad spectrum insecticides. They are nerve poisons but specific action is unknown. They seem to work on CNS. They all produce similar symptoms and that is why they tend to be regarded as having the same mode of action. The symptoms include tremors, convulsions and paralysis.

Hexachlorocyclohexanes. These compounds are related to cyclodienes and are formed by chlorination of benzene in presence of UV light. It is a mixture of compounds of 5 isomers which have lost the aromatic character of the ring. Alfa-isomer of hexachlorocyclohexane (HCH) is the most common(+80%). Gamma-isomerconstitutes about 10-15% in the crude mixture. Gamma- isomerwaspurified and called lindane which has insecticidal properties. In the literature, it is wrongly called benzene hexachloride. Benzenehexa-chloride (BHC) is used as a fungicide.

Lindane: It is odorless, white crystalline solid, stable, and is soluble in organic solvents. Lindane (gamma-isomer of HCH) has acute toxic effect greater than other isomers. The average single oral lethal dose of HCH (12% gamma-isomer) is said to have about 19m/kg. The minimum toxic dose of lindane was found to be 5 mgl kg in baby calves, and 25 mg in older cattle and sheep. Young animals are more susceptible to HCH. It is excreted in milk. Mode of action. The mode of action of lindane and its isomers is not well understood, some say it apparently attacks the ganglia of the nervous system, while others claim that it is a neurotoxic agent whose action is similar to that of DDT. In acute toxicity, hypersensitivity, tremors and convulsions are seen. Chronic exposure results liver enlargement in mammals.

Mirex and Kepone. Mirex has been used extensively in the south eastern U. S. for control of fire ant. Dietary feeding of mirex to rats resulted in hepatocellular carcinoma. It is degraded to kepone in body. Kepone is more toxic than mirex and is known to cause tremors, liver injury and reproductive dysfunction in exposed population.

Poisoning by organic phosphorus compounds

Organophosphates (OPs) and carbamates (CMs) are commonly used as pesticides in agriculture, industry, and around the home/garden throughout the world. In addition, these chemicals are used as parasiticides in veterinary medicine. Both types of chemicals produce their toxicity by virtue of inhibition of acetylcholinesterase (AChE) enzyme, which terminates the action of the neurotransmitter acetylcholine (ACh) at the synapses in nervous tissue and at the neuromuscular junctions. These chemicals are referred to as "anticholinesterases". Some of the OPs with strong AChE inhibiting potential are also used as nerve agents or nerve gases. Many compounds of both classes are extremely toxic and lack species selectivity, and therefore their inadvertent/ accidental use continues to pose a threat to the environment, human and animal health, wildlife, and aquatic systems. Small animals often encounter poisoning with these insecticides by malicious activity, while livestock by ingesting freshly sprayed crop or contaminated feed. Although these compounds are neurotoxicants, they produce a variety of cholinergic and non-cholinergic effects. Latest evidence suggests that while cholinergic mechanisms play a critical role in the initial stage of toxicity, neuronal damage/death appears to occur through noncholinergic mechanisms. OPs and CMs are discussed here together because they produce similar toxic effects in poisoned animals.

The first OP compound, tetraethyl pyrophosphate, was synthesized in 1854 by Philipe de Clermont. In 1932, Lange and Kruger described the synthesis of dimethyl and diethyl phosphorofluoridate. Based on the chemistry of these compounds, Gerhard Schrader (a chemist at the I.G. Farbenindustrie) led the exploration of OP class of compounds that could be used as insecticides. One of the earliest OP insecticides synthesized by Schrader was parathion, which is still used worldwide. Prior to World War II (WWII), the German Ministry of Defense developed highly toxic OP compounds of G series (tabun, sarin, and soman) and diisopropyl phosphorofluoridate. In the 1950s, OP compounds with super toxicity, such as VX and VR, were synthesized in the United Kingdom and Soviet Union. After WWII, thousands of OPs have been synthesized in the search for compounds with species selectivity, i.e. more toxic to insects and less toxic to mammals. Malathion is an example. This compound has been used for about half- acenturyas the most popular insecticide. Today, more than 100 OPs are in use for a variety of purposes, such as protection of crops, grains, gardens, homes, and public health.

The first CM compound, physostigmine (eserine alkaloid), was isolated from calabar beans (ordeal poison) of a perennial plant Physostigma venenosum in the mid-1860s. The compound was used to treat glaucoma. About 50 years later, an aromatic ester of carbamic acid, neostigmine, was synthesized and used in the treatment of myasthenia gravis. Most of the CMs (esters of carbamic acid) that are used as pesticides were synthesized in the 1960s and 1970s. Carbaryl was the first CM compound used as an insecticide. To mimic the structure of acetylcholine (ACh), aldicarb was synthesized which has a toxicity greater than any other compounds of this class. Like OPs, thousands of CMs have been synthesized, but less than two dozen compounds have been used practically. Today, CMs are preferred for pesticide use over OPs because some OPs have been found to be extremely toxic, whereas others cause delayed neuropathy in animals as well as in humans. In essence, both OPs and CMs have broad applications in agriculture and veterinary medicine and as a result of their indiscriminate use acute poisonings often result in animals, birds, fish, and wildlife.

Types of OPs and CMs. Basic structures of organophosphorus and CM compounds are shown in Figure x. There are at least 13 types of OPs. Despite differences in chemical structures, all OPs share one thing in common: they all have a pentavalent phosphorus atom and a characteristic phosphoryl bond (P–O) or thiophosphoryl bond (P–S). Essentially, OPs are esters of phosphoric acid with varying combinations of oxygen, carbon, sulfur, and/or nitrogen attached. Of course, the chemistry of these compounds is much more complex. The OPs that are derivatives of phosphoric or phosphonic acid possess anticholinesterase activity, unlike those that are derivatives of phosphinic acid. Usually, OP compounds have two alkyl substituents and an additional substituents group (leaving group, which is more labile to hydrolysis than the alkyl group. Basically, some OPs (such as dichlorvos, monocrotophos, and trichlorfon) are direct AChE inhibitors, while those of phosphorothioates type (such as bromophos, diazinon, fenthion, and parathion) possess minimal or no anticholinesterase (anti- AChE) activity and require desulfuration to the analoguos oxon before acquiring anti-AChE activity. Also, OPs which are used as defoliants (S,S,S-tributyl phosphorotrithioate and S,S,S-tributyl phosphorotrithioite) and herbicides (glyphosate and gluphosinate) are of very low mammalian toxicity.

OP pesticides. The majority of OP compounds are used as pesticides.

OP nerve gases/agents. OP nerve agents include tabun (GA), sarin (GB), soman (GD), cyclosarin (GF), VX, and VR. These compounds are highly toxic and pose continuous threats for the lives of humans as well as animals, because they can be used as chemical weapons of mass destruction. So far these agents have been used by dictators and terrorists. These compounds produce toxicity by directly inhibiting AChE, and are much more potent than OP pesticides, as they cause lethality to animals in the submilligram range.

Carbamates. The CM compounds are esters of carbamic acid. Unlike OPs, CM compounds are not structurally complex. Currently, CMs are used as pesticides in agricultural crops and gardens and in veterinary medicine as parasiticides.

CM pesticides. Currently, the volume of CMs used exceeds OPs because they are comparatively safer than OPs.

Pharmacokinetics deals with the processes of absorption, distribution, metabolism, and excretion (ADME). The ADME of OP and CM insecticides have been studied in animals. These insecticides gain entry into the body mainly through oral, dermal, or inhalation exposure.

Ingestion encounters with contaminated feed/food with pesticides residue, while dermal exposure is more relevant when these insecticides are used as ectoparasiticides in the form of dust, dip, or oily solution. Inhalation of airborne insecticides occurs during or soon after aerial spray, particularly due to chemical drift. Once the insecticide reaches a portal of entry, it is available for absorption. It is established that following absorption, these insecticides are well distributed in tissue throughout the body. OP insecticides may follow either activation or detoxification, or both. Activation implies that the metabolite is more toxic than the parent compound, e.g. the conversion of malathion to malaoxon. This process is often referred to as "lethal synthesis". On the other hand, detoxification implies that the metabolite is less toxic than the parent compound, e.g. the conversion of malathion to malathion monoacid and malathion diacid. Unlike OPs, CMs are metabolized to less toxic or non-toxic metabolites, though some of the metabolites of CMs are quite toxic. For example, the two major metabolites of carbofuran (3-hydroxycarbofuran and 3- ketocarbofuran) have a significant impact on overall toxicity of carbofuran.

A bulk of the metabolic activation and detoxification reactions occur within the liver. Finally, due to extensive metabolism in the body, only few metabolites are excreted in the urine that can be used as biomarkers of insecticides exposure . Residues of some OPs and CMs can also be detected in the feces, saliva, and milk.

Mechanism of action. OP and CM insecticides share a common mode of insecticidal and toxicological action associated with their ability to inhibit the enzyme AChE within nerve tissue and at the neuromuscular junctions. Both types of insecticides have a high affinity for binding to and inhibiting the enzyme AChE, an enzyme specifically responsible for the destruction of the neurotransmitter ACh. Since the cholinergic system is widely distributed within both the central and peripheral nervous systems, chemicals that inhibit AChE are known to produce a broad range of wellcharacterized symptoms of anticholinesterases. Agraphic representation for the comparison of the AChE inhibition dynamics for the interaction of ACh, carbaryl (CM), or chlorpyrifos-oxon (OP) with AChE is shown in Figure x. The cholinesterases (ChE) are serine hydrolases that catalyze the breakdown of ACh through an acyl-transfer, where water is the acceptor molecule to which the substrate acyl moiety is transferred. A serine oxygen of the active site gorge in ChEs carries out a nucleophilic attack on the electrophilic carbon of the carbonyl group of ACh, resulting in an acetylated enzyme intermediate and the release of choline. Deacetylation occurs when an attacking water molecule (hydroxyl ion) acts as a more effective nucleophile, thereby releasing acetate. The molecular interactions between OPs and AChE have been studied in much more detail than between CMs and AChE. The rates of hydrolysis and reactivation of AChE following carbamylation and phosphorylation of the active site appear to be drastically slower than for the hydrolysis of the acetylated enzyme. The turnover time for

ACh is of the order of ~150 μs, whereas the carbamylated enzyme t1D 2 for hydrolysis is substantially slower (~15–30 min). The phosphorylated enzyme is highly stable (t1D 2 days), and further dealkylation of the phosphorylation group produces an "aged" AChE that is irreversibly inhibited. In general, OPs and CMs are considered as irreversible and reversible AChE inhibitors, respectively. Details of ChEs, interaction of OPs and CMs with ChEs, and reactivation/regeneration of ChEs are described elsewhere.

Toxicity. Most animal poisoning cases in the field are acute in nature. Onset of clinical signs usually occurs within 15 min to 1 h, which is soon followed by the signs of maximal severity, although these timings tend to vary depending upon the OP compound and its dose, and species. For example, onset of clinical signs is delayed with chlorpyrifos (Dursban) and dimethoate (Rogor). Clinical signs observed in poisoned animals can be divided into local and systemic effects. The local effects involve the eyes and the lungs, owing to their exposure to vapors or droplets of the insecticides. These effects, however, are of significance in the case of animals only when exposure is via spraying. The systemic effects are primarily on the brain, skeletal muscles, lungs, heart, and other organs. The clinical signs can also be classified as muscarinic, nicotinic, and central. Muscarinic receptor-associated effects are manifested by vomiting, abdominal and chest pain, salivation, lacrimation, urination, diarrhea (SLUD), miosis (pinpoint pupils), tracheobronchial secretion, lung edema, and cyanosis. The nicotinic receptor-associated effects are produced on autonomic ganglia and skeletal muscles, and the affected animals show twitching of muscles, tremors, followed by convulsions, and seizures. This condition may lead to paralysis. The central effects include apprehension, stimulation, followed by depression. The affected animals may also show restlessness, ataxia, stiffness of the neck, and coma. Death occurs due to respiratory failure and cardiac arrest. It is important to mention that all poisoned animals may not show all the clinical signs (as described above) with every OP or CM compounds. In other words, there is a great variation in symptomatology among the individual cases of poisoning. Surviving animals usually recover within 3–6 h with CMs and within 24 h with OPs.

Poisoning cases of OP or CM are usually diagnosed by determining the level of inhibition of AChE activity in blood from a live animal and brain from a dead animal. Inhibition of Ache activity _70% is considered positive case of poisoning. It should be noted that great species variability exists in normal values of AChE activity. Also, analyzing the cortex of the brain and not the striatum for AChE analysis is preferred, since more than 6-fold variability exists (Gupta, 2004). Therefore, interpretation should be made with great caution. Residue analysis of an insecticide and/or its metabolite(s), followed by confirmation with GC/MS or LC/MS, seems an ideal approach for diagnosis.

Treatment of acute poisoning. Before instituting antidotal therapy, monogastric animals, such as dog, should be given gastric lavage. Animals of any species can be given activated charcoal to stop further absorption of insecticides. Animals should be washed thoroughly with water if they are exposed to insecticides dermally. Intravenous (IV) fluid therapy is always beneficial. In the case of OP poisoning, antidotal treatment requires the combined use of atropine sulfate and pyridine-2- aldoxime methochloride (2-PAM). Atropine sulfate acts by blocking the muscarinic receptors from ACh. In ruminants, one-fourth of the total recommended dose (0.5 mg/kg) can be given as a slow IV injection, and the remainder through intramuscular (IM) or subcutaneous (SC) injection. The total dose of atropine sulfate for an average size horse is about 65 mg, and for a dog is about 2 mg. Atropine sulfate treatment can be repeated at an interval of every hour until all hyper-secretory signs have subsided. 2-PAM reactivates the AChE inhibited by OPs. The recommended therapeutic dose of 2-PAM is 20 mg/kg, IV. The injection of 2-PAM can be repeated once after 1 h at half of its initial dose. Care should be taken that only a freshly prepared solution of 2-PAM be used. It needs to be emphasized that the combined therapy of atropine sulfate and 2-PAM is superior to any other treatment till today in the case of OP poisoning. Although many other oximes have been tested against many OPs, none has been proven to be better than 2-PAM. Furthermore, depressant drugs, such as morphine and barbiturates, are contraindicated, since they aggravate the condition. Diazepam without atropine sulfate also accentuates the toxicity of OPs. Unlike with OP poisoning, 2-PAM and other oximes are ineffective in CM poisoning cases. In fact, in the case of some CMs, such as carbaryl and carbofuran, 2-PAM therapy accentuates the toxicity. Some anticonvulsant drugs, such as barbiturates and diazepam, also aggravate the toxicity of CMs. Therefore, atropine sulfate, with doses as described for OPs, is the only preferred antidote. When the animals are exposed to very higher doses of CMs, atropine sulfate does not appear to be a life saving antidote.

OP-induced delayed polyneuropathy. OP compounds that produce delayed neurotoxic effects are esters of phosphorus-containing acids. Over 35 years ago, tri-o-cresyl phosphate (TOCP) was known to produce delayed neurotoxic effects in man and chicken, characterized by ataxia and weakness of the limbs, developing 10–14 days after exposure (Johnson, 1969). This sindrome was called OP-induced delayed neuropathy (OPIDN). In recent literature, the syndrome has been renamed as OP-induced delayed polyneuropathy (OPIDP). OPIDP is characterized by distal degeneration of long- and largediameter motor and sensory axons of both peripheral nerves and spinal cord. Among all animal species hen appears to be the most sensitive and therefore used as an animal model. TOCP and certain other compounds have minimal or no anti-AChE property, however they cause phosphorylation and aging (dealkylation) of a protein in neurons called neuropathy target esterase (NTE), and subsequently lead to OPIDP. Studies on the sensitivity of the target enzymes of a variety of OPs showed that the comparative inhibitory power of OPs against hen AChE and NTE in vitro correlates with their comparative effects in vivo (i.e. delayed neuropathy or death). The relationship between the degree of NTE inhibition and the severity of OPIDP changes according to the compound involved. For example, certain compounds cause OPIDP with a minimum of 70% NTE inhibition, while others require almost complete inhibition to cause OPIDP. However, the cascade of events from NTE inhibition/aging to impairment of retrograde axonal transport and axonal degeneration is yet to be explained. Today, many compounds, such as DFP, N,N_ -diisopropyl phosphorodiamidic fluoride (mipafox), tetraethyl pyrophosphate (TEPP), paraoxon, parathion, o-cresyl saligenin phosphate, and haloxon, are known to produce this sindrome. Treatment of this syndrome is symptomatic.

OP-induced intermediate syndrome. OP insecticide-induced intermediate syndrome (IMS) was reported for the first time in human patients in Sri Lanka in 1987 (Senanayake and Karalliedde 1987). Thereafter this syndrome has been diagnosed in OP-poisoned patients in South Africa (1989), Turkey (1990), Belgium (1992), the United States (1992), Venezuela (1998), France (2000), and elsewhere. To date, OPs that are known to cause IMS include bromophos, chlorpyrifos, diazinon, dicrotophos, dimethoate, disulfoton, fenthion, malathion, merphos, methamidophos, methyl parathion, monocrotophos, omethoate, parathion, phosmet, and trichlorfon. IMS is usually observed in individuals who have ingested a massive dose of an OP insecticide either accidentally or in a suicide attempt. IMS is clearly a separate clinical entity from acute toxicity and delayed neuropathy. Clinically, IMS is characterized by acute paralysis and weakness in the areas of several cranial motor nerves, neck flexors, facial, extraocular, palatal, nuchal, proximal limb, and respiratory muscles 24–96 h after poisoning. Generalized eakness, depressed deep tendon reflexes, ptosis, and diplopia are also evident. These symptoms may last for several days or weeks depending on the OP involved. A similar syndrome has also been observed in dogs and cats poisoned maliciously or accidentally with massive doses of certain OPs. It should be noted that despite Severe AChE inhibition, muscle fasciculations and muscarinic receptor-associated hyper-secretory activities are absent. Although the exact mechanism involved in pathogenesis of IMS is unclear, studies suggest that decrease of AChE and nicotinic ACh receptor mRNA expression occurs after oral poisoning with disulfoton in rats. Based on electromyographic (EMG) findings from OP-poisoned patients and experimental studies on laboratory animals, it has been found that the defect in IMS is at the neuromuscular endplate and postsynaptic level, but the effects of neural and central components in producing muscular weakness have not been ruled out. It seems clear that some OPs are greatly distributed to muscles and have higher affinity for nicotinic ACh receptors. Currently, very little is known about the type of damage at the motor endplate or about risk factors contributing to its development. There is no specific treatment, and therapy relies upon atropine sulfate and 2-PAM. The administration of atropine sulfate and 2-PAM should be continued for a long period, even if efficacy of these drugs on the development of IMS appears to be limited.

Tolerance development. Tolerance development to the toxicity of OPs was noted more than 50 years ago. Following prolonged exposure to an OP, the physiological effects often diminish more than expected from the degree of AChE inhibition or repeated additions of OP give lower responses with time. Tolerance to AChE inhibiting OPs (such as DFP, disulfoton, methyl parathion, and others) has been observed using different forms of administration and in different species, such as mice, rats, guinea pig, and man. Tolerance to OP toxicity can develop in several ways. Most often, it occurs due to receptor changes either in the number of receptors or by decreased affinity of the receptor molecule. However, it can also occur due to the presence of other proteins that can bind or inactivate the inhibitor and thereby make it less readily available. Some of the examples for binding to the OPs are carboxylesterases (CarbEs), butyrylcholinesterases (BuChEs), or other binding proteins such as albumin. In addition, tolerance can be achieved through more rapid metabolism of the OP compounds by OP-hydrolyzing enzymes such as paraoxonases (PONs) and somanases. ACh receptors (both mAChRs and nAChRs) are involved in the development of tolerance to OP toxicity. Treatment with a cholinergic agonist for a prolonged time leads to a decrease in the muscarinic ACh receptors (mAChRs). This is common for G protein-linked receptors. In some studies, OPs have been found to cause decrease in the numbers of mAChRs in the brain, while in others both the number of mAChRs and the affinity to the ligand in ileum and striatum. Significant reductions in nAChRs numbers (Bmax), without change in affinity (KD), have been found in the brain of tolerant rats treated with disulfoton in skeletal muscle of rats treated with DFP. In tolerant rats, significant recovery of CarbEs and BuChEs has also been found. In essence, tolerance development following subchronic or chronic treatment with AChE inhibiting OPs occurs through multiple mechanisms.

Conclusions and future directions. OPs and CMs constitute a large number of chemicals that are used in agriculture primarily as insecticides and in veterinary medicine as parasiticides. These chemicals exert a broad range of toxic effects, varying from mild effects as salivation and tremors to as serious as convulsions, seizures, paralysis, and death. Basically, OPs and CMs are neurotoxicants, but directly or indirectly several vital organs are affected, as these chemicals produce a variety of toxicological effects on the central nervous system, peripheral nervous system, cardiovascular, pulmonary, ocular, neurobehavioral, immunological, reproductive, placental, cutaneous, and other body systems. In addition, these insecticides cause neurodegeneration, oxidative stress, endocrine disruption, and many other complications. In general, OPs produce more serious and lingering health effects than CMs. For example, some of the complex syndromes like OPIDP and IMS have devastating effects, and have yet to be thoroughly characterized mechanistically and need to be well defined. It is expected that newer compounds of both OP and CM classes will be developed with greater selective toxicity. Also, newer antidotes need to be developed that can be effective in patients with OPIDP or IMS, or against CMs.

Poisoning by mercury and mercury organic compounds

Mercury (Hg) is a naturally occurring element that is found in the environment. It exists in several forms, such as elemental (metallic), inorganic, and organic. About 80% of the mercury released in the environment is metallic mercury, and comes from human activities, such as fossil fuel combustion, mining, smelting, and from solid waste incineration. Human activity can lead to mercury levels in soil as much as 200,000 times higher than natural levels. Metallic mercury in a pure form looks like a shinywhite liquid substance at room temperature. It is commonly used in thermometers, barometers, blood pressure devices, batteries, electric switches, dental fillings (amalgams), etc. Inorganic mercury compounds, or mercury salts, occur when it combines with other elements, such as chlorine, sulfur, and oxygen. Most of these compounds are white, except mercuric sulfide or cinnabar ore (i.e. red, but it turns black after exposure to light). Some of mercury compounds are used as fungicides, while others are used for medicinal purposes, e.g. laxatives, deworming agents, antiseptics, and disinfectants. When mercury combines with carbon, it is called organic mercury (organomercurials). Methylmercury, ethylmercury, and phenylmercury are a few examples. In the environment (by bacteria and fungi) and mammalian systems, various forms of mercury are interchangeable. For example, inorganic mercury can be methylated to methylmercury, and methylmercury can change to inorganic or elemental mercury. The animals at the top of the food chain tend to bioaccumulate methylmercury in their bodies.

Therefore, poisoning by mercury is due to consumption of meat or grain contaminated with mercury. Poisoning can also result from excessive exposure to inorganic and organic mercury compounds from misuse or overuse of mercury-containing products. Much of the information presented in this chapter is from experimental studies conducted in laboratory animals and poisoning incidences.

Mercury exists naturally in the environment, and as a result everyone is exposed to very low levels. Aristotle named it "Quicksilver". Animal poisoning by mercury is rare because of strict federal, state, and local regulations. Most of the mercury found in the environment is in the form of metallic mercury and inorganic compounds. The most common natural forms of mercury found in the environment are metallic mercury, mercury sulfide (cinnabar ore), mercuric chloride, and methylmercury. Methylmercury is of particular concern because it can bioaccumulate in certain edible freshwater and saltwater fish and marine mammals to levels that are many times greater than levels in the surrounding water. As a result, the larger and older fish living in contaminated water build up levels of mercury in their bodies. Inorganic mercury does not bioaccumulate in the food chain to any extent. Cultivation of edible mushrooms, where waste as compost material with high levels of mercury is used, can also accumulate high levels of mercury. The release of methylmercury into an ocean bay (Minamata) in Japan in the 1950s led to a massive health disaster, and the clinical syndrome was named Minamata disease. Thousands of people were poisoned, and hundreds of them had severe brain damage.

The Food and Drug Administration (FDA) estimates that on average most people are exposed to about 50 ng mercury/kg of body weight/day in the food they eat. This level is not enough to cause any harmful effects. A large part of this mercury is in the form of methylmercury and the majority of that comes from eating fish. Fish for food consumption are not allowed to have more than 1 ppm. This level is below a level that can be associated with adverse effects. Foods other than fish that may contain higher levels of mercury include wild animals, birds, and mammals (bears) that eat large amounts of contaminated fish. Meat and/or fat from fish, marine mammals, fisheating wildlife, and birds; and mercury-based fungicidestreated grains have the highest-mercury levels. Certain species of commercially available saltwater fish, such as shark, swordfish, kingfish, and tilefish, can contain high levels of methylmercury. In addition, edible mushrooms can have unsafe levels of mercury. These are the potential sources of mercury poisoning. In horses, mercury toxicity occurs from wound dressings (blisters) when dimethyl sulfoxide (DMSO) is applied simultaneously, because DMSO enhances absorption of mercury.

Toxicokinetics. Absorption of mercury from oral ingestion depends on the form of mercury. Metallic mercury is maximally absorbed (about 80%) from the lungs, while very little is absorbed from the gastrointestinal (GI) tract. Once mercury enters the circulation, it is rapidly distributed to other tissues, but more so in the kidneys, where it accumulates. Metallic mercury can stay in the body for weeks and months. Due to its high lipophilicity, metallic mercury can readily cross the blood–brain barrier and placental barrier. When metallic mercury enters the brain, it is readily converted to an inorganic divalent mercury (oxidized by the hydrogen peroxidase–catalase pathway), and it gets trapped there for a long time. The inorganic divalent cation can, in turn, be reduced to metallic mercury. Most of the absorbed metallic mercury excretes in the urine and feces, some amount passes in the milk, and very little in the exhaled air.

Inorganic mercury compounds (e.g. mercurous chloride and mercuric chloride) are absorbed 10– 40% from the GI tract on ingestion, distributed to different organs, and mainly accumulate in the kidneys. In an experimental study, female Sprague–Dawley rats given a single dose of mercuric chloride (7.4 or 9.2mgHg/kg, p.o.) showed 12.6 and 18.9 ppm mercury, respectively, in the kidneys when sacrificed 14 days post-exposure. Trace amounts were also detected in the liver, brain, and serum. These compounds do not readily cross the blood– brain barrier or placental barrier. Inorganic mercury excretes in the urine and feces, and only detectable levels pass through the milk. Organic mercury (e.g. methylmercury) readily gets absorbed from the GI tract (about 95%). From circulation it gets distributed to other organs. The distribution of methylmercury is similar to that of metallic mercury, i.e. a relatively large amount of mercury can accumulate in the brain and fetus (compared to inorganic mercury) because of its ability to penetrate the blood–brain and placental barriers and its conversion in the brain and fetus to the inorganic divalent cation mercury. However, the extent of conversion is less than with metallic mercury. In the brain, methylmercury can be changed to inorganic mercury and can remain in this tissue for a long time. Organic mercury excretes in the form of inorganic mercury in the feces over a period of several months. Some organic mercury also excretes in the urine and milk.

Depending on the route of exposure, dose, and single versus repeat exposure, toxicokinetics of mercury can follow one- or two-compartment model. Studies have shown that repeat or continuous exposure to any form of mercury can result in the accumulation of mercury in the body. Retention of mercury in the brain may persist long after cessation of short- and long-term exposures. Blood levels of mercury are closely related to whole-body retention of mercury during the first 3 days after administration. After the initial 3 days, the amount of mercury in the blood declines more rapidly than the whole-body burden.

Evidence suggests that the metabolism of all forms of mercury is similar for humans and animals. Mercury is metabolized through the oxidation–reduction cycle that takes place in intestinal microflora and after absorption in many tissues and in the red blood cells (RBCs). Elimination rates for methylmercury appear to vary with species, dose, sex, and strain. The elimination half- life in the blood of monkeys receiving inorganic and organic mercury was found to be 26 days. In a study of organs from sled dogs fed methylmercury-laden meat and organs from predatory marine animals, the highest concentration of total mercury was found in the mesenterial lymph nodes, followed by liver and kidneys, indicating that the lymphatic system may play an important role in the transport of mercury to target organs. The tissue concentrations of mercury observed in this study were found to be age related, and the results sugest that demethylation takes place in all organs, except the skeletal muscles. Demethylation of methylmercury was found to be lower in the brain than in other organs.

Mechanism of action. Mechanism of action for the toxic effects of organic and inorganic mercury is similar. Toxicities of the different forms of mercury are related, in part, to its differential accumulation in sensitive tissues. This theory is supported by the observation that mercury rapidly accumulates in the kidneys and specific areas of the brain (the two major target organs). High- affinity binding of the divalent cationic mercury to thiol or sulfhydryl (SH) groups of proteins is believed to be a major mechanism involved in the toxicity of mercury. As a result, mercury can cause inactivation of various enzymes, structural proteins, transport proteins, and alteration of cell membrane permeability by the formation of mercaptides. In addition, mercury may induce one or more of the following effects: increased oxidative stress, mitochondrial dysfunction, changes in heme metabolism, glutathione depletion, increased permeability of the blood–brain barrier; and disruption of microtubule formation, protein synthesis, DNA replication, DNA polymerase activity, calcium homeostasis, synaptic transmission, and immune response (ATSDR, 1999). The nervous system is especially sensitive to mercury. The degree of damage depends on the form of mercury and its dose. Metallic mercury at high doses causes irreparable damage to the brain. In many poisoning incidents, permanent damage to the brain and kidneys occurred by methylmercury. Since inorganic mercury does not readily cross the blood–brain barrier, it is highly unlikely that inorganic mercury may cause any damage to the brain or nerves. Most of the information concerning neurotoxicity in humans following oral exposure to organic mercury comes from reports describing the effects of ingesting contaminated fish or fungicide-treated grains, or meat from animals fed such grains. Studies conducted in experimental animals strongly indicate that organic mercury is a potent neurotoxicant.

Evidence suggests that a single dose of mercuric chloride (0.74 mg/kg) caused disruption of the blood–brain barrier in rats. These investigators also administered mercuric chloride to rats at the same dose daily for 11 weeks. Within 2 weeks, there were coagulative or lucid changes in cerebellar granule cells and fragmentation, vacuolation, and cytoplasmic lesions in the neurons of dorsal root ganglia. Neurological disturbances consisted of severe ataxia and sensory loss, with an accompanying loss in body weight. It is expected that mercuric chloride administered via subcutaneous route would be much more toxic than that administered orally because of the poor absorption of inorganic forms of mercury from the GI tract.

Neurotoxic effects seen in the Minamata (Japan) and Eraqi poisonings have been associated with neuronal degeneration and glial proliferation in the cortical and cerebellar gray matter and basal ganglia, and derangement of basic developmental processes such as neuronal migration and neuronal cell division. In the brain, Purkinje, basket, and stellate cells were severely affected. In the brain, methylmercury selectively damages specific focal areas such as the granule cells of the cerebellum and the neurons in the interstices of the visual cortex. Methylmercury selectively inhibits protein synthesis in the brain (reversibly in neurons from the cerebrum and Purkinje cells; and irreversibly in granule cell of cerebellum), and this effect usually precedes the appearance of clinical signs. This selective action on the brain may be due to the fact that certain cells are susceptible because they cannot repair damage from methylmercury. Cheung and Verity (1985) identified the most sensitive step in the protein synthesis, i.e. the peptide elongation can be affected by the high concentrations of mercury, but the first stage of synthesis associated with tRNA may be the most sensitive. Methylmercury inhibits one or more of the amino acyl tRNA synthetase enzymes. Microtubules are essential for cell division (main component of the mitotic spindle), and methylmercury reacts with the SH groups on tubulin monomers, and thereby disrupts the assembly process. The dissociation process continues, and that leads to depolymerization of the tubule. In all forms, mercury accumulates in the kidneys, and thereby it causes greater damage to this organ. The kidney damage appears to be dose dependent, and that means recovery can occur if exposure is at low level. Following entry of the mercuric or methylmercuric ion into the proximal tubular epithelial cells via transport across the brushborder or basolateral membrane, mercury interacts with thiol-containing compounds, such as glutathione and metallothionein. This interaction initially produces alterations in membrane permeability to calcium ions and inhibition of mitochondrial function. Subsequently, by unknown signaling mechanisms, mercury induced the synthesis of glutathione, glutathione-dependent enzymes, metallothionein, and several stress proteins. Furthermore, in the kidney, epithelial cell damage occurs as a result of excess free radical formation and lipid peroxidation.

Toxicity. In general, toxic effects of mercury depend on the form of mercury, the amount and duration of exposure, and the route of exposure. Mercury, in all forms, has been found to be toxic to both man and animals. There are many similarities in the toxic effects of the various forms of mercury, but there are also differences. Exposure to mercury, however, does not necessarily mean that adverse health effects will occur. Each form and route leads to different effects. Practically, it is organic mercury, which is more toxic and often encountered in poisonings. In both humans and animals, exposure to mercury is more frequently encountered by oral ingestion. The major targets of toxicity following oral exposure to inorganic and organic mercury are the kidneys and the central nervous system (CNS), respectively. In humans, deaths resulting from organic mercury ingestion have been documented following outbreaks of poisoning (Minamata disease) after consumption of methylmercury-contaminated fish in Minamata Bay, Japan and after consumption of grains contaminated with methylmercury and ethylmercury in Iraq. Noninflammatory damage to the brain was the primary cause of deaths, while pneumonia and nonischemic heart damage were reported as secondary cause of death. The signs of mercury acute toxicity in animals are similar to those described in humans. Signs and symptoms associated with short-term exposure to metallic mercury may include nausea, vomiting, diarrhea, increase in blood pressure or heart rate, skin rashes, and eye irritation. Inorganic mercury, if swallowed in large quantities, may cause damage to the kidney, and also in stomach and intestine, including nausea, diarrhea, and ulcers.

Animal studies revealed that long-term oral exposure to inorganic mercury salts causes kidney damage, increase in blood pressure and heart rate, and effects on the stomach. Studies also show that nervous system damage occurs after long-term exposure to high levels of inorganic mercury. Short-term, highlevel exposure of laboratory animals to inorganic mercury has been shown to affect the developing fetus and may cause termination of the pregnancy.

Laboratory animals exposed to long term, high levels of methylmercury or phenylmercury showed damage to the kidneys, stomach, and large intestine, changes in blood pressure and heart rate, and adverse effects on the developing fetus, sperm, and male reproductive organs; and increases in the number of spontaneous abortions and stillbirths. In livestock animals, clinical signs of mercury poisoning vary greatly. In cattle, toxicity signs include ataxia, neuromuscular incoordination, and renal failure, followed by convulsions and a moribund state. Average time from ingestion to death is reported to be about 20 days. Ingestion of phenylmercuric acetate may cause sudden death with massive internal hemorrhage, without other signs of toxicity. In horses, signs of acute toxicity include severe gastroenteritis and nephritis. In chronic cases, signs may include neurological dysfunction, laminitis, in addition to renal disease which is characterized by glycosuria, proteinuria, phosphaturia, reduced urine osmolarity, reduced glomerular filtration rate, azotemia, and elevated creatinine and blood urea nitrogen. In sheep, the poisoning is characterized by severe neurological symptoms, and tetraplegia. Pigs show incoordination, unstable gait, lameness, recumbency and death. Some of the toxic effects are described in detail for each organ/system affected by mercury exposure.

Nervous system. Adverse effects on the nervous system of animals occur at lower doses than do harmful effects to most other systems of the body. This difference indicates that the nervous system is more sensitive to mercury than are other organs in the body. Animal studies also provide evidence of damage to the nervous system from exposure to methylmercury during development, and evidence suggests that the effects worsen with age, even after the exposure stops. The reason for this greater susceptibility is that mercury affects processes unique to the developing nervous system, namely cell migration and cell division. Both human epidemiology and experimental animal studines indicate that organic mercury is a potent neurotoxicant. Studies suggest that cats and monkeys are more sensitive than rodents to the neurotoxic effects of mercury (especially methylmercury). In several animal species, the major effects that are seen across the studies include motor disturbances, such as ataxia and tremors, as well as signs of sensory dysfunction, such as impaired vision. The predominant pathological feature is degenerative changes in the cerebellum, which is likely to be the mechanism involved in many of the motor dysfunctions. In a chronic study, cats fed tuna contaminated with methylmercury, showed degenerative changes in the cerebellum and the cortex. Neonatal monkeys exposed to methylmercuric chloride at 0.5 mgHg/kg/day for 28–29 days exhibited stumbling, and falling, blindness, crying, tember tomtrums, and coma. Histopathological analysis revealed diffuse degeneration in the cerebral cortex, cerebellum basal ganglia, thalamus, amygdala, and lateral geniculate nuclei. Rats acutely intoxicated with methylmercury (19.9mg Hg/kg, oral gavage) showed signs of lethargy and ataxia, which was not accompanied by histopathological changes. Symptoms disappeared within 2–3 h. Administration of a single dose of methylmercuric chloride (0.8mgHg/kg) produced blood–brain barrier dysfunction in rats similar to that described for inorganic mercury.

Following inhalation exposure to metallic mercury vapors, the CNS has been found to be the most sensitive organ in guinea pigs, rats, and mice. With increasing concentrations of mercury, damage to CNS becomes irreversible. Rabbits appear to be less sensitive. In rabbits given 5.5 mgHg/kg as methylmercuric acetate for 1–4 days, widespread neuronal degenerative changes in cervical ganglia cells, cerebellum, and cerebral cortex have been observed without accompanying behavioral changes. In similar studies, mice exposed to 1.9 or 9.5 mgHg/kg/day as methylmercury in the drinking water for 28 weeks exhibited degeneration of Purkinje cells and loss of granular cells in the cerebellum. At higher doses, hind limb paralysis was observed. Neuronal degeneration and microgliosis were observed in the corpus striatum, cerebral cortex, thalamus, and hypothalamus, accompanied by hind leg weakness, in mice given with 1 or 4mgHg/kg/day as methylmercuric chloride by gavage for 60 days. Neurotoxic signs observed in rats exposed to methylmercury (4 mgHg/kg/day for 8 days) include muscle spasms, gait disturbances, flailing, and hind limb crossing. Histopathological examination of the nervous system of affected rats has shown degeneration of cerebellar granule cells and dorsal root ganglia and degenerative changes in peripheral nerves.

Renal system. Mercury, in all forms, has been shown to cause renal toxicity (structural and functional damage) in humans and animal species that are tested. Renal toxicity has been observed in rats and mice (B6C3F1) following acute, intermediate, and chronic exposures to mercuric chloride. In a 14-day study, male and female rats were exposed by gavage to 0.93–14.8 mgHg/kg/day as mercuric chloride for 5 days a week. There was a significant increase in the absolute and relative kidney weights of males beginning at the 1.9 mgHg/kg/day dose level. An increased incidence of tubular necrosis was observed in rats exposed to at least 3.7mgHg/kg/day. Severity was dose dependent. In chronic studies, mercuric chloride produced a variety of pathological changes in kidneys. Degenerative effects have been found in the kidneys of animals exposed to moderateto- high levels of metallic mercury vapors following acute or subacute exposures. Effects ranging from marked cellular degeneration to tissue destruction and widespread necrosis were observed in rabbits exposed to mercury vapor at a concentration of 28.8mg/m3 for 2–3 h. In rats, slight degenerative changes (i.e. dense deposits in tubule cells and lysosomal inclusions) in the renal tubular epithelium were evident following exposure to 3mg/m3 mercury vapor for 3h/day, 5 days a week, for 12–42 weeks. Low-level, long-term exposure to mercury (0.1mg/m3) has not been found toxic to kidneys of rats, rabbits, and dogs. Exposure to organic mercury via inhalation is extremely rare.

Cardiovascular system. Mercury has been shown to have adverse effects on cardiovascular system. Adecrease in heart rate was observed in male rats given two gavage doses of 2 mgHg/kg as methylmercuric chloride. An increase in systolic blood pressure was observed in male rats after daily oral gavage doses of 0.4 mgHg/kg/day as methylmercuric chloride for 3–4 weeks. This effect began approximately 60 days after initiation of exposure and persisted for at least 9 months. GI tract. Ingestion of mercuric chloride is highly irritating to the tissues of GI tract. Inflammation and necrosis of the glandular stomach were observed in mice that were given oral doses of 59 mg/kg as mercuric chloride 5 days a week for 2 weeks (NTP, 1993). In a 2-year gavage study, an increased incidence of forestomach hyperplasia was observed in male rats exposed to 1.9 or 3.7mgHg/kg/day as mercuric chloride compared to the control group. Mice showed ulceration of the glandular stomach compared to the control group. It showed ulceration of the glandular stomach after 2 years of dietary exposure to methylmercuric chloride at 0.69mgHg/kg/day. In experimental studies, methylmercury has been found very strainand sex-specific in mice. Asingle oral dose of methylmercuric chloride at 16 mgHg/kg resulted in the death of 4 of 6 male mice (C57BL/6N Jc1 strain) but no death was noted in females. Mortality in female mice (4 of 6) was noted at 40mgHg/ kg dose. In a chronic study, 26 weeks of dietary exposure to methylmercuric chloride resulted in increased mortality in both male and female mice (ICR strain) at 3.1 mgHg/kg/day.

Hematopoietic system. In general, acute mercury toxicity does not produce any characteristic hematological changes. In a chronic study conducted in rats, phenylmercuric acetate given in water at a dose of 4.2 mgHg/kg/day caused decreases in hemoglobin, hematocrit, and RBC counts. The anemia observed in this study may have been secondary to blood loss associated with the ulcerative lesions in the large intestine. However, methylmercuric chloride at low dose (0.1 mgHg/kg/day for 2 years) given in diet for 2 years caused no changes in hematological parameters.

Other effects. In laboratory animals, mercury has been found to have potential for inducing genotoxicity, carcinotoxicity, reproductive and developmental toxicity, and immunotoxicity. Diagnosis. Presently there are reliable and accurate ways to measure mercury levels in the body. These tests involve mercury analysis in blood, urine, milk, hair, and liver and kidney. Levels found in blood, urine, and hair may be used together to predict possible health effects that may be caused by the different forms of the mercury. Mercury levels in the blood provide more useful information after recent exposures than after long-term exposures. Mercury in urine is determined to test for exposure to metallic or inorganic mercury, while whole blood or hair is used to determine exposure to methylmercury. Kidney is an ideal specimen for mercury analysis from dead animals.

Treatment. Activated charcoal (1–3 g/kg body weight, p.o.) is very effective in reducing further absorption of mercury. Specific treatment of mercury poisoning rests with the use of chelators along with protein solutions to bind and neutralizme mercury compounds. The use of a particular chelator is dependent on the type of mercury exposure. Among several chelators, dimercaprol (British anti-Lewisite (BAL), 3 mg/kg, i.m.) has been found to be the most effective against mercury poisoning. However, chelation releases mercury from soft tissues which can be redistributed to the brain. Oral administration of sodium thiosulfate (1 g/kg) can assist in eliminating mercury. Animal studies sugest that antioxidants (particularly vitamin E) may be useful for decreasing the toxicity of mercury. Improved chelation and drug therapies for treating acute and chronic mercury poisonings are greatly needed.

Conclusion. Toxicity by mercury depends on the form of mercury, dose, duration, and route of exposure. Organic mercury tends to bioaccumulate in the higher food chain, and as result, the maximum concentrations are found in the meat of fish, marine mammals, and fish-eating birds and wildlife. Methylmercury is the most toxic among the mercury species because of its volatility and its ability to pass through biological membranes such as the blood–brain barrier and the placental barrier. Nervous system and kidneys are the two major target organs. Not all forms of mercury cross the blood–brain barrier (e.g. inorganic mercury), but in all forms it accumulate in the kidney and thereby causes damage to this organ. Chelation therapy appears to be the best treatment.

Poisoning phenoxy acid compounds

Chemically, 2,4-D is 2-chloro-4-phenoxyacetic acid. It is usually formulated as salts, esters, or amine derivatives.2 2,4-D is used for control of broadleaf weeds on residential and commercial properties and in some areas on roadside rights-of-way. Residues on treated turf are in the range of 35 to 75 ppm and dissipate rapidly in the first several days following the application. A residue tolerance of 300 ppm has been established on pasture grasses. Poisoning is almost always due to accidental ingestion of concentrates or sprays. Technical-grade phenoxy herbicides are irritating to the eye and mucous membranes and somewhat less irritating to skin and are also phytotoxic to most plants. Dogs appear to be somewhat more sensitive to phenoxy herbicides than other species of domestic animals. The approximate oral median lethal dose (LD50) for 2,4-D in the dog has been reported to be 100 mg/kg.3 However, Beasley and colleagues4 orally dosed English pointers with 8.8, 43.7, 86.7, 175, and 200 mg/kg 2,4-D, and all survived. Doses of 175 or 220 mg/kg of body weight produced overt signs of toxicosis characterized by myotonia, vomiting, and weakness. The lower doses did not produce overt clinical signs, but electromyographic abnormalities were detectable at exposures of 8.8 mg/kg. Multiple dosages of 20 mg/kg daily for approximately 3 weeks or 25 mg/kg for 6 days were lethal for dogs.3,5 Even at an exposure of 20 mg/kg of body weight, it is not likely that dogs will be significantly poisoned by exposure to properly treated lawns. The greatest hazard to dogs is ingestion of undiluted product, discarded or excess spray that had been previously mixed, or pools of spray that have collected in low spots or in containers. Arnold and colleagues6 attempted to produce

2,4-D toxicosis by placing English pointers on enclosed turf plots to confine the animals for controlled periods of continuous exposure. One enclosure was sprayed with 2,4-D at a rate of 168 mg/square meter, which is the maximum recommended rate for lawns, and another enclosure was sprayed at four times the maximum recommended rate. The dogs were placed in the enclosures within 30 minutes of spraying and were observed five times each day for a period of 7 days. Detailed clinical examinations included electromyograms, which were performed on days 1 and 7 after exposure. No adverse effects were detected in any of the clinical, hematological, biochemical, electrophysiological, or postmortem examinations. A 2,4-D concentration of 500 ppm (25 mg/kg of body weight) in the diet caused no ill effects in dogs during a 2-year study. In a more recent study, the level of 2,4-D at which no observable effects were noted in chronic toxicity in dogs was determined to be 1.0 mg/kg/day. Orally administered 2,4-D is rapidly and extensively absorbed by the gastrointestinal tract. The extent of dermal absorption varies according to the chemical form of the product and the species of animal, varying from about 5% for the acid in humans to 85% for the ester in rats. Absorbed 2,4-D salts and esters are rapidly converted to 2,4-D acid and excreted by the renal anion transport system. The renal anion transport system is saturable and appears to account for the longer half-life and greater sensitivity to toxicity in the dog. 2,4-D concentrations of 718 μg/mL and 1075 μg/mL were present in the serum of dogs dosed with 175 or 220 mg 2,4- D/kg of body weight, respectively. The peak serum concentration was 121 μg/mL following an oral exposure of 8.8 mg/kg of body weight. 2,4-D is widely distributed in tissues with little accumulation in fat. Plasma or serum appears to be the best specimen to use for laboratory confirmations of 2,4-D poisoning. Kidney tissue is an alternative sample. Pharmacokinetic data suggest that kidney to plasma ratios approach unity as the renal organic anion system becomes saturated. Data also sugest that plasma and kidney concentrations of up to 100 ppm may be present in animals that do not show signs of intoxication. The clinical signs in dogs are characteristic and include vomiting and an initial disinclination to move and a passivity that gradually becomes worse as a pattern of myotonia develops. This rigidity of skeletal muscles is combined with ataxia, progressive apathy, depression, and muscular weakness, particularly of the posterior limbs. Myotonia has been produced by exposure to 2,4-D and 2,4,5-T in dogs. At high doses, the condition can be induced in less than 1 hour after administration. Spontaneous movement ceases, and when startled, animals make sudden spastic movements and sometimes lose the ability to stand or rise. Opisthotonos may also occur. A potential biochemical lesion associated with myotonia is an increase in basic paranitrophenyl phosphatase related to increased passive flux of potassium. This may lead to myotonia through a compensatory decrease in chloride conductance. Periodic clonic spasms and finally coma are the typical sequelae of phenoxy herbicide poisoning in dogs. During the clinical course of poisoning there is marked anorexia; there may be vomiting and occasionally passage of blood-tinged feces. Postmortem examination often reveals necrotic ulcers of the oral mucosa, signs of irritation in the gastrointestinal tract, and sometimes necrosis of the small intestine, and focal necrosis in the liver and degeneration of renal tubules. However, there are no reports of renal failure in dogs from any exposure to 2,4-D. There are no specific antidotes. Since 2,4-D is excreted almost quantitatively in urine as the free acid, forced alkaline diuresis should enhance excretion. Unless there is severe central nervous system (CNS) depression (rare), recovery should be rapid. An association between 2,4-D and canine malignant lymphoma in dogs was reported in the Journal of the National Cancer Institute (NCI) in 1991.8 This report not only raised concern by homeowners and veterinarians, but also in some instances was used to indict lawn care in a generic sense. The NCI reported a twofold increase in risk of canine malignant lymphoma associated with four or more yearly homeowner applications of 2,4-D. It is unusual for any homeowner (or commercial lawn care companies) to apply 2,4-D four or more times per year; thus the validity of pet owners' responses to the NCI questionnaire or interviews concerning the application of lawn care products is questionable. Two critiques of the NCI report have been published. A panel of experts concluded that because of numerous limitations in the design of the study an association was not established between 2,4-D and canine malignant lymphoma. Kaneene and Miller did not confirm a dose- response relationship between 2,4-D use and canine malignant lymphoma and concluded that the occurrence of canine malignant lymphoma was not significantly associated with the use of 2,4-D. An increased risk of transitional cell carcinoma of the urinary bladder of Scottish Terriers exposed to lawns or gardens treated with phenoxy herbicides was reported by Glickman et al.11 The authors proposed a gene-environment interaction to the development of the bladder tumors. The cause-effect relationship was based on information obtained by questionnaires completed by owners of dogs in the case and control groups. Additional studies are needed to replicate the results and to more specifically confirm exposures.

Toxicity of phenolic compounds

Several substituted dinitrophenols alone or as salts such as DNP (2,4-dinitrophenol), DNOC (dinitro-o-cresol), and dinoseb {2-(1-methylpropyl)-4,6-dinitro} are used as insecticides, fungicides, acaricides, and herbicides. The main source of poisoning in animals are human negligence in removing the preparation if it spills, in disposing off the containers and preventing animals access to treated fields. The structural formulae of DNP, DNOC, and dinoseb are given as under. In general, the dinitro compounds are not very water-soluble and are highly hazardous to animals. The oral acute LD50 of DNOC in mice, guinea pigs, rabbits, hens, dogs, pigs, and goats ranges from 25 to 100 mg/kg b.w. In sheep, dosage of 25 mg/kg/day causes toxicosis in 2–4 days. Clinical signs include fever, dyspnea, acidosis, oliguria, muscular weakness, tachycardia and convulsions, followed by coma and death with a rapid onset of rigor mortis. Abortions have been reported in sows. In cattle and ruminants, methemoglobinemia, intravascular hemolysis, and hemoproteinemia have been observed. Cataract can occur with chronic dinitrophenol intoxication. Exposure to these compounds may cause yellow staining of skin, conjunctiva, or hair.

POISONING BY TRIAZINE COMPOUNDS

Triazines and triazoles have been used extensively as selective herbicides for more than 40 years. These herbicides are inhibitors of photosynthesis and include both the asymmetrical and symmetrical triazines. Examples of symmetrical triazines are chloro-S-triazines (simazine, atrazine, propazine, and cyanazine); the thiomethyl-S-triazines (ametryn, prometryn, terbutryn), and the methoxy-S-triazine (prometon). The commonly used asymmetrical triazine is metribuzin. The structures of symmetrical and asymmetrical triazines are given as below. These herbicides have low oral toxicity and are unlikely to pose acute hazards in normal use, except ametryn and metribuzin, which may be slight to moderately hazardous. They are neither irritant to the skin or eye, nor are skin sensitizers. The exceptions are atrazine, which is skin sensitizer and cyanazine, which is toxic by the oral route. But sensitivity of sheep and cattle to these herbicides is appreciably high. The oral LD50 for rats is ~5 g/kg b.w. The main symptoms are anorexia, hemotoxia, hypothermia, locomotor disturbances, irritability, tachypnea, and hypersensitivity.

Doses of 500 mg/kg of simazine or 30 mg/kg atrazine for 30–60 days are lethal to sheep. Deaths have been reported in sheep and horses grazing triazine-treated pasture 1–7 days after spraying. Cumulative effects are not seen. Metribuzin is slightly more toxic than simazine but does not produce any harmful effects in dogs fed at 100 ppm in the diet. Simazine is excreted in milk, so it is of public health concern. Atrazine is more toxic to rats, but comparatively less toxic to sheep and cattle than simazine. When cultured human cells are exposed to atrazine, splenocytes are damaged, bone marrow cells are not affected. This class of herbicides is liver microsomal enzyme inducer and is converted to N-dealkylated derivatives. In contrast to simazine, it is not excreted in milk. Triazines seem to have no potential to be mutagenic or to produce carcinogenicity in animals. The exception is cyanazine, which is more acutely toxic, week mutagenic, and results in developmental toxicity, presumably because of the presence of cyano moiety.

Cyanide poisoning

Cyanide inhibits cytochrome oxidase and causes death from histotoxic anoxia.

Etiology. Cyanides are found in plants, fumigants, soil sterilizers, fertilizers, and rodenticides (eg, calcium cyanomide). Toxicity can result from improper or malicious use, but in the case of livestock, the most frequent cause is ingestion of plants that contain cyanogenic glycosides. These include Triglochin maritima (arrow grass), Hoecus lunatus (velvet grass), Sorghum spp (Johnson grass, Sudan grass, common sorghum), Prunus spp (apricot, peach, chokecherry, pincherry, wild black cherry), Sambucus canadensis (elderberry), Pyrus malus (apple), Zea mays (corn), and Linum spp. (flax). The seeds (pits) of several plants such as the peach have been the source of cyanogenic glycosides in many cases. Eucalyptus spp , kept as ornamental houseplants, have been implicated in deaths of small animals. The cyanogenic glycosides in plants yield free hydrocyanic acid (HCN), otherwise known as prussic acid, when hydrolyzed by â-glycosidase or when other plant cell structure is disrupted or damaged, eg, by freezing, chopping, or chewing. Microbial action in the rumen can further release free cyanide.

Apple and other fruit trees contain prussic acid glycosides in leaves and seeds but little or none in the fleshy part of the fruits. In Sorghum spp forage grasses, leaves usually produce 2-25 times more HCN than do stems; seeds contain none. New shoots from young, rapidly growing plants often contain high concentrations of prussic acid glycosides. The cyanogenic glycoside potential of plants can be increased by heavy nitrate fertilization, especially in phosphorus-deficient soils. Spraying of cyanogenic forage plants with foliar herbicides such as 2,4-D can increase their prussic acid concentrations for several weeks after application.

The cyanogenic glycoside potential is slow to decrease in drought-stricken plants containing mostly leaves. Grazing stunted plants during drought is the most common cause of poisoning of livestock by plants that produce prussic acid.

Frozen plants may release high concentrations of prussic acid for several days. After wilting, release of prussic acid from plant tissues declines. Dead plants have less free prussic acid. When plant tops have been frosted, new shoots may regrow at the base; these can be dangerous because of glycoside content and because livestock selectively graze them.

Ruminants are more susceptible than monogastric animals, and cattle slightly more so than sheep. Hereford cattle have been reported to be less susceptible than other breeds.

Clinical Findings. Signs can occur within 15-20 min to a few hours after animals consume toxic forage. Excitement can be displayed initially, accompanied by rapid respiration rate. Dyspnea follows shortly, with tachycardia. Salivation, excess lacrimation, and voiding of urine and feces may occur. Vomiting may occur, especially in pigs. Muscle fasciculation is common and progresses to generalized spasms before death. Animals stagger and struggle before collapse.

Mucous membranes are bright red but may become cyanotic terminally. Death occurs during severe asphyxial convulsions. The heart may continue to beat for several minutes after struggling and breathing stops. The whole syndrome usually does not exceed 30-45 min. Most animals that live e"2 hr after onset of clinical signs recover, unless continuous absorption of cyanide from the GI tract occurs.

Lesions. In acute or peracute cyanide toxicoses, blood may be bright cherry red initially but can be dark red if necropsy is delayed; it may clot slowly or not at all. Mucous membranes may also be pink initially, then become cyanotic after respiration ceases. The rumen may be distended with gas, and the odor of "bitter almonds" may be detected after opening. Agonal hemorrhages of the heart may be seen. Liver, serosal surfaces, tracheal mucosa, and lungs may be congested or hemorrhagic; some froth may be seen in respiratory passages. Neither gross nor histologic lesions are consistently seen.

Multiple foci of degeneration or necrosis may be seen in the CNS of dogs chronically exposed to sublethal amounts of cyanide. These lesions have not been reported in livestock.

Diagnosis. Appropriate history, clinical signs, postmortem findings, and demonstration of HCN in rumen (stomach) contents or other diagnostic specimens support a diagnosis of cyanide poisoning. Specimens recommended for cyanide analyses include the suspected source (plant or otherwise), rumen or stomach contents, heparinized whole blood, liver, and muscle. Antemortem whole blood is preferred; other specimens should be collected as soon as possible after death, preferably within 4 hr. Specimens should be sealed in an airtight container, refrigerated or frozen, and submitted to the laboratory without delay. When cold storage is unavailable, immersion of specimens in 1-3% mercuric chloride has been satisfactory.

Hay, green chop, silage, or growing plants containing >220 ppm cyanide as HCN on a wet-weight (as is) basis are very dangerous as animal feed. Forage containing <100 ppm HCN, wet weight, is usually safe to pasture. Analyses performed on a dry-weight basis have the following criteria:

>750 ppm HCN is hazardous, 500-750 ppm HCN is doubtful, and <500 ppm HCN is considered safe.

Normally expected cyanide concentrations in blood of most animal species are usually <0.5 μg/mL. Minimal lethal blood concentrations are ~3.0 μg/mL or less. Cyanide concentrations in muscle are similar to those in blood, but concentrations in liver are generally lower than those in blood.

Differential diagnoses include poisonings by nitrate or nitrite, urea, organophosphate, carbamate, chlorinated hydrocarbon pesticides, and toxic gases (carbon monoxide and hydrogen sulfide), as well as infectious or noninfectious diseases that cause sudden death.

Treatment, Control, and Prevention. Immediate treatment is necessary. Sodium nitrite (10 g/100 mL of distilled water or isotonic saline) should be given IV at 20 mg/kg body wt, followed by sodium thiosulfate (20%), IV, at e"500 mg/kg; the latter may be repeated as needed with little hazard. Sodium nitrite therapy may be carefully repeated at 10 mg/kg, every 2-4 hr or as needed. In one study investigating cyanide poisoning treatment in dogs, either dimethylaminophenol (DMAP) IM at 5 mg/kg or hydroxylamine hydrochlorine IM at 50 mg/kg were as effective as nitrite and thiosulfate.

Sodium thiosulfate alone is also an effective antidotal therapy at e"500 mg/kg, IV, plus 30 g/cow, PO, to detoxify any remaining HCN in the rumen. Oxygen may be helpful in supplementing nitrite or thiosulfate therapy, especially in small animals. Hyperbaric oxygen therapy (100% oxygen breathed intermittently at a pressure >1 atmosphere absolute) causes an above normal partial pressure of oxygen (PO2) in arterial blood and markedly increases the amount of oxygen dissolved in plasma. Oxygen-dependent cellular metabolic processes benefit from heightened oxygen tension in capillaries and enhanced oxygen diffusion from capillaries to critical tissues. Activated charcoal is not efficacious in absorbing cyanide and thus is not recommended PO for antidotal therapy.

Caution is indicated in treatment. Many clinical signs of nitrate and prussic acid poisoning are similar, and injecting sodium nitrite induces methemoglobinemia identical to that produced by nitrate poisoning. If in doubt of the diagnosis, methylene blue, IV, at 4-22 mg/kg, may be used to induce methemoglobin. Because methylene blue can serve as both a donor and acceptor of electrons, it can reduce methemoglobin in the presence of excess methemoglobin or induce methemoglobin when only hemoglobin is present (but sodium nitrate is the more effective treatment for cyanide poisoning if the diagnosis is certain).

Pasture grasses (eg, Sudan grass and sorghum-Sudan grass hybrids) should not be grazed until they are 15-18 in. tall to reduce danger from prussic acid poisoning. Forage sorghums should be several feet tall. Animals should be fed before first turning out to pasture; hungry animals may consume forage too rapidly to detoxify HCN released in the rumen. Animals should be turned out to new pasture later in the day; prussic acid release potential is reported to be highest during early morning hours. Free-choice salt and mineral with added sulfur may help protect against prussic acid toxicity. Grazing should be monitored closely during periods of environmental stress, eg, drought or frost. Abundant regrowth of sorghum can be dangerous; these shoots should be frozen and wilted before grazing.

Green chop forces livestock to eat both stems and leaves, thereby reducing problems caused by selective grazing. Cutting height can be raised to minimize inclusion of regrowth.

Sorghum hay and silage usually lose e"50% of prussic acid content during curing and ensiling processes. Free cyanide is released by enzyme activity and escapes as a gas. Although a rare occurrence, hazardous concentrations of prussic acid may still remain in the final product, especially if the forage had an extremely high cyanide content before cutting. Hay has been dried at oven temperatures for up to 4 days with no significant loss of cyanide potential. These feeds should be analyzed before use whenever high prussic acid concentrations are suspected.

Potentially toxic feed should be diluted or mixed with grain or forage that is low in prussic acid content to achieve safe concentrations in the final product.

Poisoning by carbamic acid compounds

The carbamic acid class of fungicides includes dithiocarbamates (ferbam, thiram, ziram, propamocarb, etc.) and EDBCs (maneb, mancozeb, zineb, nabam, metiram, etc.). In general, carbamic acid derivatives have low or moderate acute toxicity by the oral, dermal and respiratory routes, except nabam. The main features of toxicity include anorexia, diarrhea and flatulence followed by neurological effects, ataxia, muscular contractions and prostration. With repeated ingestion, there is a possibility of cutaneous effects, alopecia, risk of antithyroid effects specially with maneb. Certain compounds inhibit ovulation and egg laying (thiram, ziram). On histopathology, hepatic, renal and pulmonary congestion is common. Occasionally hepatic degeneration, ascites, enteritis and hydrothorax have been observed. Propamocarb is non-irritating to the eye or skin. It induces sensitization in a Magnusson–Kligman maximization test. The signs of toxicity include hypokinesia, lethargy, hunched posture, body tremors, clonic convulsions, nasal hemorrhages, piloerection, staggering gait and ataxia. Vacuolar changes in various tissues including choroid plexus in the brain and reduction in organ weights have been observed in the rat and dog. The common development and reproductive abnormalities include reduction in copulation index (female rats) and body weight, retardation in ossification (rat) and increased post-implantation loss (rabbit). The principal target organ upon repeated exposure to EBDCs is thyroid. These fungicides alter thyroid hormone levels and/or weights. The developmental toxicity includes malformations and embryo-fetotoxic effects at maternally toxic dose levels with EBDCs in rats.

Poisoning by urea and urea compounds

The ureas and thioureas (polyureas) are available under different names such as diuron, fluometuron, isoproturon, linuron, buturon, chlorbromuron, chlortoluron, chloroxuron, difenoxuron, fenuron, methiuron, metobromuron, metoxuron, monuron, neburon, parafluron, siduron, tebuthiuron, tetrafluron, and thidiazuron. Of these, diuron and fluometuron are the most commonly used in the United States, whereas isoproturon is mostly used in other countries including India. In general, polyureas have low acute toxicity and are unlikely to present any hazard in normal use, except tebuthiuron, which may be slightly hazardous. In general, the cattle are more sensitivity to polyurea herbicides than sheep, cats, and dogs. Diuron and Monuron are potent inducers of hepatic metabolizing enzymes compared to those polyurea herbicides with one or no halogen substitutions (chlortoluron and isoproturon).

Male rats are more sensitive than females to the enzyme-inducing activity of diuron and can lead to detoxication of EPN and O- demethylation of p-nitro anisole. N-demethylation of aminopyrine increases for 1–3 weeks, then return to normal. Recovery from diuron intoxication is quick (within 72 h) and no signs of skin irritation or dermal sensitization have been reported in guinea pigs. After repeated administration, hemoglobin levels and erythrocyte counts are significantly reduced, while methemoglobin concentration and white blood cell counts are increased. Increased pigmentation (hemosiderin) in the spleen is seen histopathologically. Linuron in sheep causes erythrocytosis and leucocytosis with hypohemoglobinemia and hypoproteinemia, hematuria and ataxia, enteritis, degeneration of the liver, muscular dystrophy.

In chickens it leads to loss of weight, dyspnea, cyanosis, and diarrhea. It is non-toxic to fish. Fluometuron is less toxic than diuron. In sheep, depression, salivation, grinding of teeth, chewing movements of the jaws, mydriasis, dyspnea, in-coordination of movements, and drowsiness is commonly seen. On histopathology severe congestion of red pulp with corresponding atrophy of the white pulp of spleen and depletion of the lymphocyte elements have been reported. The acute LD50 of isoproturon in rats is almost similar to diuron and does not produce any overt signs of toxicity, except at very high doses. Single oral dose of isoproturon in mice may produce some neurotoxic effects at very high doses and may reduce spontaneous and forced locomotor activity. Polyurea herbicides have been suspected to have some mutagenic effects but do not have carcinogenic potential. In general, the compounds do not cause developmental and reproductive toxicity, except monolinuron, linuron, and buturon are known to cause some teratogenic abnormalities in experimental animals.

CHLORATE POISONING

This is now seldom used as a herbicide but remains registered. Treated plants and contaminated clothing are highly combustible and constitute fire hazards. In addition, many cases of chlorate poisoning of livestock have occurred both from ingestion of treated plants and from accidental consumption of feed to which it was mistakenly added as salt. Cattle sometimes are attracted to foliage treated with sodium chlorate. Considerable quantities must be consumed before signs of toxicity appear. The minimum lethal dose is 1.1 g/kg body wt for cattle, 1.54-2.86 g/kg for sheep, and 5.06 g/kg for poultry. Ingestion results in hemolysis of RBC and conversion of Hgb to methemoglobin. Treatment with methylene blue (10 mglkg) must be repeated frequently because, unlike the nitrites, the chlorate ion is not inactivated during the conversion of Hgb to methemoglobin and is capable of producing an unlimited quantity of methemoglobin as long as it is present in the body. Blood transfusions may reduce some of the tissue anoxia caused by methemoglobin; isotonic saline can hasten elimination of the chlorate ion. Mineral oil containing 1% sodium thiosulfate will inhibit further absorption of chlorate in monogastric animals.

Poisoning by rodenticides

Many poisons have been used against rodent pests. Farm animals, pets, and wildlife often gain access to these poisons via the baits or the poisoned rodents or by malicious intent. This discussion covers the most commonly used rodenticides. Strychnine poisoning is discussed separately.

Anticoagulant Rodenticides (Warfarin and congeners). Potentially dangerous to all mammals and birds, anticoagulant rodenticides are the most frequent cause of poisoning in pets. Pets and wildlife may be poisoned directly from baits or indirectly by consumption of poisoned rodents. Intoxications in domestic animals have resulted from contamination of feed with anticoagulant concentrate, malicious use of these chemicals, and feed mixed in equipment used to prepare rodent bait.

All anticoagulants have the basic coumarin or indanedione nucleus. The "first-generation" anticoagulants (warfarin, pindone, coumafuryl, coumachlor, isovaleryl indanedione, and others less frequently used) require multiple feedings to result in toxicity. The "intermediate" anticoagulants (chlorophacinone and in particular diphacinone) require fewer feedings than "first- generation" chemicals, and thus are more toxic to nontarget species. The "second-generation" anticoagulants (brodifacoum and bromadiolone) are highly toxic to nontarget species (dogs, cats, and potentially livestock) after a single feeding.

The anticoagulants antagonize vitamin K, which interferes with the normal synthesis of coagulation proteins (factors I, II, VII, IX, and X) in the liver; thus, adequate amounts are not available to convert prothrombin into thrombin. A latent period, dependent on species, dose, and activity, is required, during which clotting factors already present are used up. New products have a longer biologic half-life and therefore prolonged effects (which require prolonged treatment).

For example, the half-life in canine plasma of warfarin is 15 hr, diphacinone is 5 days, and bromadiolone is 6 days, with maximum effects estimated at 12-15 days. Brodifacoum may continue to be detectable in serum for up to 24 days.

Clinical signs generally reflect some manifestation of hemorrhage, including anemia, hematomas, melena, hemothorax, hyphema, epistaxis, hemoptysis, and hematuria. Signs dependent on hemorrhage, such as weakness, ataxia, colic, and polypnea, may be seen. Depression and anorexia occur in all species even before bleeding occurs.

Anticoagulant rodenticide toxicosis is usually diagnosed based on history of ingestion of the substance. Differential diagnoses when massive hemorrhage is encountered include disseminated intravascular coagulation, congenital factor deficiencies, von Willebrand's disease, platelet deficiencies, and canine ehrlichiosis. A prolonged prothrombin, partial thromboplastin, or thrombin time in the presence of normal fibrinogen, fibrin degradation products, and platelet counts is strongly suggestive of anticoagulant rodenticide toxicosis, as is a positive therapeutic response to vitamin K1.

Vitamin K1 is antidotal. Recommended dosages vary from 0.25-2.5 mg/kg in warfarin (coumarin) exposure, to 2.5-5 mg/kg in the case of long-acting rodenticide intoxication (diphacinone, brodifacoum, bromadiolone). Vitamin K1 is administered SC (with the smallest possible needle to minimize hemorrhage) in several locations to speed absorption. IV administration of vitamin K1 is contraindicated, as anaphylaxis may occasionally result. The oral form of K1 may be used daily after the first day, commonly at the same level as the loading dose (divided bid). Fresh or frozen plasma (9 mL/kg) or whole blood (20 mL/kg) IV is required to replace needed clotting factors and RBC if bleeding is severe. One week of vitamin K1 treatment is usually sufficient for first- generation anticoagulants. For intermediate and second-generation anticoagulants or if anticoagulant type is unknown, treatment should continue for 2-4 wk to control longterm effects. Administration of oral vitamin K1 with a fat-containing ration, such as canned dog food, increases its bioavailability 4-5 times as compared with vitamin K1 given PO alone.

Coagulation should be monitored weekly until values remain normal for 5-6 days after cessation of therapy. Vitamin K3 given as a feed supplement is ineffective in the treatment of anticoagulant rodenticide toxicosis. Additional supportive therapy may be indicated, including thoracocentesis (to relieve dyspnea due to hemothorax) and supplemental oxygen if needed.

ANTU (á-Naphthylthiourea). ANTU causes local gastric irritation; when absorbed, it increases permeability of the lung capillaries in all animals, although species variability in dose response is marked. Properties of ANTU, when compared with those of warfarin, have led to near abandonment of its use. Dogs and pigs are occasionally poisoned; ruminants are resistant.

Animals with an empty stomach readily vomit after ingestion of ANTU; however, food in the stomach decreases the stimulation to vomit, and fatal quantities may be absorbed. Signs include vomiting, hypersalivation, coughing, and dyspnea. Animals prefer to sit. Severe pulmonary edema, moist rales, and cyanosis are present. Dependent signs include weakness; ataxia; rapid, weak pulse; and subnormal temperature. Death from hypoxia may occur within 2-4 hr of ingestion, while animals that survive 12 hr may recover.

The lesions are suggestive. The most striking findings are pulmonary edema and hydrothorax. Hyperemia of the tracheal mucosa; mild to moderate gastroenteritis; marked hyperemia of the kidneys; and a pale, mottled liver are found in most cases. Tissue for chemical analysis must be obtained within 24 hr.

Emetics should be used only if respiratory distress is not evident. Prognosis is grave when severe respiratory signs occur. Agents providing sulfhydryl groups, eg, n-amyl mercaptan, sodium thiosulfate (10% solution), or n-acetylcysteine are beneficial. Positive-pressure oxygen therapy, an osmotic diuretic (eg, mannitol), and atropine (0.02-0.25 mg/kg) may relieve the pulmonary edema.

Bromethalin. This nonanticoagulant, single-dose rodenticide is a neurotoxin that appears to uncouple oxidative phosphorylation in the CNS. CSF pressure increases, which places pressure on nerve axons and results in decreased conduction of nerve impulses, paralysis, and death. In dogs, a dose of 1.67 mg/kg is toxic, and 2.5 mg/kg (25 g of bait/kg body wt) is lethal.

Bromethalin can cause either an acute or a chronic syndrome. The acute effects follow consumption of e"5 mg/kg bromethalin. Signs, which include hyperexcitability, muscle tremors, grand mal seizures, hindlimb hyperreflexia, CNS depression, and death, may appear ~10 hr after ingestion. Chronic effects are seen with lower dosages and may appear 24-86 hr after ingestion. This syndrome is characterized by vomiting, depression, ataxia, tremors, and lateral recumbency.

The effects may be reversible if exposure to bromethalin is discontinued. Bromethalin toxicosis should be considered when cerebral edema or posterior paralysis is present.

Treatment should be directed at blocking absorption from the gut and reducing cerebral edema. Use of mannitol as an osmotic diuretic and corticosteroids have been suggested but have shown little effect in dogs poisoned by bromethalin. Use of activated charcoal for several days may improve the recovery rate.

Cholecalciferol. Although this rodenticide was introduced with claims that it was less toxic to nontarget species than to rodents, clinical experience has shown that rodenticides containing cholecalciferol are a significant health threat to dogs and cats. Cholecalciferol produces hypercalcemia, which results in systemic calcification of soft tissue, leading to renal failure, cardiac abnormalities, hypertension, CNS depression, and GI upset.

Signs generally develop within 18-36 hr of ingestion and can include depression, anorexia, polyuria, and polydipsia. As serum calcium concentrations increase, clinical signs become more severe. Serum calcium concentrations >16 mg/dL are not uncommon. GI smooth muscle excitability decreases and is manifest by anorexia, vomiting, and constipation. Hematemesis and hemorrhagic diarrhea may develop as a result of dystrophic calcification of the GI tract and should not lead to a misdiagnosis of anticoagulant rodenticide toxicosis. Loss of renal concentrating ability is a direct result of hypercalcemia. As hypercalcemia persists, mineralization of the kidneys results in progressive renal insufficiency.

Diagnosis is based on history of ingestion, clinical signs, and hypercalcemia. Other causes of hypercalcemia, such as hyperparathyroidism, normal juvenile hypercalcemia, paraneoplastic hypercalcemia, hemoconcentration (hyperproteinemia), and diffuse osteoporosis should be ruled out. Gross lesions associated with hypercalcemia include pitted, mottled kidneys; diffuse hemorrhage of the GI mucosa; and roughened, raised plaques on the great vessels and on the surface of the lungs and abdominal viscera.

Recommended therapy includes gastric evacuation, generally followed by administration of activated charcoal at 2-8 g/kg body wt in a water slurry. Calciuresis is accomplished with 0.9% sodium chloride solution and administration of furosemide (initial bolus of 5 mg/kg, IV, followed by a constant rate IV infusion of 5 mg/kg/hr) and corticosteroids (prednisolone, 1-2 mg/kg, bid). Furosemide and prednisolone should be continued for 2-4 wk, and the serum calcium concentration monitored at 24 hr, 48 hr, and 2 wk after cessation of treatment. Additionally, calcitonin may be used at 4-6 IU/kg, SC, every 2-3 hr, until the serum calcium stabilizes at <12 mg/dL. The IV use of calcium chelators such as Na-EDTA has been used in severe cases, but this use is experimental and requires close monitoring of blood calcium to prevent hypocalcemia. The dose of prednisolone should be tapered if it is administered for >2 wk to prevent acute adrenocortical insufficiency. Continuous peritoneal dialysis may be considered if the animal is in renal failure. A low-calcium diet should be provided in all cases of significant exposure to cholecalciferol rodenticides.

Recently, pamidronate disodium, a specific inhibitor of bone resorption used for the treatment of hypercalcemia of malignancy and Paget's disease in humans, has shown promise in the treatment of cholecalciferol toxicosis in dogs. It is given slowly IV at 1.3-2.0 mg/kg in saline solution over 2-4 hr. Two infusions are given 4 days apart. Pamidronate disodium has a long-lasting inhibitory action on bone resorption, thus requiring only limited infusions. Total serum calcium and BUN should be monitored 2 and 4 days after the last infusion.

Metaldehyde. This polymer of acetaldehyde is used as a snail or slug bait, to which dogs and livestock may be exposed. (See also metaldehyde poisoning, Metaldehyde Poisoning: Introduction.) Toxic effects are due to absorption of limited acetaldehyde from metaldehyde hydrolysis in the stomach, but primarily to the metaldehyde itself. Signs range from salivation and vomiting to anxiety and incoordination with muscle tremors, fasciculations, and hyperesthesia leading to continuous muscle spasms, prostration, and death. Generally, the muscle spasms are not initiated by external stimuli, but excessive muscular activity is common, often producing high body temperatures. Differential diagnoses include strychnine poisoning and anticholinesterase insecticide toxicity. The finding of metaldehyde bait or pellets in the vomitus and the possible odor of acetaldehyde from stomach contents or on the animal's breath may assist in diagnosis. Treatment is most effective if initiated early. Further toxicant absorption should be prevented by induced emesis, gastric lavage, and oral dosing with activated charcoal. Hyperesthesia and muscle activity may be controlled with diazepam at 2-5 mg, IV, or light barbiturate anesthesia and muscle relaxants as needed. IV fluid therapy with lactated Ringer's solution or 5% glucose should be aggressive to promote toxin excretion and to combat dehydration and the acidosis induced by the excessive muscle activity. Continuous supportive care is important.

Prognosis is heavily determined by the exposure dose, but if death does not occur earlier, animals poisoned by metaldehyde may show clinical improvement 24-36 hr after initial onset of signs.

Phosphorus. In its white (or yellow) form, phosphorus is hazardous to all domestic animals and is locally corrosive and hepatotoxic when absorbed. Phosphorus is infrequently used as a rodenticide today, but dogs occasionally become exposed through ingestion of fireworks that contain white phosphorus. The onset of signs of poisoning is sudden. Early signs include vomiting, severe diarrhea (often hemorrhagic), colic, and a garlic-like odor to the breath. Apparent recovery can occur up to 4 days after ingestion, but additional signs of acute liver damage may develop, including hemorrhages, abdominal pain, and icterus. Hepatic encephalopathy is followed by convulsions and death. Lesions include severe gastroenteritis; fatty liver; multiple hemorrhages; and black, tarry blood that fails to clot. Body tissues and fluids may be phosphorescent, and the gastric contents have a garlic odor. Death is due to hepatic and renal failure.

Prognosis is grave unless treatment is instituted early. A 1% solution of copper sulfate is an effective emetic and also forms a nonabsorbable copper phosphide complex. Gastric lavage with a 0.01-0.1% potassium permanganate solution or a 0.2-0.4% copper sulfate solution should be followed by activated charcoal adsorbent and 30 min later by a saline cathartic. Any fat in the diet must be avoided for 3-4 days or longer because fats favor additional absorption of phosphorus.

Mineral oil orally has been recommended because it dissolves phosphorus and prevents absorption.

Red Squill. This rodenticide is a cardiac glycoside derived from the plant Urginea maritima . It is of limited current use. Because rats are incapable of vomiting, red squill is more toxic to that species. It is unpalatable to domestic animals but, when eaten, usually induces vomiting in dogs and cats. Large quantities are required for toxicity in farm animals. It is considered relatively safe, but dogs, cats, and pigs have been poisoned. Signs are vomiting, ataxia, and hyperesthesia followed by paralysis, depression, or convulsions. Bradycardia and cardiac arrhythmias may end in cardiac arrest. The clinical course seldom is longer than 24-36 hr.

Treatment consists of supportive therapy and evacuation of the GI tract using gastric lavage and saline cathartics. Atropine sulfate SC at 6- to 8-hr intervals may prevent cardiac arrest. Phenytoin at 35 mg/kg, tid, should be given to dogs to suppress arrhythmias.

Sodium Monofluoroacetate (1080). 1080 is a colorless, odorless, tasteless, water-soluble chemical that is highly toxic (0.1-8 mg/kg) to all animals, including humans. Its use is restricted to certain commercial applications. Fluoroacetate is metabolized to fluorocitrate, which blocks the tricarboxylic acid cycle—a mechanism necessary for cellular energy production. It causes toxic effects by overstimulating the CNS, resulting in death by convulsions, and by causing alteration of cardiac function that results in myocardial depression, cardiac arrhythmias, ventricular fibrillation, and circulatory collapse. CNS stimulation is the main effect in dogs, while the cardiac effects predominate in horses, sheep, goats, and chickens. Pigs and cats appear about equally affected by both. A characteristic lag phase of e"30 min after ingestion occurs before the onset of nervousness and restlessness. Marked depression and weakness follow in all species except dogs and pigs.

Affected animals rapidly become prostrate, and the pulse is weak and 2-3 times normal rate. Death is due to cardiac failure. Usually, dogs and pigs rapidly develop tetanic convulsions similar to those of strychnine poisoning. Many exhibit severe pain. Vomiting is prominent in pigs. Dogs usually have urinary and fecal incontinence and exhibit frenzied running. The course is rapid; affected animals die within hours after signs appear. Few animals that develop marked signs recover. Congestion of organs, cyanosis, subepicardial hemorrhages, and a heart stopped in diastole are common necropsy findings.

Emetics are contraindicated if clinical signs are present. Gastric lavage and adsorbents (activated charcoal, 0.5 g/kg) are recommended. Prognosis is grave if clinical signs are severe. Barbiturates are preferred for controlling seizures. Glyceryl monoacetate (monacetin) has been used with inconsistent results as a competitive antagonist of fluoroacetate. The recommended dose is 0.55 mL/kg, IM, or IV in 5 parts of sterile saline solution, every 30 min for several hours.

The danger of secondary poisoning due to ingestion of rodents killed with 1080 is high and has led to restrictions in its use (and use of fluoroacetamide) in the USA. Only certified, insured exterminators can purchase 1080, and a black dye must be mixed with it for identification.

Sodium Fluoroacetamide (1081). 1081 causes signs similar to those of 1080 (see above) and requires the same treatment.

Thallium Sulfate. This general cellular poison can affect all species of animals. It has been banned for use as a rodenticide. Onset of clinical signs may be delayed 1-3 days and, although all body systems are affected, the most prominent signs are of the GI, respiratory, integumentary, and nervous systems. Signs include gastroenteritis (occasionally hemorrhagic), abdominal pain, dyspnea, blindness, fever, conjunctivitis, gingivitis, and tremors or seizures. After 4-5 days and an apparent recovery, or after repeated small doses, a chronic dermatitis characterized by alopecia, erythema, and hyperkeratosis occurs. Necrosis of many tissues is a common necropsy finding.

Treatment of the acute phase of thallium poisoning includes emetics, gastric lavage with a 1% sodium iodide solution, and IV administration of 10% sodium iodide. Diphenylthiocarbazone (dithizone, 70 mg/kg, PO, tid) is antidotal but must be given within

24 hr of exposure. At the same time and for 14 days thereafter, Prussian blue 100 mg/kg should be given bid in oral aqueous suspension to stop enterohepatic recirculation of the thallium and to enhance its excretion in the feces. Symptomatic treatment of the diarrhea and convulsions is needed with particular attention to fluid and electrolyte balance, nutrient needs, prevention of secondary infection, and good nursing care.

Zinc Phosphide and Aluminum Phosphide. Zinc phosphide has been used extensively around farms and barns because affected rats tend to die in the open. Toxicity is due to liberation of phosphine gas at the acid pH in the stomach. The gas results in direct GI tract irritation along with cardiovascular collapse. The toxic dose is ~40 mg/kg, and onset is rapid in animals with a full stomach. Clinical signs include vomiting, abdominal pain, and aimless running and howling, followed by depression, dyspnea, and convulsions (which may resemble those seen in strychnine or fluoroacetate poisoning). Death is due to respiratory arrest. The odor of acetylene is present in vomitus or stomach contents. Less frequent lesions include visceral congestion and pulmonary edema. Diagnosis is based on history of exposure to zinc phosphide, suggestive clinical signs, and detection of zinc phosphide in stomach contents. Zinc levels in the blood, liver, and kidneys may be increased. Treatment must include supportive therapy, calcium gluconate, and appropriate fluids to reduce acidosis. Sodium bicarbonate (in cattle, 2-4 L of 5%), PO, to neutralize stomach acidity is recommended.

Toxicity of nitrates and nitrites

Nitrate poisoning can occur commonly in animals. Poisoning is usually associated with animals ingesting forage or feed with a high nitrate content. Sheep and cattle are more susceptible to poisoning than non-ruminant species, because microbes in their digestive tracts favor the conversion of nitrate to nitrite.

What is nitrate poisoning? Nitrate in itself is not toxic to animals, but at elevated levels it causes a disease called nitrate poisoning. Nitrates are normally found in forages and are converted by the digestion process to nitrite, and in turn the nitrite is converted to ammonia. The ammonia is then converted to protein by bacteria in the rumen. If cattle rapidly ingest large quantities of plants that contain high levels of nitrate, nitrite will accumulate in the rumen. Nitrite is ten times as toxic to cattle as nitrate.

Nitrite is absorbed into red blood cells and combines with hemoglobin (oxygen-carrying molecule) to form methemoglobin. Methemoglobin cannot transport oxygen as efficiently as hemoglobin, so the animal's heart rate and respiration increases, the blood and tissues of the animal take on a blue to chocolate brown tinge, muscle tremors can develop, staggering occurs, and the animal eventually suffocates.

What plant factors favor nitrate poisoning? The majority of nitrate poisoning cases in North Dakota occur with drought-stressed oats, corn and barley. However, a number of other plants can also accumulate nitrate, including sudangrass, sorghum-sudan hybrids, and pearl millet. Table 1 lists common plants known to accumulate nitrate if conditions are favorable. Plants that have been fertilized have higher nitrate levels than non-fertilized plants.

The abnormal accumulation of nitrate in plants is influenced by various factors such as moisture conditions, soil conditions and type of plant. Plant stresses such as drought are associated with increased levels of nitrate in plants. Soils high in nitrogen readily supply nitrate to plants. Acidity, sulfur or phosphorus deficiencies, low molybdenum, and low temperatures are known to increase nitrate uptake by plants.

Plant parts closest to the ground (stalks) contain the highest concentrations of nitrates. Leaves contain less than stalks or stems, and the seed (grain) and flower usually contain little or no nitrate. Most of the plant nitrate is in the bottom third of the stalk. Research from Oklahoma has shown that the lower 6 inches of the stem in pearl millet contains three times more nitrate than the top part of the plant. While difficult to do with drought-stressed forages, raising the cutter bar above 6 inches can reduce nitrate content of forages.

Nitrate decreases as plants mature. Young plants have higher nitrate concentrations than mature plants. However, mature plants can still have excessive nitrate concentrations if environmental and soil conditions are favorable.

Weather conditions that favor nitrate accumulation by plants. Not all drought conditions cause high nitrate levels in plants. Some moisture must be present in the soil for the plant to absorb and accumulate nitrate. If the major supply of nitrates for the plant is in the dry surface soil, very little nitrate will be absorbed by the roots. In plants that survive drought conditions, nitrates are often high for several days following the first rain (as the plant regrows following drought). Frost, hail and low temperatures all interfere with normal plant growth and can cause nitrates to accumulate in the plant. Frost and hail may damage, reduce or completely destroy the leaf area of the plant. A decrease in leaf area limits the photosynthetic activity of the plant, so nitrates absorbed by the roots are not converted to plant proteins but are accumulated in the stem or stalk instead. Most plants require temperatures above 55°F for active growth and photosynthesis. Nitrates can be absorbed quickly by plants when temperatures are low, but conversion to amino acids and protein occurs very slowly in plants during periods of cool weather. This allows nitrate to accumulate in the plant.

Water may be a source of toxic levels of nitrate for livestock. Water may become contaminated by fertilizer, animal wastes or decaying organic matter. Shallow wells with poor casings are susceptible to contamination. Marginally toxic levels of nitrate in water and feed may together cause nitrate toxicity in animals. Remember to consider both sources of nitrate.

Acute nitrate poisoning may occur if livestock consume nitrate fertilizer. Avoid grazing immediately after spreading fertilizer. Areas where the fertilizer spreader turns or areas where filling (and consequently spilling) take place may have excessive quantities of nitrate freely available to animals.

What are the clinical signs of toxicity? Clinical signs of nitrate poisoning are related to the lack of oxygen in the blood. Acute poisoning usually occurs between a half hour to four hours after consuming toxic levels of nitrate. Onset of symptoms are rapid and include:

- bluish/chocolate brown mucous membranes
- rapid/difficult breathing
- noisy breathing
- rapid pulse (150+/min)
- salivation, bloat, tremors, staggering
- weakness, coma, death
- dark "chocolate-colored" blood

Pregnant females that survive nitrate poisoning may abort due to lack of oxygen to the fetus.

Abortions generally occur approximately 10-14 days following exposure to nitrates.

How can you diagnose nitrate toxicity? Diagnosis of nitrate intoxication is based on observed clinical signs and the possibility of exposure to toxic plants or water. Consult a veterinarian for a definitive diagnosis. Laboratory analysis can be performed on suspected plants, water, stomach contents, blood, urine, and aqueous humor of the eye of dead cattle to confirm the diagnosis.

Postmortem specimens of rumen contents are of little value for nitrate determination because most nitrate in the rumen is reduced by anaerobic fermentation to ammonia.

Samples from fresh grass or dry forages need to be representative of the field or bales in question. Package these samples in a clean plastic bag and ship them to the laboratory for analysis. Collect water samples in a sterile bottle. When collecting from a water system, let the water flow for a couple of minutes before collecting the sample. Results of chemical analysis are interpreted according to guidelines in Table 5. These guidelines apply to livestock only.

Is there a treatment for nitrate toxicity? Animals can be treated by intravenous injections of methylene blue. Commercial preparations intended for treatment of prussic acid poisoning only should not be used to treat nitrate poisoning. Note that methylene blue is not approved by the FDA for use in food-producing animals. Consult your veterinarian before using this treatment. Prevention of nitrate poisoning is best achieved by controlling type and quantity of forage offered to livestock. Avoid forages with potentially toxic levels of nitrate or at least dilute them with feeds low in nitrate. When in doubt, have feeds and forages analyzed for nitrate before grazing or feeding them. Forages with sub-lethal nitrate levels can be fed to livestock with appropriate precautions. No single level of nitrate is toxic under all conditions. When grazing, feed a dry roughage first to reduce the amount of affected plants ingested by hungry animals. Harvested forages that are high in nitrate can often be safely fed by mixing with other feeds to reduce the total dietary intake of nitrate. Contact your veterinarian or extension personnel if you need assistance in determining the correct ratios of high and low nitrate forages to blend to develop a ration for a particular class of livestock.

Management guidelines for dealing with nitrate toxicity:

- Test drought-stressed small grain forages and other forages suspected of being high in nitrates before feeding.
- Dilute high nitrate forages with other forages or feedstuffs which are low in nitrates. This can bring the nitrate level of the diet down low enough where it is safe to feed.
- Frequent intake of small amounts of high nitrate feed helps adjust livestock to high nitrate feeds and increases the total amount of nitrate that livestock can consume daily without adverse effects.
- Allow cattle time to adapt to increased nitrate in the diet. If nitrate levels are not excessively high (9000 ppm nitrate) the animals can adapt to increasing amounts in the feed.
- Allow livestock access to fresh, nitrate-free water at all times.
- Be sure you don't overstock pastures when grazing high nitrate forages. Overstocking increases the amount of high nitrate plant parts (stems and stalks) that are consumed by livestock.
- Do not strip graze high nitrate forages. Strip grazing also increases the amount of stem and stalk material consumed by livestock.
- Do not allow hungry cattle access to high nitrate forages or pastures. Feed cattle hays or forages low in nitrates before turning them onto high nitrate pastures.
- Supplement cattle grazing high nitrate forages with other low-nitrate feedstuffs such as low nitrate forages, feed grains, or byproducts.
- If possible, graze cattle on high nitrate pastures during the day and remove them at night for the first week of grazing. This reduces the amount of high nitrate forage consumed and helps acclimate cattle to the high nitrate levels.

- If possible, don't graze high nitrate pastures until one week after a killing frost.
- Observe cattle frequently when you turn into a suspected field or pasture in order to detect any signs of toxicity.
- Cattle in poor health and condition, especially cattle suffering from respiratory disease, are more susceptible to nitrate poisoning.
- Consider harvesting and feeding high nitrate forages as silages. Nitrate levels are reduced by the fermentation process that occurs when feeds are ensiled.
- Do not allow cattle access to areas where fertilizers are stored.
- Do not feed green chop which has heated after cutting or which has been held overnight. Heating favors the formation of nitrite which is more toxic than nitrate.

Many species are susceptible to nitrate and nitrite poisoning, but cattle are affected most frequently. Ruminants are especially vulnerable because the ruminal flora reduces nitrate to ammonia, with nitrite (-10 times more toxic than nitrate) as an intermediate product. Nitrate reduction (and nitrite production) occurs in the cecum of equids but not to the same extent as in ruminants. Young pigs also have GI microflora capable of reducing nitrate to nitrite, but mature monogastric animals (except equids) are more resistant to nitrate toxicosis because this pathway is age-limited.

Acute intoxication is manifested primarily by methemoglobin formation (nitrite ion in contact with RBC oxidizes ferrous iron in Hgb to the ferric state, forming stable methemoglobin incapable of oxygen transport) and resultant anoxia. Secondary effects due to vasodilatory action of the nitrite ion on vascular smooth muscle may occur. The nitrite ion may also alter metabolic protein enzymes. Ingested nitrates may directly irritate the GI mucosa and produce abdominal pain and diarrhea. Although usually acute, the effects of nitrite or nitrate toxicity may be subacute or chronic and are reported to include retarded growth, lowered milk production, vitamin A deficiency, minor transitory goitrogenic effects, abortions and fetotoxicity, and increased susceptibility to infection. Chronic nitrate toxicosis remains a controversial issue and is not as yet well characterized, but most current evidence does not support allegations of lowered milk production in dairy cows due to excessive dietary nitrate exposure alone.

Etiology. Nitrates and nitrites are used in pickling and curing brines for preserving meats, certain machine oils and antirust tablets, gunpowder and explosives, and fertilizers. They may also serve as therapeutic agents for certain noninfectious diseases, e.g., cianide poisoning. Toxicoses occur in unacclimated domestic animals most commonly from ingestion of plants that contain excess nitrate, especially by hungry animals engorging themselves and taking in an enormous body burden of nitrate. Nitrate toxicosis can also result from accidental ingestion of fertilizer or other chemicals. Nitrate concentrations may be hazardous in ponds that receive extensive feedlot or fertilizer runoff; these types of nitrate sources may also contaminate shallow, poorly cased wells. Although nitrate concentrations are increasing in groundwater in the USA, well water is rarely the sole cause of excess nitrate exposure. Water with both high nitrate content and significant coliform contamination has greater potential to affect health adversely and lower productivity than do either nitrate or bacteria alone. Livestock losses have occurred during cold weather due to the concentrating effect of freezing, which increases nitrate content of remaining water in stock tanks. Crops that readily concentrate nitrate include cereal grasses (especially oats, millet, and rye), com (maize), sunflower, and sorghums. Weeds that commonly have high nitrate concentrations are pigweed, lamb's quarter, thistle, Jimson weed, fireweed (Kochia), smartweed, dock, and Johnson grass. Anhydrous ammonia and nitrate fertilizers and soils naturally high in nitrogen tend to increase nitrate content in forage. Excess nitrate in plants is generally associated with damp weather conditions and cool temperatures (13°C]), although high concentrations are also likely to develop when growth is rapid during hot, humid weather. Drought conditions, particularly if occurring when plants are immature, may leave the vegetation with high nitrate content.

Decreased light, cloudy weather, and shading associated with crowding conditions can also cause increased concentrations of nitrates within plants. Well-aerated soil with a low pH, and low or deficient amounts of molybdenum, sulfur, or phosphorus in soil tend to enhance nitrate uptake, whereas soil deficiencies of copper, cobalt, or manganese tend to have opposing effects. Anything that stunts growth increases nitrate accumulation in the lower part of the plant. Phenoxy acid derivative herbicides, e.g., 2, 4-D and 2, 4, 5-T, applied to nitrate-accumulating plants during early stages, cause increased growth and a high nitrate residual (10-30%) in surviving plants, which are lush and eaten with apparent relish even though previously avoided. Nitrate, which does not selectively accumulate in fruits or grain, is found chiefly in the lower stalk with lesser amounts in the upper stalk and leaves. Nitrate in plants can be converted to nitrite under the proper conditions of moisture, heat, and microbial activity after harvesting.

Clinical Findings. Signs of nitrite poisoning usually appear suddenly due to tissue hypoxia and low blood pressure as a consequence of vasodilation. Rapid, weak heartbeat with subnormal body temperature, muscular tremors, weakness, and ataxia are early signs of toxicosis when methemo- globinemia reaches 30-40%. Brown, cyanotic mucous membranes develop rapidly as methemoglobinemia exceeds 50%. Dyspnea, tachypnea, anxiety, and frequent urination are common. Some monogastric animals, usually because of excess nitrate exposure from nonplant sources, exhibit salivation, vomiting, diarrhea, abdominal pain, and gastric hemorrhage. Affected animals may die suddenly without appearing ill, in terminal anoxic convulsions within 1 hr, or after a clinical course 'of

12-24 hr or longer. Acute lethal toxicoses almost always are due to development of 380% methemoglobinemia. Under certain conditions, adverse effects may not be apparent until animals have been eating nitrate-containing forages for days to weeks. Some animals that develop marked dyspnea recover but then develop interstitial pulmonary emphysema and continue to suffer respiratory distress; most of these recover fully within 10-14 days. Abortion and stillbirths may be seen in some cattle 5-14 days after excessive nitrate/nitrite exposure, but likely only in cows that have survived a 350% methemo-globinemia for 6-12 hr or longer.

Prolonged exposure to excess nitrate coupled with cold stress and inadequate nutrition may lead to the alert downer cow syndrome in pregnant beef cattle; sudden collapse and death can result.

Lesions. Blood that contains methemoglobin usually has a chocolate-brown color, although dark red hues may also be seen. There may be pinpoint or larger hemorrhages on serosal surfaces. Dark brown discoloration evident in moribund or recently dead animals is not pathognomonic, however, and other methemoglobin inducers must be considered. If necropsy is postponed too long, the brown discoloration may disappear with conversion of methemoglobin back to Hgb.

Diagnosis. Excess nitrate exposure can be assessed by laboratory analysis for nitrate in both preand postmortem specimens. High nitrate and nitrite values in postmortem specimens may be an incidental finding, indicative only of exposure and not toxicity. Plasma is the preferred premortem specimen, because some plasma-protein-bound nitrate could be lost in the clot if serum was collected. Nitrite present in whole blood also continues to react with Hgb in vitro, so these specimens must be centrifuged immediately. Plasma separated to prevent erroneous values of both. Additional postmortem specimens from either toxicoses or abortions include ocular fluids, fetal pleural or thoracic fluids, fetal stomach contents, and maternal uterine fluid. All specimens should be frozen in clean plastic or glass containers before submission, except when whole blood is collected for methemoglobin analysis. Because the amount of nitrate in rumen contents is not representative of concentrations in the diet, evaluation of rumen contents is not indicated.

Bacterial contamination of postmortem specimens, especially ocular fluid, is likely to cause conversion of nitrate to nitrite at room temperature or higher; such specimens may have abnormally high nitrite concentrations with reduced to aBsent nitrate concentrations. Endogenous biosynthesis of nitrate and nitrite by macrophages stimulated by lipopolysaccharide or other bacterial products may also complicate interpretation of analytical findings; this should be considered as a possible maternal or fetal response to an infectious process. Methemoglobin analysis alone is not a reliable indicator of excess nitrate or nitrite exposure except' in acute toxicosis, because 50% of methemoglobin present will be converted back to Hgb in -2 hr, and alternate forms of nonoxygenated Hgb that may be formed by reaction with nitrite are not detected by methemoglobin analysis. Nitrate and nitrite concentrations >20 flg N03 ImL and >0.5 flg N02 ImL, respectively, in maternal and perinatal serum, plasma, ocular fluid, and other similar biologic fluids are usually indicative of excessive nitrate or nitrite exposure in most domestic animal species; nitrate concentrations of up to 40 flg NO/mL have been present in the plasma of healthy calves at birth, but are reduced rapidly as normal neonatal renal function eliminates nitrate in the urine. Normally expected nitrate and nitrite concentrations in similar diagnostic specimens are usually <10 flg NO/ml and <0.2 flg NO/ml, respectively. Nitrate and nitrite concentrations >10 but <20 f-Lg NO/ml and >0.2 but <0.5 f-Lg NO/ml, respectively, are suspect and indicate nitrate or nitrite exposure of unknown duration, extent, or origin. The possible contribution of endogenous nitrate or nitrite synthesis by activated macrophages must also be considered. The biologic half-life of nitrate in beef cattle, sheep, and ponies was determined to be 7.7, 4.2, and 4.8 hr, respectively, so it will be at least 5 biologic half-lives (24-36 hr) before elevated nitrate concentrations from excessive nitrate exposure diminish to normally expected values, allowing additional time for valid premortem specimen collection. A latent period may exist between excessive maternal dietary nitrate exposure and equilibrium in perinatal ocular fluids. Aqueous humor is actively secreted into the anterior chamber at a rate of 0.1 ml/hr, and nitrate and nitrite are thought to enter the globe of the eye by this mechanism. Equilibrium between aqueous and vitreous humor is by passive diffusion rather than by active secretion, so nitrate or nitrite may be present in comparatively lesser concentrations in vitreous humor after acute exposure. Field tests for nitrate are presumptive and should be confirmed by standard analytical methods at a qualified laboratory. The diphenylamine blue test (1% in concentrated sulfuric acid) is more suitable to determine the presence or absence of nitrate in suspected forages. Nitrate test strips (dipsticks) are effective in determining nitrate values in water supplies and can be used to evaluate nitrate and nitrite content in serum, plasma, ocular fluid, and urine. Differential diagnoses include poisonings by cyanide, urea, pesticides, toxic gases (e.g., carbon monoxide, hydrogen sulfide), chlorates, aniline dyes, aminophenols, or drugs (e.g., sulfonamides, phenacetin, and acetaminophen), as well as infectious or noninfectious diseases (e.g., grain overload, hypocalcemia, hypomagnesemia, pulmonary adenomatosis, or emphysema) and any sudden unexplained deaths.

Treatment. Slow IV injection of 1% methylene blue in distilled water or isotonic saline should be given at 4-22 mg/kg body wt, or more, depending on severity of exposure. Lower dosages may be repeated in 20-30 min if the initial response is not satisfactory. Lower dosages of methylene blue can be used in all species, but only ruminants can safely tolerate higher dosages. If additional exposure or absorption occurs during therapy, retreating with methylene blue every 6-8 hr should be considered. Rumen lavage with cold water and antibiotics may stop the continuing microbial production of nitrite.

Control. Animals may adapt to higher nitrate content in feeds, especially when grazing summer annuals such as sorghum-Sudan hybrids. Multiple, small feedings help animals adapt. Trace mineral supplements and a balanced diet may help prevent nutritional or metabolic disorders associated with long term excess dietary nitrate consumption. Feeding grain with high-nitrate forages may reduce nitrite production. Forage nitrate concentrations >1% nitrate dry-weight basis (la, 000 ppm N03) may cause acute toxicoses in unacclimated animals, and forage nitrate concentrations £5, 000 ppm N03 (dry-weight basis) are recommended for pregnant beef cows.

However, even forage concentrations of I, 000 ppm N03 dry-weight basis have been lethal to hungry cows engorging themselves in a single feeding within an hour, so the total dose of nitrate ingested is a deciding factor. High-nitrate forages may also be harvested and stored as ensilage rather than dried hay or green chop; this may reduce the nitrate content in forages by up to 50%. Raising cutter heads. of machinery during harvesting operations selectively leaves the more hazardous stalk bases in the field. Hay appears to be more hazardous than fresh green chop or pasture with similar nitrate content. Heating may assist bacterial conversion of nitrate to nitrite; feeding highnitrate hay, straw, or fodder that has been damp or wet for several days, or stockpiled, green-chopped forage should be avoided. Large round bales with excess nitrate are especially dangerous if stored uncovered outside; rain or snow dm leach and subsequently concentrate most of the total nitrate present into the lower third of these bales. Water transported in improperly cleaned liquid fertilizer tanks may be extremely high in nitrate. Young unweaned livestock, especially neonatal pigs, can be more sensitive to nitrate in water.

Poisoning by sodium chloride

Salt toxicity (sodium chloride, NaCl), which is more appropriately called "water deprivation sodium ion toxicosis" can result when excessive quantities of salt are ingested and intake of potable water is limited. Salt toxicity is unlikely to occur as long as salt-regulating mechanisms are intact and fresh drinking water is available. It has been reported in virtually all species of animals all over the world. It is more common in swine (the most sensitive species), cattle, and poultry. Sheep are relatively resistant. The acute oral lethal dose of salt is 2.2 g/kg in swine and 6.0 g/kg in sheep.

Etiology. Salt toxicity is directly related to water consumption. Water intake in animals can be reduced significantly or abolished completely due to factors such as mechanical failure of waterers, overcrowding, unpalatable medicated water, new surroundings, or frozen water. With water deprivation, sodium propionate, acetate, or carbonate can produce the same toxicosis as sodium chloride.

Feeder pigs on feed containing only 0.25% salt have had salt poisoning when water intake was limited, yet even 13% salt in feed may not produce poisoning when adequate fresh water is consumed. Swine feed should contain 0.5-1% salt, and fresh drinking water should always be available. Feeding whey or brine containing 3-4% salt can result in toxicosis in most livestock and poultry species. Similarly, ingestion of 1-3 kg of salt in deprived animals can result in salt toxicosis even when water is available, especially in cattle.

Chickens can tolerate up to 0.25% salt in drinking water but are susceptible to sodium ion toxicosis when water intake is restricted. Wet mash containing 2% salt caused poisoning in ducklings. High salt content in wet mash is more likely to cause poisoning than in dry feed, probably because birds eat more wet mash.

Cattle and sheep on range can develop salt poisoning when a high percentage of mineral supplement is provided, and the water supply is limited or saline. Sheep can tolerate 1% salt in drinking water; however, 1.5% may be toxic. It is generally recommended that drinking water should contain <0.5% total salt for any species of livestock. Chronic salt poisoning in cattle can cause gastroenteritis, depressed appetite, weight loss, and dehydration.

Clinical Findings. In pigs, early signs (rarely seen) may be increased thirst, pruritus, and constipation. Affected pigs may be blind, deaf, and oblivious to their surroundings; they will not eat, drink, or respond to external stimuli. They may wander aimlessly, bump into objects, circle, or pivot around a single limb. After 1-5 days of limited water intake, intermittent seizures occur with the pig sitting on its haunches, jerking its head backward and upward, and finally falling on its side in clonic-tonic seizures and opisthotonos. Terminally, pigs may lie on their sides, paddling in a coma, and die within a few to 48 hr.

In cattle, signs of acute salt poisoning involve the GI tract and CNS. Salivation, increased thirst, vomiting (regurgitation), abdominal pain, and diarrhea are followed by ataxia, circling, blindness, seizures, and partial paralysis. Cattle sometimes manifest belligerent and aggressive behavior. A sequela of salt poisoning in cattle is dragging of hindfeet while walking or, in more severe cases, knuckling of the fetlock joint. In poultry, increased thirst, dyspnea, fluid discharge from the beak, weakness, diarrhea, and leg paralysis are some of the common signs of salt poisoning.

Lesions. During the first 48 hr, swine develop eosinopenia, eosinophilic cuffs around vessels in the cerebral cortex and adjacent meninges, and cerebral edema or necrosis. After 3-4 days, eosinophilic cuffs are usually no longer present. The GI mucosa may be inflamed and congested and may have pinpoint, blood-filled ulcers. Cattle do not have eosinophilic cuffs; they have gastric inflammation or ulceration (or both), edema of skeletal muscles, and hydropericardium. Chickens have hydropericardium. In acute cases, no gross lesions may be present in any species. Diagnosis. Serum and CSF concentrations of sodium >160 mEq/L, especially when CSF has a greater sodium concentration than serum, are indicative of salt poisoning. In brain (cerebrum), >1,800 ppm of sodium (wet wt) is compatible with toxicosis. Characteristic brain lesions and analyses of feed or water for sodium content are useful for establishing a diagnosis.

In swine, differential diagnoses include insecticide poisoning (organochlorine, organophosphorous, and carbamate), phenylarsonic poisoning, and pseudorabies. In cattle, differential diagnoses include insecticide and lead poisoning, polioencephalomalacia, hypomagnesemic tetany, and the nervous form of ketosis.

Treatment. There is no specific treatment. Immediate removal of offending feed or water is imperative. Fresh water must be provided to all animals, initially in small amounts at frequent intervals. Ingestion of large amounts of water may exacerbate neurologic signs due to brain edema. Severely affected animals should be given water via stomach tube. The mortality rate may be >50% in affected animals regardless of treatment. In small animals, slow administration of hypertonic dextrose or isotonic saline may be useful.

Poisoning by smoke compounds (carbon monoxide)

Smoke inhalation is the leading cause of death from fires for both humans and animals. More than 80% of fire-related deaths are the result of smoke inhalation and not from surface burns.1 Smoke itself is the complex mixture of vapors, gases, fumes, heated air and particulate matter, and liquid and solid aerosols produced by thermal decomposition. Thermal decomposition can result from flaming combustion or from pyrolysis, which is the application of intense heat. These thermal decompositions can result in the rapid oxidation of a substance by heat. Pyrolysis occurring with high heat and relatively low oxygen concentration is known as smoldering. Although flaming combustion generates light (flame), heat, and smoke, smoke can be produced in the absence of flames. Thus flames are not a prerequisite for smoke production, and furthermore, the gaseous product of combustion (smoke) is not always visible. Combustion products are difficult to predict in fires. Even within the same fire, the concentration of the smoke may vary. Temperature, oxygen concentration, and the chemical composition of the burning material determine the combustion products. In recent years, the use of newer synthetic building materials and furnishings has led to an increase in inhalational injuries caused by fires. Although more rigorous building codes make new structures less likely to burn, the materials used to make them have become more dangerous when they do catch fire through their production of more toxic smoke. At about the time of the World War II, differences were noted between natural materials and synthetics in terms of their combustion products and relative toxicity when burning. It is now recognized that compared with natural materials (e.g., cotton, wood, wool) plastics generate more heat more swiftly, spread flames faster, generate larger amounts of denser visible smoke, and release more toxic and greater concentrations of invisible products of thermal decomposition.

Despite testimonials to the contrary, plastics are neither nonburning nor self-extinguishing and, like many other synthetic substances, burn hotter and smokier than wood or other natural substances.

The majority of fires in the United States (more than 70%) occur in residential homes. Carelessness with cigarettes, heating devices, matches, flammable liquids, and malfunctioning electrical appliances is overwhelmingly the most common initiatory cause of fires. Every year there are nearly 3600 human deaths in the United States. Deaths in companion animals as a consequence of fire are harder to quantify, but certainly thousands of animals suffer fire-related injuries and smoke inhalation each year.

Toxic dose. There is no standard toxic or lethal dose for smoke inhalation in animals. The composition of smoke can vary tremendously even from the same fire. Combustion products and their concentrations are difficult to predict, and the relative toxicity of smoke produced depends upon the composition of the substance burning, amount of oxygen available, the temperature of the fire, the length of exposure, and the size of the animal involved. In addition, the incredible variety of materials currently used in an animal's environment and their wide spectrum of toxic combustion products ensure that there is no such thing as "typical" smoke. If burns are present and respiratory tract tissue displays burn edema, the episode becomes much more serious and much more likely to be life threatening. Increased vascular permeability of burned, edematous respiratory tissue greatly enhances the toxic effects of smoke inhalation. In one study in humans, mortality as a result of smoke inhalation alone was 12%;where smoke inhalation was also associated with burns, 61% were fatal. Thus mortality from smoke inhalation is dramatically increased in animals with concomitant thermal burns.

Toxicokinetics and mechanism of toxicity. The pathophysiology of smoke inhalation can be traced to the mechanism of action of the individual toxins involved, their subsequent physiological effects, and the cause of clinical toxicity after exposure. Toxic combustion products are classified as simple asphyxiants, irritant toxins, and chemical asphyxiants. Simple asphyxiants are space occupying and fill enclosed spaces at the expense of oxygen. In addition to this effect, combustion uses oxygen and creates an oxygen-deprived environment. The net effect is less oxygen available to the animal. Irritant toxins are chemically reactive substances. They produce local effects on the tissue or the respiratory tract. Ammonia is produced by burning wool, silk, nylon, and synthetic resins. Ammonia has high water solubility and dissolves in moist membranes of the upper respiratory tract, resulting in nasopharyngeal, laryngeal, and tracheal inflammation. Acrolein is lipid soluble and penetrates cell membranes. It denatures nucleic acid and intracellular proteins and results in cell death. Acrolein is a very common irritant gas generated by combustion. Sulfur dioxide is found in more than 50% of smoke from fires. Sulfur dioxide reacts with the moist respiratory membrane mucosa, producing the potent caustic, sulfurous acid. Polyvinyl chloride is ubiquitously found in floor coverings, office and home furniture, electrical insulation, and clothing. The resultant combustion products phosgene, chlorine, and hydrogen chloride are produced in many residential fires.8 Together with water in the mucosa, chlorine produces hydrogen chloride free oxygen radicals and is very damaging to tissue. Phosgene descends and produces more delayed alveolar injuries. Isocyanates are produced from burning and smoldering upholstery, and intense irritation of both upper and lower respiratory tissue results. Organic material produces finely divided carbonaceous particulate matter upon combustion. This particulate matter or soot is suspended in the gases and hot air of smoke. Not only just carbon, soot has aldehydes, acids, and reactive radicals that adhere to its surface. The inhalation of soot and associated aerosols heightens the effect of other irritant toxins. Soot binds with respiratory mucosal surfaces, allowing other irritant chemicals to adhere and react with adjacent tissue. The penetrance and deposition of these particles within the respiratory tract is dependent on size.

Small particles (1 to 3 μm) reach the alveoli. In various animals, lung injury is decreased when smoke is filtered to remove particulate matter. Sulfur dioxide shows a high propensity to adhere to soot. In addition, polyvinyl chloride combustion produces a large amount of soot-containing smoke coated with its particular combustion products phosgene, chloride, and hydrogen chloride. In addition to soot and related particles, irritant gases, acids, and other combustion products can also adhere to aerosol droplets.

The most important determining factor in predicting the level of respiratory injury is the water solubility of the toxin. Water-soluble chemicals injure the mucosa of upper respiratory airways by releasing the mediators of inflammation and deleterious free radicals. This type of inflammation increases microvascular membrane permeability and results in a net influx of fluid from intravascular spaces into the upper respiratory tissue. The underlying tissue of the supraglottic larynx may become terrifically swollen and edematous. This edematous reaction can result in minutes to hours postexposure, continue to progress, and close off upper airways completely. Low water-soluble molecules react with the lung parenchyma. They react more slowly and produce delayed toxic effects. Concentration of the toxic element inhaled, particle size, duration of exposure, respiratory rate, absence of protective reflexes, preexisting disease, and size and age of the animal also contribute to the level and degree of respiratory injury in addition to the water solubility of toxic products. An intense inflammatory reaction develops secondary to the initial injury to respiratory mucosal cells by toxic combustion products. Inhaled soot and toxic gases generate increased airway resistance caused by inspissated secretions, increased mucosal airway edema, and associated bronchospasm. Damaged mucosal cells stimulate copious exudates rich in protein, inflammatory cells, and necrotic debris. If this reaction continues, mucosal sloughing ensues. The degenerative exudates, bronchorrhea, and extensive sloughing produce casts of the airways. In animal victims of smoke inhalation, these casts increase airway resistance by blocking major airways and prevent oxygen passage to the alveoli. In addition, increased vascular permeability of respiratory tissue contributes to airway blockage. Bronchoconstriction and reflexive wheezing follow in response to inflammation and the toxic mucosal injury. Chemical asphyxiants produce toxic systemic effects at tissue distant from the lung. Carbon monoxide is generated during incomplete combustion and is regarded as the most serious systemic agent to smoke inhalation victims. Carbon monoxide prevents oxygen binding to hemoglobin, thereby producing a functional anemia. Furthermore, carbon monoxide inhibits release of oxygen, thereby shifting the oxyhemoglobin dissociation curve to the left. Carbon monoxide itself has other toxic effects that cause lipid peroxidation and directly damage cellular membranes. Carbon monoxide is invariably present in smoke from fires and is thought to be the cause of most immediate deaths from smoke inhalation. Nitrogencontaining products, such as wool, silk, nylon, plastics, paper, rubber, pyroxylin, polyurethanes, and polyacrylonitriles, all produce cyanide upon their combustion. Cyanide has been detected in samples from many other types of fires as well.

Together with carbon monoxide, cyanide has at least an additive and perhaps synergistic toxic effect in victims of smoke inhalation. Nitrogen-containing compounds produce oxides of nitrogen on their burning, which are potent respiratory irritants. Other combustion products can cause systemic and local toxicity. Metal oxides, hydrogen fluoride, hydrogen bromide, and various hydrocarbons can all be retrieved from toxic smoke. Benzene can be detected in the smoke of plastic and petroleum fires. Antimony, cadmium, chromium, cobalt, gold, iron, lead, and zinc have all been recovered from smoke samples during fires. Natural disasters, accidents at illegal drug labs, transportation accidents, industrial fires, and acts of terrorism are situations where unusual types of toxic smoke combustion products may be encountered. In fact the entire spectrum of potentially toxic combustion products from fires is endless, and we must remain vigilant. Super-heated air and steam in smoke results in thermal burns to tissue of the respiratory tract. In animals the higher the air temperature and humidity, the greater is the mortality in affected individuals. Exposure to dry air heated to 200 °C for 5 minutes or to 125 °C for 15 minutes is potentially lethal in mammals. Shorter exposure to dry air at temperatures of 350° C to 500 °C results in tracheitis in dogs. Exposure to steam alone results in tracheitis, bronchitis, and pulmonary parenchymal damage. Respiratory tract injury secondary to steam or heat alone is relatively uncommon in animals. Combustion progressively consumes oxygen. This decrease in oxygen concentration produces hypoxic asphyxia. The normal oxygen fraction at sea level is roughly 21%. Acute reductions in ambient oxygen fractions to 15% result in dyspnea. A reduction to 10% produces dyspnea and altered mentation, and fractions from 8% to 6% cause loss of consciousness followed by death in less than 8 minutes. It is noteworthy to examine the dynamics of smoke dispersal from fires. Spreading smoke initially accumulates and forms a hot layer mainly at the ceiling, which gradually descends to the floor. The main toxic combustion agent threats (e.g., heat, irritants, asphyxiants, noxious gases, and particulate material) are found in this ceiling layer. Depending on the size of the enclosed room, the amount of smoke produced, and duration of time, the toxic products will eventually disperse to the floor. Thus at least initially, animals at the floor are breathing cooler and much less contaminated air and are receiving less radiant heat. Because of this pattern of dispersal, the chance for survival exists for limited periods. Carcinogens are also some of the toxic products of thermal decomposition. All fires produce benzopyrene, the classic initiator of carcinogenesis. Plastic fires, particularly those involving polyvinyl chloride, produce arsenic, benzene, chromium, and acrylonitrile, all of which are suspected human and animal carcinogens. Smoke from wood and plastic produces the potent carcinogen, formaldehyde. Soot, so long known to cause cancer in chimney sweeps and tobacco smokers, is a principal product of most fires. The exact association of smoke inhalation and the development of cancer are unknown for animals at present. Smoke inhalation causes progressive physiological dysfunction and ultimately can lead to death. Irrespective of cause, asphyxia is the underlying mechanism. This asphyxia may be due to inhibition of cellular respiration, impaired oxygen transport and delivery, central respiratory depression, direct or indirect occlusion of airways, or a decreased supply of oxygen. For smoke inhalation, there is a direct correlation between the duration of exposure and the severity of effects. Finally the greater the exposure the more rapid and pronounced are the effects observed.

Clinical signs. Smoke inhalation victims generally have signs of respiratory compromise, systemic toxicity, or a combination of both. Concomitant surface burns may be noted and cutaneous burns of more than 15% body surface area, a history of exposure in an enclosed space, altered mentation and carbonaceous sputum or saliva production are all associated with a high incidence of bronchopulmonary injury. The presence of extensive body burns indicates greater exposure and a potentially larger, hotter fire and is typically associated with a higher incidence of both upper and lower respiratory tract injury and a worse prognosis. A respiratory abnormality may worsen with time after presentation, and signs of systemic toxicity are maximal at the time of exposure. Signs of smoke inhalation may be notoriously nonspecific and include cough, dyspnea, tachycardia, tachypnea, and hypoxemia. Almost always, signs of lacrimation, conjunctivitis, pharyngitis, and rhinitis are present. Erythema, edema, and soot may be evident upon examination of the nose, mouth, and throat. Corneal abrasions are common, and exposure to fire or prolonged heat can produce corneal burns. Mucosal ulcerations and hemorrhagic areas may be present.

Drooling, dysphagia, hoarseness, and stridor are all signs of laryngotracheal involvement and injury. In severe cases, copious exudates combined with severe laryngeal edema can result in complete upper airway obstruction. Blistering, erythema, ulcerations, mucosal sloughing, mucosal edema, hemorrhage, and laryngospasm may all be evident with the aid of a laryngoscope. The increased secretions may contain soot and carbonaceous particulate matter. Laryngeal edema and tracheal narrowing may be visible on radiographs. Auscultation of the lungs may reveal rales, rhonchi, and wheezing. In bronchospasm breath sounds can be virtually inaudible. Rales or crackles may be localized (atelectasis) or diffuse (pneumonitis). Fever and leukocytosis may accompany atelectasis and pneumonitis. Radiographs obtained soon after fire exposure may be normal. After progression, chest radiographs can reveal peribronchial cuffing (caused by airway edema) and diffuse infiltrates (caused by atelectasis, pneumonitis, or pulmonary edema). Crusts, casts of debris and exudates, and plugs of mucus and soot may develop and block airways. Central nervous system (CNS) signs and cardiovascular dysfunction signs reveal systemic toxicity caused by hypoxia and hypercapnia. Further CNS effects can include agitation, confusion, ataxia, abnormal posturing, transient loss of consciousness, and seizures. Cardiovascular signs include hypotension, dysrhythmias, and cardiac arrest. As with any severe shock or prolonged period of tissue hypoxia, elevated plasma lactate concentration (0.1 mEq/L) may be present. However, this finding is by no means specific to smoke inhalation. The methemoglobin fractions are elevated in virtually all animals with significant signs of either smoke inhalation or systemic toxicity. Notable methemoglobinemia appears to be a rare finding.

Minimum database and confirmatory tests. Details of fire and smoke exposure, current medications, past medical treatment, and any therapy before hospital arrival should all be obtained in a good history from family members of the victim. Of particular importance is what substance generated the smoke (e.g., wood, plastic, polyvinyl chloride), a description of the smoke (odor, intensity, and color), the duration of the exposure, and whether the exposure took place indoors or in the open. The nature and type of signs displayed by the animal at the time of the exposure and at the time of hospital arrival are helpful in determining the severity of the exposure. Altered mentation, ataxia, collapse, and syncope at the time of exposure all suggest carbon monoxide or cyanide intoxication. These substances can be easily missed if delayed presentation has allowed clinical improvement by the time of hospital presentation or if oxygen was administered in the field. Vital signs cannot be overlooked. Physical examination should first focus on determining the patency and condition of airways, adequacy of respirations, and assessment of ventilation. The respiratory rate of the animal involved is critical. The heart rate, mucous membrane color, CNS function, body temperature, and skin turgor should all be closely monitored. Animals showing respiratory signs must be assessed for hypoxia, hypercapnia, and upper airway obstruction. All fire victims should receive a thorough ophthalmic evaluation, including retinal examination, checking for particulate matter under the lids, and fluorescein stain evaluation of the cornea for ulcerations. In addition to the heart rate, respiratory rate, body temperature, and central nervous function, pulse oximetry (SpO_2) should be obtained in all smoke inhalation animals. Oxygen saturation obtained through this method may be falsely elevated and near normal when methemoglobinemia and carboxyhemoglobinemia are present. Cyanosis that is unresponsive to oxygen in an animal without respiratory distress is suggestive of methemoglobinemia. Any animal with cyanosis with normal vital signs and a SpO2 greater than 90% should have a methemoglobin fraction measured. Animals with altered mental status, respiratory distress, and atypical chest auscultations should be assessed for hypercapnia and acid-base imbalances by formal blood gas analysis. Decreased oxygen saturation measured by co-oximetry in conjunction with a normal PO2 (and hence normal calculated oxygen) suggests a diagnosis of either carbon monoxide poisoning or methemoglobinemia. The difference between calculated and measured oxygen saturation can be used to estimate either the fraction of carbon monoxide or methemoglobin. If a metabolic lactic acidosis is present, and carboxyhemoglobin and methemoglobin and PO2 are all normal, then cyanide poisoning should be suspected. Unresponsive hypotension and coma are likewise suggestive of cyanide poisoning.

Diagnostic tests must focus on the animal's oxygenation and ventilation. Thus for the suspected smoke inhalation victim, arterial blood gas analysis, carboxyhemoglobin concentration, methemoglobin levels, and chest radiographs are the most valuable diagnostic tools to be used. Establishing whether a metabolic acidosis is present can help in diagnosing underlying tissue hypoxia. The presence of elevated carboxyhemoglobin concentrations in smoke victims indicates substantial exposure to toxic combustion products has occurred, and the potential for ongoing, developing smoke-related pathological conditions exists. However, carboxyhemoglobin concentrations alone are poor predictors of severity of exposure since low or nondetectable concentrations do not rule out the possibility of significant underlying tissue damage and potential for progressive pathological conditions. Admission carboxyhemoglobin levels do not reflect peak blood concentrations and are usually significantly decreased by the time the animal reaches the veterinary hospital. Furthermore, it is important to note that transcutaneous measurement of oxygen saturation (pulse oximetry) is unreliable in smoke inhalation patients since it overestimates actual oxygen saturation in the presence of methemoglobin. Lactic acidosis seen in animals suffering from smoke inhalation is a result of tissue hypoperfusion, carbon monoxide poisoning, and cianide poisoning resulting in pulmonary dysfunction. Nevertheless, lactic acidosis also is an insensitive indicator of smoke inhalation since hypoxia from any cause impedes aerobic metabolism and generates lactic acid. Chest radiographs are most commonly normal at the time of and for the first few hours following smoke inhalation. Thus early radiographs are another inaccurate predictor of pulmonary injury. Within 24 to 36 hours of exposure, radiographic changes ranging from patchy atelectasis to diffuse interstitial and alveolar involvement may be evident. Subtle radiographic findings within the first 24 hours of exposure can include perivascular haziness, peribronchial cuffing, bronchial wall thickening, and subglottic edema. For radiographic studies to be helpful in diagnosis of smoke inhalation, serial chest radiographs must be obtained over time. The hallmark pulmonary injuries secondary to smoke inhalation usually develop more than 24 hours after smoke inhalation and can include acute lung tissue injury, aspiration, infection, volume overload, respiratory distress, and cardiogenic pulmonary edema. Animals demonstrating abnormal cardiac rhythms, hypertension, and those suffering from underlying cardiovascular disease should receive an electrocardiogram after episodes of smoke inhalation. Sedation and bronchoscopy can be performed in larger animals, but is rarely done in veterinary medicine for diagnosis alone. Finally, effective, aggressive therapy for smoke inhalation victims should never be postponed awaiting blood results. Blood results (e.g., carboxyhemoglobin levels, methemoglobin concentrations, confirmation of blood cyanide levels) may take hours or longer to obtain. Initiation of treatment for acutely smoke-poisoned animals must never await results of laboratory analysis.

Treatment. Successful management of smoke inhalation begins with prompt and safe removal of the animal from the smoke-filled environment. Care must be taken and, unless rescuers possess skin, respiratory, and eye protection, removing animals from fires is best left to professional firefighters. Never enter smoky environments without adequate protection. Basic emergency support measures can be instituted as necessary at the scene of the exposure. Decontamination measures, such as irrigation of eyes and skin, can be initiated immediately. The cornerstones of smoke inhalation therapy are maintenance of airway patency, adequate ventilation and oxygenation, aggressive measures for countering pulmonary debris, and stabilization of hemodynamic imbalances. A major concern in managing smoke inhalation patients is failing to appreciate their potential for rapid deterioration. Critical airway compromise can develop suddenly and insidiously in these animals. The patency of upper airways must be rapidly ascertained and established if compromised. Upper airway injury is always almost certain if obvious oropharyngeal burns are present. If telltale burns are not present, it is very easy to underestimate the degree of injury after episodes of smoke inhalation. If evidence of upper airway injury is present, endotracheal intubation should be undertaken rather than waiting for the injured animal to decompensate and deteriorate. Animals displaying coma, visible burns, full-thickness neck burns, edema of the oropharynx, and stridor should all be swiftly intubated. Fluid administration to burn victims contributes to the formation of upper airway edema. As a result, burned animals receiving aggressive fluid therapy also require intubation. Inhalant ß2-adrenergic agonists are the first line of defense for acute reversible bronchoconstriction (e.g., asthma and chronic obstructive pulmonary disease). Although effective for these conditions, their efficacy has not been established in smoke inhalation victims.18 Since pathophysiological changes induced by irritant toxins in smoke are partially reversible, ß2-adrenergic agonists should have some beneficial effects on airway obstruction. Corticosteroids are effective in treatment of refractory acute asthma, but mortality and infection rates are increased in animals with smoke inhalation that receive steroids. Furthermore, benefits of corticosteroid treatment of smoke injury have not been demonstrated in clinical animal studies. Mammalian lungs may show progressive pathophysiological changes over time (hours to days) after exposure to smoke. Counters for progressive respiratory compromise and failure include mechanical ventilation techniques, continuous positive airway pressure, positive end-expiratory pressure, and vigorous clearing of pulmonary secretions and debris. Frequent airway suctioning may be necessary to clear plugs, casts, inspissated secretions, and necrotic debris. Toxic combustion products damage the respiratory tract but also potentially the skin, eyes, and other mucous membranes. The extent of chemical injury to eyes, skin, and other membranes is largely determined by the duration of contact between the irritant or toxin and the animal's tissue. The eyes of all animals suffering from smoke inhalation must be thoroughly examined for corneal burns caused by thermal, chemical, or irritant injury. Animals with signs of ocular injury should have the eyes irrigated copiously with artificial tears or normal saline. Dermal decontamination should be initiated if necessary to prevent ongoing dermal burns from toxin-laden soot adherent to the skin. Rapid removal of soot and smoke debris from skin may prevent continued injury and burns. Candidates for immediate endotracheal intubation include animals with respiratory distress and signs of upper airway obstructions and animals that are cyanotic or hypoxic (SpO_2 <90%) despite aggressive maximum oxygen therapy with a nonrebreathing mask. Animals with respiratory depression (<10 to 12/min or PCO_2 >50 mm Hg), pulmonary edema, depressed mentation, and full-thickness neck or face burns also should be intubated. Any animal should be considered for intubation that does not improve with oxygen delivered by mask. Intubation should be done with the largest possible endotracheal tube so that suction can be employed, or bronchoscopy can be performed if required. The use of nasal catheters for oxygen delivery may be necessary in animals with excessive airway edema, if extensive perioral burns and constricting neck burns are present, or if direct visualization of the larynx for intubation is impossible.

Be certain to evaluate the whole animal for other injuries (e.g., neck trauma and cervical spine injury) by performing a thorough physical examination. Do not focus only upon the smoke inhalation. The potential for other injuries occurring during fires should never be overlooked. Once intubated, the airway should be suctioned regularly to remove secretions, inhaled debris, and necrotic material. Supplemental oxygen must be humidified to prevent drying of respiratory tissue and secretions. Positive and expiratory pressure should be routinely administered to prevent and to treat atelectasis and for those who remain hypoxic despite administration of 100% oxygen. Bronchoscopy can be employed to help direct effective removal of bronchial secretions. Repeated suctioning may be necessary to help break up inspissated mucus plugs, casts, and accumulated debris. Tracheostomy is reserved only for animals with complete airway obstruction, either caused by constriction, edema, or trauma, or for those that may require prolonged intubation.

Performing a tracheostomy must be carefully considered since it requires upkeep, can be associated with significant complications, and because it is usually reserved for animals that are tremendously compromised. Animals that do not require intubation nonetheless may benefit from inhalant, aerosolized, and racemic epinephrine. However, they still must be carefully monitored for sudden deterioration and may still need to be intubated. Experimental evidence exists for a variety of smoke inhalation therapies, ranging from nonsteroidal drug administration, antioxidants, and free radical scavengers to inhaled nitric oxide. Hyperbaric oxygen may be of benefit in the treatment of pulmonary edema and pneumonitis.

Hyperbaric oxygen can also be effective in treating carbon monoxide poisoning, cyanide poisoning, cerebral edema, and thermal burns associated with fires and smoke inhalation. It can also be considered in animals with refractory hypoxemia. Nevertheless, the availability of hyperbaric oxygen is still limited for veterinary medical applications. The treatment of carbon monoxide poisoning in animals associated with smoke inhalation is supplemental oxygen therapy administered through a tight-fitting mask or endotracheal tube. The amount of cianide exposure in animal victims of smoke inhalation is not predictable. However, cyanide poisoning should be suspected in animals with serious episodes of smoke exposure. Until hydroxocobalamin therapy becomes available in the United States, cyanide poisoning is treated according to the usual guidelines. Methemoglobinemia can result from inhalation of certain toxic products of combustion. Oxygen therapy alone is effective for most instances. Methylene blue should be reserved for only those cases where methemoglobin concentration is greater than 20% to 30%.

Treatment with antibiotics should only be initiated in patients with a documented infection. The prophylactic use of antibiotic therapy begun immediately in cases of smoke inhalation is of no benefit and in fact may help implement the development of infection with antibiotic-resistant organisms. Selection of antibiotics should be directed by the results from the Gram stain and culture and sensitivity of the sputum and secretions collected. Animals with fever and persistent leukocytosis following more than 2 days postexposure should be treated with antibiotics.

Empirical therapy with agents effective against Staphylococcus aureus (such as cefazolin) and gram-negative organisms like Pseudomonas (gentamycin) can be started in the absence of culture results. Because of the high incidence of sudden deterioration and decompensation in cases of smoke inhalation, it is recommended that exposed animals be observed closely for 6 to 8 hours postexposure before they are released. No smoke inhalation victims should be discharged until they are asymptomatic, are normal on physical examination and ancillary tests, and are otherwise stable. All discharged animals should be seen again within 72 hours to ensure that underlying pulmonary injuries are not progressing. Definite therapy and treatment is available for animals suffering from smoke inhalation. Early intervention and respiratory support are essential in these cases. Establishing and maintaining patent airways and the delivery of high-flow oxygen is the basis for successful treatment of smoke inhalation. The use of hyperbaric oxygen and other therapeutic regimens is being explored in hope of more effective therapy for smokeintoxicated animals.

Prevention and prognosis. Prevention of smoke inhalation injuries begins with the prompt and safe removal of animals from environments filling with smoke. However, no rescues can be attempted unless rescuers have adequate skin, eye, and respiratory protection. In many instances removing animals from fires and contact with toxic smoke is best left to professional firefighters. Common sense must outweigh emotion and hazardous heroics. The simple use of smoke alarms and sprinkler systems cannot be underestimated in reducing the hazardous effects of fires. The mere presence of a smoke alarm is a tremendous deterrent to fire-related injury simply through its early-warning merits. If sprinkler systems are in place and activated, its response to a fire is swift and unmistakable. Sprinkler systems require no action from occupants, do not depend on their presence or location, and immediately quench the toxic potential of fire and smoke. In the absence of smoke detectors and sprinkler systems, fires can progress to their most dangerous potential.

Both smoke detectors and sprinkler systems are widely available, fairly inexpensive, and relatively easy to install. Commercial fire extinguishers using a variety of retardants are also easily and inexpensively obtainable at home improvement outlets. Family members should all be well versed in where smoke alarms and fire extinguishers are located in the home and be instructed in their function. Stickers for doors are available informing rescuers how many and what type of animals live in that residence. These should be prominently placed and currently updated. Following smoke inhalation, a whole spectrum of related injuries is possible, ranging from asymptomatic, unaffected animals to rapid upper airway occlusion, to a few to several days later the appearance of delayed pulmonary edema and progressive pathological changes.

Prognosis depends upon several factors, such as duration of exposure, the concentration of the inhaled smoke, the toxic combustion products of the smoke involved, and the presence of preexisting underlying disease. Animals suffering from smoke inhalation may have a variety of complications caused by a number of pulmonary sequelae. Wheezing and chronic cough may reflect underlying chronic hyperreactive airways. Chronic bronchitis, bronchiectasis, bronchial stenosis, pulmonary fibrosis, bronchiolitis obliterans, and atelectasis may result after exposure to smoke and subsequent inflammation and scarring. Tracheal stenosis has been seen as a complication of long-term endotracheal intubation. The precise outcome of smoke inhalation exposure may not be evident for some time. As a result, these cases often require extensive follow-up, serial radiographs, bronchoscopy, and other diagnostics to document the extent and the nature of pulmonary injuries and how much normal function will be maintained. Early intervention certainly is beneficial in the prognosis and outcome of smoke inhalation cases.

Finally, we must continue to strive to identify safer, less toxic construction and furnishing materials that do not release poisonous combustion products upon burning.

Dross and histological lesions. The effects of smoke inhalation can be characterized with regard to the mechanism of action of the individual toxicants, their physiological effects, or their course of toxicity after exposure. Thermal burns are caused by hot air and steam, and chemical and thermal burns are produced by irritant gases. Target tissue for smoke are skin, mucosal surfaces (eyes and respiratory tract), and any mucous membranes. The effects of smoke irritation are mediated by polymorphonuclear leukocytes, arachidonic acid metabolites (leukotrienes, thromboxanes, and prostacyclins), cytokines (platelet activity factors), and free radicals (such as superoxides, peroxides, and hydroxyl). Minimal exposure results only in inflammation, but prolonged exposure to smoke can cause ulceration and cellular necrosis. Smoke causes inflammation of the larynx, trachea, bronchi, bronchioles, and alveoli, which then produces local edema, bronchospasm, cessation of ciliary function, loss of surfactant, and increased permeability of microvasculature membranes, upper and lower airway obstruction, atelectasis, loss of lung compliance, and ventilation-perfusion mismatch may develop. From the perspective of time, smoke inhalation produces a progressive pulmonary dysfunction that can ultimately be fatal. Asphyxia is the end result, no matter what the predisposing causes or smoke sources. Asphyxia can be due to a number of causes: inhibition of cellular respiration (carbon monoxide and cyanide), impaired oxygen transport and delivery (carboxyhemoglobin and methemoglobin), central respiratory depression (carbon monoxide, cyanide, and carbon dioxide), direct or indirect occlusion of airways (effects of irritant gases, heat, soot, and humidity), or a decreased supply of oxygen (air robbed of oxygen by combustion or containing toxic gases that lower the partial pressure of oxygen). There is a strong correlation between duration of exposure and severity of pathological injuries. Following smoke inhalation the clinical sequence of events can be divided into early, intermediate, and late phases. In the initial 24 to 36 hours, the systemic effects of carbon monoxide, cyanide, and methemoglobin and the airway effects of heat, humidity, irritant gases, and soot predominate. From 6 hours to 5 days following exposure, pulmonary edema becomes the most significant problem. It can develop swiftly, progress to severe respiratory distress, and a high mortality rate. Cerebral edema is present in animals with severe or prolonged hypoxia. In days to weeks after exposure, late manifestations include permanent CNS damage secondary to anoxia, bronchiectasis, and subglottic stenosis as a result of airway injury and prolonged endotracheal intubation, and sepsis and pneumonia secondary to edema, debris, impaired defense mechanisms, and opportunistic bacteria. Pneumonia is the most common late complication, and it can have a mortality rate approaching 50%. Early pneumonia (within 3 to days of exposure) is usually due to S.

Aureus or Escherichia coli when occurring later in the progression of pathological events. Chronic sequelae, such as asthma, pulmonary fibrosis, chronic obstructive pulmonary disease, bronchiolitis obliterans, and neoplasia, can develop months to years after the original exposure injury. Nuclear imaging, pulmonary function tests, and bronchoscopy can all be used to document chronic pulmonary lesions following smoke inhalation. These types of studies are often more sensitive than changes detectable by radiographs. The morbidity and mortality of smoke inhalation increase greatly when associated with thermal burns. Burned animals demonstrante increased vascular permeability that leads to a large fluid flux from the circulatory plasma to the interstitial spaces. The resultant lung edema formation is progressively more severe the more extensive the cutaneous burns are. Animals with combined burn and smoke inhalation injuries show an increase in transpulmonary fluid flux (lung lymph flow), an increase in lung water content, and a significant drop in the partial pressure of inspired oxygen as a result of the edema. This constriction also produces significant airway obstruction. The major cause of progressively worsening pulmonary gas exchange is the development of airway obstruction. By 48 to 72 hours postexposure, there is a progressive reduction in bronchi and bronchiolar luminal cross section. This reduction in cross section is due to the progressive development of obstructive casts and exudative material that occludes the lumen of the airway. This poor oxygenation of blood leads to hypoxemic events in tissue distant to the respiratory tract. Therapy targeting these pathological mechanisms, removing obstructive materials and cellular debris, and preventing the development of permanent pathological pulmonary changes might be a more effective way of successful airway management and smoke inhalation treatments in the future.

Differential diagnoses. The manifestations of smoke inhalation can be unfortunately nonspecific. Most of the time, smoke inhalation causes injuries that preponderantly involve the lower respiratory tract. Look-alikes of smoke inhalation include any situation leading to respiratory compromise.

Poisoning by hydrogen sulphide

Hydrogen sulfide is considered a broad-spectrum poison, meaning that it can poison several different systems in the body, although the nervous system is most affected. The toxicity of H2S is comparable with that of hydrogen cyanide. It forms a complex bond with iron in

- the mitochondrial cytochrome enzymes, thus preventing cellular respiration.
- Since hydrogen sulfide occurs naturally in the body, the environment and the gut, enzymes exist in the body capable of detoxifying it by oxidation to (harmless) sulfate. Hence, low levels of hydrogen sulfide may be tolerated indefinitely.

- At some threshold level, believed to average around 300–350 ppm, the oxidative enzymes become overwhelmed. Many personal safety gas detectors, such as those used by utility, sewage and petrochemical workers, are set to alarm at as low as 5 to 10 ppm and to go into high alarm at 15 ppm.
- A diagnostic clue of extreme poisoning by H2S is the discoloration of copper coins in the pockets of the victim. Treatment involves immediate inhalation of amyl nitrite, injections of sodium nitrite, inhalation of pure oxygen, administration of bronchodilators to overcome
- eventual bronchospasm, and in some cases hyperbaric oxygen therapy(HBO). HBO therapy has anecdotal support and remains controversial.

Exposure to lower concentrations can result in eye irritation, a sore throat and cough, nausea, shortness of breath, and fluid in the lungs. These effects are believed to be due to the fact that hydrogen sulfide combines with alkali present in moist surface tissues to form sodium sulfide, a caustic. These symptoms usually go away in a few weeks.

Long-term, low-level exposure may result in fatigue, loss of appetite, headaches, irritability, poor memory, and dizziness. Chronic exposure to low level H2S (around 2 ppm) has been implicated in increased miscarriage and reproductive health issues among Russian and Finnish wood pulp workers, but the reports have not (as of circa 1995) been replicated.

- 0.00047 ppm or 0.47 ppb is the odor threshold, the point at which 50% of a human panel can detect the presence of the compound.
- OSHA has established a permissible exposure limit (PEL)(8 hour time-weighted average(TWA)) of 10 ppm.
- 10–20 ppm is the borderline concentration for eye irritation.
- 20 ppm is the acceptable ceiling concentration established by OSHA.
- 50 ppm is the acceptable maximum peak above the ceiling concentration for an 8 hour shift, with a maximum duration of 10 minutes.
- 50–100 ppm leads to eye damage.
- At 100–150 ppm the olfactory nerve is paralyzed after a few inhalations, and the sense of smell disappears, often together with awareness of danger.
- 320–530 ppm leads to pulmonary edema with the possibility of death.
- 530–1000 ppm causes strong stimulation of the central nervous system and rapid breathing, leading to loss of breathing.
- 800 ppm is the lethal concentration for 50% of humans for 5 minutes exposure (LC50).

- Concentrations over 1000 ppm cause immediate collapse with loss of breathing, even after inhalation of a single breath.

Although respiratory paralysis may be immediate, it can also be delayed up to 72 hours. Hydrogen sulfide was used by the British Army as a chemical agent during World War I. It was not considered to be an ideal war gas, but, while other gases were in short supply, it was used on two occasions in 1916.

Poisoning by petroleum products

Ingestion or inhalation of or skin contact with petroleum, petroleum condensate, gasoline, diesel fuel, kerosene, crude oil, or other hydrocarbon mixtures may cause illness and occasionally death in domestic and wild animals. Both dogs and cats may ingest petroleum products during grooming if their fur becomes contaminated. Dogs may ingest these products directly when they are left in open containers. Inhalation may occur when animals are confined in poorly ventilated areas where these chemicals have been used or stored. Cattle, and less frequently sheep or goats, may ingest such products because they are curious or seeking salt or other nutrients, water is not available, or food or water is contaminated. A cow may consume several gallons at one time.

Petroleum fractions have been used as insecticides and acaricides for many years, either alone or as part of formulations. Small quantities of these may be applied to the skin with few or no harmful effects, but large quantities and prolonged exposure can induce severe reactions. Pipeline breaks, accidental release from storage tanks, and tank car accidents may contaminate land and water supplies. Animals may have access to open or leaky containers of fuel or other hydrocarbon materials. The lower the molecular weight and the higher the degree of unsaturation or aromaticity, the greater the volatility. More volatile hydrocarbons are more lipid soluble and therefore more readily absorbed by inhalation or ingestion. Crude petroleum that has lost much of its lighter, more volatile components through weathering may still be hazardous.

Crude oil and gasoline contain varying amounts of aromatic hydrocarbons including benzene, toluene, ethyl-benzene, and xylene. For example, gasoline in the USA typically contains up to 2% benzene. Gasoline in some other countries may contain up to 5% benzene. These compounds, if ingested or inhaled in sufficient amounts, can have acute and chronic effects different from the other hydrocarbons that make up the majority of oil and gas products. Benzene, for example, is a known carcinogen at high levels of exposure and has a variety of hemotoxic properties. Toluene can cause profound neurologic signs and damage at sufficient doses.

Variation in composition of petroleum and petroleum-derived hydrocarbon mixtures explains some of the differences in toxic effects. Mixtures of low viscosity (eg, gasoline, naphtha, kerosene) have a high aspiration hazard and irritant activity on pulmonary tissues. Gasoline and naphtha fractions may induce vomiting, which contributes to aspiration hazard. Fractions more viscous than kerosene are less likely to be inhaled and, even if aspirated, are somewhat less damaging to lung tissue. Older formulations of lubricating oils and greases can be particularly hazardous because of toxic additives or contaminants (eg, lead).

Clinical Findings. Petroleum hydrocarbon toxicity may involve the respiratory, GI, or integumentary systems or the CNS. In most cases of ingestion, no clinical signs are observed. Pneumonia due to aspiration of hydrocarbons into the lungs is usually the most serious consequence of ingestion of these materials. Aspiration can occur during vomition or eructation of rumen contents. Acute bloat is not a consistent finding but has been reported to cause death very shortly after consumption of highly volatile hydrocarbons such as gasoline or naphtha. CNS effects are usually associated with aspiration. CNS signs may be a result of the anesthetic-like action of low-molecular-weight aliphatic hydrocarbons and/or cerebral anoxia that can result from lung damage or displacement of oxygen by the more volatile hydrocarbons. Some compounds when absorbed in high doses may sensitize the myocardium to endogenous catecholamines.

Anorexia, decreased rumen motility, and mild depression may begin in ~24 hr and last 3-14 days depending on dose and content. Hypoglycemia may be seen several days after ingestion. These signs and weight loss may be the only responses seen in animals that do not bloat or aspirate oil. Some animals fail to reestablish normal rumen function after ingestion and can develop a chronic wasting condition.

After ingestion of oil, the feces may not be affected until several days later, at which time they become dry and formed in the case of kerosene or lighter hydrocarbon fractions; in contrast, heavier hydrocarbon mixtures tend to be cathartic. Oil may be found in feces and rumen contents up to 2 wk following ingestion. Regurgitated or vomited oil may be seen on the muzzle and lips. Signs attributable to pulmonary adsorption of hydrocarbons or cerebral anoxia include excitability (associated with aromatic fractions—benzene, toluene, etc), depression (aliphatic or saturated low-molecular-weight hydrocarbons), shivering, head tremors, visual dysfunction (sometimes associated with lead contamination), and incoordination. Acute pneumonia and possibly pleuritis (coughing, tachypnea, shallow respiration, reluctance to move, head held low, weakness, oily nasal discharge, dehydrated appearance) are seen in some animals that aspirate highly volatile mixtures; death usually are seen within days. Respiratory signs may be limited to dyspnea shortly before death in animals that aspirate heavier hydrocarbons. Increased PCV, Hgb, and BUN, indicating mild to moderate hemoconcentration, are associated with development of pneumonia. Neutropenia, lymphopenia, and eosinopenia occur initially and are followed by a relative increase in neutrophils.

There are a few anecdotal reports of abortion following exposure. Laboratory data in rodents support the occurrence of increased fetal loss and decreased fetal growth. However, the doses necessary to affect the fetus were also sufficient to profoundly affect maternal health and weight. Lesions. Aspiration pneumonia is the most consistent postmortem finding in animals that did not die of bloat. This may be accompanied by tracheitis, pleuritis, and hydrothorax if highly volatile fractions such as gasoline or naphtha are involved. Lung lesions are usually bilateral and found in the caudoventral apical, cardiac, cranioventral diaphragmatic, and intermediate lobes. Affected portions are dark red and consolidated and may contain multiple abscesses. Encapsulated pulmonary abscesses may be found in cattle surviving up to several months after aspiration. Skin lesions may be obvious after repeated topical application or severe exposure and include drying, cracking, or blistering.

Diagnosis. A hydrocarbon odor may be detected in lungs, ruminal contents, and feces. Even if ingested in large doses, hydrocarbons may not be visible in ruminal contents after ~4 days.

Adding warm water to the GI contents may cause any oily contents to collect at the surface, but finding oil in the GI tract does not in itself justify a diagnosis of poisoning; most oils have low toxicity if not aspirated. Samples of GI contents, lung, liver, kidney, and the suspected source should be collected for chemical analysis to demonstrate presence of hydrocarbons in tissue (particularly lung) and GI contents and to match those found in tissues and ingesta with the suspected source. Samples must be carefully protected from cross-contamination during necropsy and transportation to the laboratory. Check with the diagnostic laboratory to ensure collection equipment and transport containers are appropriate to prevent evaporative loss of important components and contamination. Positive chemical findings together with appropriate clinical and pathologic findings are confirmatory. Diagnosis in oil-field situations has historically been complicated by involvement of other toxicants, eg, explosives, lead from grease and "pipe dope," arsenicals, organophosphate esters, caustics (acids or alkalis), and saltwater.

Treatment. Bloat pressure should be released by passing a stomach tube if absolutely necessary to save the life of the animal; using a trocar risks forcing oil into the peritoneal cavity, which results in peritonitis. Passing a stomach tube dramatically increases the risk of aspiration and extreme caution is necessary. In the absence of bloat, the prime objectives are to prevent aspiration and to mitigate GI dysfunction. Rumenotomy to remove ruminal contents and replace them with healthy ruminal material is safer. More chronic cases involving primarily hypofunction of the rumen may also respond to this procedure. Cathartics, if used, should be of the saline type; however, there is no evidence that they improve prognosis. Activated charcoal has occasionally been suggested for use in small animals. Although it does not effectively adsorb petroleum distillates, it may be given if necessary to adsorb additives and other contaminants. Care should be taken to avoid inducing vomiting and aspiration.

Animals with evidence of respiratory involvement may require broad-spectrum antibiotic treatment. Pathogens can be introduced into the lungs from aspirated rumen contents mixed with the hydrocarbons. The use of steroids in hydrocarbon aspiration may further reduce the chance for recovery. Treatment of aspiration pneumonia is rarely effective, and the prognosis is poor.

However, because signs of aspiration may not appear for several days, prognosis based on initial clinical findings may be misleading.

Most high-molecular-weight compounds pass through the digestive tract unchanged. Most of the petroleum hydrocarbons are highly lipophilic and will be stored for varying times in tissues with high lipid content including fat, nervous tissue, and the liver. Some of the absorbed compounds are metabolized into more toxic byproducts (eg, benzene, toluene, n-hexane). Although most of these compounds do not remain in the body for prolonged periods, little is known about exactly how long tissue levels persist in highly exposed animals. The potential for tissue residues must be considered prior to the slaughter of animals intended for human consumption.

In poisoning or damage due to cutaneous exposure, the material should be removed from the skin with the aid of soap or mild detergents and copious amounts of cool water. The skin should not be brushed or abraded. Further treatment depends on the clinical signs and is largely restricted to supportive therapy.

Petroleum hydrocarbon poisoning can be avoided only by preventing access to these materials through proper storage of home and farm chemicals and well maintained fencing around high-risk petroleum facilities.

Effects of Oil and Gas Fields on Cattle Health and Production. Anecdotal reports in the literature have documented producer concerns about the effect of oil fields on cattle health and production.

Some recent observational studies have suggested that exposure to emissions from sour gas processing plants and sour gas flares (natural gas containing hydrogen sulfide) may be associated with an increased risk of certain reproductive losses in cattle. Current research is re-examining these findings and exploring the impact of oil and gas field emissions on the immune system.

Poisoning by detergents, solvents, corrosives and other household preparations

Several daily use products (soaps, detergents, drain cleaners; acids, alkalies, etc.) available in homes and other work places may cause dermal irritation. There is scarce informatikon in literature regarding toxicosis due to these daily use products in animals. We will discuss briefly the possibilities of dermal irritation in animals based on wellknown scientific principles and published reports. The evaluation of cleansing products depends upon their cleansing properties (detergent or adsorptive), on the rinsability of the product, on the amount and nature of the additives and on the general and specifically epidermal toxicity of the components. In this context, one such preparation is denture cleaner in which the major toxic component is sodium perborate which decomposes to form hydrogen peroxide and sodium borate. These products are strongly alkaline and very irritating to the skin and mucous membranes thus can produce dermal toxicity in exposed animals. Similarly, constant or repeated exposure of skin to anionic detergents (sulfomated or phosphorylated hydrocarbons) causes irritation with the removal of natural oils and can result in thickening of skin along with weeping, cracking, scaling and blistering. This is evident that detergent substances are important components of cleansing materials. But their detergent cleansing action may also result in skin toxicity. Non-ionic detergents such as alkyl and aryl polyether sulfates, alcohols or sulfonates are comparatively less irritating than ionic ones.

The acute skin irritation potential of various detergent formulations can be assessed with patch test. In this test, the time of exposure required for 50% of subjects to show a positive skin reaction (TR50 value) is calculated for each product. Using this approach, 24 detergent preparations were tested in 7 individual studies. The dermal irritation profiles could be categorized as follows (by decreasing irritancy): mold/ mildew removers (average TR50 ~ 0.37 h) _ disinfectants/ sanitizers (0.64 h) _ fabric softener concentrate (1.09 h) _aluminum wash (1.20h) _ 20% SDS (1.81h) _ liquid laundry detergents (3.48 h) _ liquid dish detergents (4.16 h) _ liquid fabric softeners (4.56 h) _ liquid hand soaps (4.58 h) _ shampoos (5.40 h) _ hard surface cleaners (6.34 h) _ powder automatic dish detergents (~16 h) _powder laundry detergents (~16 h). Cleansing can also be accomplished with non-detergent containing cleaners. Drain cleaners consisting of high concentrations (25–36%) of sodium hydroxide and sodium hypochlorite are extremely toxic and caustic to the skin. Direct skin contact to these agents causes coagulative to liquefaction necrosis by dissolving proteins and saponifying lipids. Soaps have been used for thousands of years as part of daily life. Soaps are limited by their irritancy to the skin due to their alkaline nature and their tendency to form insoluble and inactive salts when combined with either hard water or sea water. Calcium cyanide, which is a special constituent of fertilizers, may act as contact irritant causing skin ulcers. Chlorobromomethane (a fire extinguisher liquid) causes intense skin irritation from exposure to either liquid or its vapors, although this chemical has low dermal absorption. Skin contact to another chemical methyl bromide (used in fire extinguishers, refrigerants or fumigants) can result in irritation and vesiculation. Clinical poisoning may occur from its skin absorption.

Liquid soldering fluxes mainly contain zinc chloride and a high proportion of hydrochloric acid. These components are caustic or corrosive and acute exposure results in direct irritation to skin. Glues and adhesives are often supplied in hydrocarbon solvents. Dermal exposure to these preparations may also produce irritation. The corrosive activity of hypochlorite (a common component of laundry bleaches) on skin is due to oxidizing potency or available chlorine.

Alkalinity of some preparations may contribute to tissue injury. Metal cleaners and oven cleaners contain components like soda/potash and KOH/NaOH in high concentrations, respectively, in addition to petroleum-based solvents. All these components would be expected to cause dermal irritation. Ethylene glycol and propylene glycol (PG) are used in most of the topical pharmaceutical and cosmetic preparations as solvents. The localized dermal effects from both solvents are mild, while the published data suggested that PG might have a skin contact sensitization potential. The alkalies present in oven cleaners may cause Severe necrotic lesions similar to drain cleaners. Domestic animals, especially pets, may be exposed dermally to paints and varnish removers containing mixtures of benzene, methanol, acetone and toluene. These solvents can absorb through the skin in considerable amounts. Certain perfumes can cause local irritation of skin. This is probably due to alcohol serving as the primary vehicle. Rubbing alcohol (ethyl alcohol) may sometimes cause cutaneous hyperemia. Rust removers whose major toxic components are hydrochloric acid, phosphoric acid, hydrofluoric acid, etc. can have direct corrosive and necrotizing action like other acids. Most shampoos would not be expected to cause severe dermal irritation. Generally, skin exposures to acids, alkalies and phenols may lead to lesions that can vary from mild dermatitis to severe corrosion of the skin.

9

MYCOTOXICOLOGY GENERAL MYCOTOXICOSIS PROFILE

Introduction. Acute or chronic toxicoses can result from exposure to feed or bedding contaminated with toxins that may be produced during growth of various saprophytic or phytopathogenic fungi or molds on cereals, hay, straw, pastures, or any other fodder. A few principles characterize mycotoxic diseases: 1) the cause may not be immediately identified; 2) they are not transmissible from one animal to another; 3) treatment with drugs or antibiotics has little effect on the course of the disease; 4) outbreaks are usually seasonal because particular climatic sequences may favor fungal growth and toxin production; 5) study indicates specific association with a particular feed; and 6) although large numbers of fungi found on examination of feedstuff does not necessarily indicate that toxin production has occurred.

Confirmation of diagnosis of mycotoxic disease requires a combination of information. Detection of fungal spores alone, even at high concentrations, is not sufficient for diagnosis; fungal spores or even mold growth may be present without formation of mycotoxins. Especially important in diagnosis is the presence of a disease documented to be caused by a known mycotoxin, combined with detection of the mycotoxin in either feedstuffs or animal tissues.

Sometimes more than one mycotoxin may be present in feedstuffs, and their different toxicologic properties may cause clinical signs and lesions that are not consistent with those seen when animals are dosed experimentally with pure, single mycotoxins. Several mycotoxins are immunosuppressive, which may allow viruses, bacteria, or parasites to create a secondary disease that is more obvious than the primary.

In reaching a diagnosis of mycotoxicosis characterized by reduced feed intake, reproductive failure, or increased infectious disease due to immunosuppression, differential diagnoses must be carefully established and eliminated by a combination of thorough clinical and historical evaluation, examination of production records, and close attention to appropriate diagnostic testing.

There are no specific antidotes for mycotoxins; removal of the source of the toxin (ie, the moldy feedstuff) eliminates further exposure. The absorption of some mycotoxins (eg, aflatoxin) has been effectively prevented by aluminosilicate. If financial circumstances do not allow for disposal of the moldy feed, it can be blended with unspoiled feed just before feeding to reduce the toxin concentration or fed to less susceptible species. When contaminated feed is blended with good feed, care must be taken to prevent further mold growth by the toxigenic contaminants. This may be accomplished by thorough drying or by addition of organic acids (eg, propionic acid) to prevent mold growth. Important mycotoxic diseases occur in domestic animals worldwide.

Sampling and Submitting Feeds for Laboratory Analysis. Much of the error in detecting mycotoxins in feed results from sampling (or subsampling) rather than from analytical methodology. Samples can be taken at various stages—from growing crops or during transport or storage. Whenever possible, samples should be taken after particulate size has been reduced (eg, by shelling or grinding) and soon after blending has occurred (as in harvesting, loading, or grinding). Sampling is most effective if small samples are taken at periodic, predetermined intervals from a moving stream of grain or feed. These individual stream samples should be combined and mixed thoroughly, after which a subsample of 10 lb (4.5 kg) should be taken.

Probe sampling is acceptable when grain has been recently blended but is less reliable because different microenvironments within the storage facility may cause areas of mold or mycotoxin concentration. A suggested method of probe sampling is to sample at 5 locations, each 1 ft (30 cm) from the periphery of a bin, plus once in the center. This should be done for each 6 ft (2 m) of bin depth. Thus, taller bins would require more samples, and the total weight should be >10 lb.

Dry samples are preferable for transport and storage. Samples should be dried at 80-90°C for ~3 hr to reduce moisture to 12-13%. If mold studies are to be done, drying at 60°C for 6-12 hr should preserve fungal activity.

Containers should be appropriate for the nature of the sample. For dried samples, paper or cloth bags are recommended. Plastic bags should be avoided unless grain is dried thoroughly. Plastic bags are useful for high-moisture samples only if refrigeration, freezing, or chemicals are used to retard mold growth during transport and storage. Once a sample has been cooled or frozen, warming may induce condensation and allow mold growth.

AFLATOXICOSIS

Aflatoxins are produced by toxigenic strains of Aspergillus flavus and A parasiticus on peanuts, soybeans, corn (maize), and other cereals either in the field or during storage when moisture content and temperatures are sufficiently high for mold growth. Usually, this means consistent day and night temperatures >70°F. The toxic response and disease in mammals and poultry varies in relation to species, sex, age, nutritional status, and the duration of intake and level of aflatoxins in the ration. Earlier recognized disease outbreaks called "moldy corn toxicosis," "poultry hemorrhagic syndrome" and "Aspergillus toxicosis" may have been caused by aflatoxins.

- Aflatoxicosis occurs in many parts of the world and affects growing poultry (especially ducklings and turkey poults), young pigs, pregnant sows, calves, and dogs. Adult cattle, sheep, and goats are relatively resistant to the acute form of the disease but are susceptible if toxic diets are fed over long periods. Experimentally, all species of animals tested have shown some degree of susceptibility. Dietary levels of aflatoxin (in ppb) generally tolerated are d"50 in young poultry,
- d"100 in adult poultry, £50 in weaner pigs, d"200 in finishing pigs, <100 in calves, and <300 in cattle. Dietary levels as low as 10-20 ppb may result in measurable metabolites of aflatoxin
- (aflatoxin M1 and M2) being excreted in milk; feedstuffs that contain aflatoxins should not be fed to dairy cows.
- Aflatoxins bind to macromolecules, especially nucleic acids and nucleoproteins. Their toxic effects include mutagenesis due to alkylation of nuclear DNA, carcinogenesis, teratogenesis, reduced protein synthesis, and immunosuppression. Reduced protein synthesis results in reduced production of essential metabolic enzymes and structural proteins for growth. The liver is the principal organ affected. High doses of aflatoxins result in severe hepatocellular necrosis; prolonged low dosages result in reduced growth rate and liver enlargement.

Clinical Findings. In acute outbreaks, deaths occur after a short period of inappetence. Subacute outbreaks are more usual, and unthriftiness, weakness, anorexia, and sudden deaths can occur. Generally, aflatoxin concentrations in feed >1,000 ppb are associated with acute aflatoxicosis. Frequently, there is a high incidence of concurrent infectious disease, often respiratory, that responds poorly to the usual chemotherapy.

Lesions. In acute cases, there are widespread hemorrhages and icterus. The liver is the major target organ. Microscopically, the liver shows marked fatty accumulations and massive centrilobular necrosis and hemorrhage. In subacute cases, the hepatic changes are not so pronounced, but the liver is somewhat enlarged and firmer than usual. There may be edema of the gallbladder. Microscopically, the liver shows proliferation and fibrosis of the bile ductules; the hepatocytes and their nuclei (megalocytosis) are enlarged. The GI mucosa may show glandular atrophy and associated inflammation. In the kidneys, there may be tubular degeneration and regeneration. Prolonged feeding of low concentrations of aflatoxins may result in diffuse liver fibrosis (cirrhosis) and carcinoma of the bile ducts or liver.

Diagnosis. Disease history, necropsy findings, and microscopic examination of the liver should indicate the nature of the hepatotoxin, but hepatic changes are somewhat similar in Senecio poisoning (Plants Poisonous to Animals). The presence and levels of aflatoxins in the feed should be determined. Aflatoxin M1 can be detected in urine or kidney or in milk of lactating animals if toxin intakes are high.

Control. Contaminated feeds can be avoided by monitoring batches for aflatoxin content. Young, newly weaned, pregnant, and lactating animals require special protection from suspected toxic feeds. Dilution with noncontaminated feedstuff is one possibility. Ammoniation of grain reduces contamination but is not currently approved for use in food animals. Hydrated sodium calcium aluminosilicates (HSCAS) have shown promise in reducing the effects of aflatoxin when fed to pigs or poultry; at 10 lb/ton (5 kg/tonne), they provided substantial protection against dietary aflatoxin. HSCAS reduced, but did not eliminate, residues of aflatoxin M1 in milk from dairy cows fed aflatoxin B1.

Ochratoxicosis

Ochratoxins and citrinin are produced by several species of genera Aspergillus and Penicillium. The two most common species that produce ochratoxin A (OTA) are Aspergillus ochraceus and Penicillium verrucosum. These fungi are ubiquitous and the potential for contamination of animal feed and human food is widespread. Aspergillus spp. appears to produce ochratoxins at conditions of high humidity and temperature, whereas some Penicillium spp. may produce ochratoxins at temperatures as low as 5 °C. OTA has been found in a variety of food/feed, with levels in commodities used as feed ranging up to 27 ppm, and with levels in foodstuffs for human consumption in the range of trace to about 100 ppb. Unlike OTA, the occurrence of ochratoxin B is rare. Chemical structures of OTA and ochratoxin B are shown below. Ochratoxin B lacks chlorine and thereby it is less toxic than OTA. Citrinin was first isolated as a pure compound from a culture of P. citrinum in 1931. Later, it was also isolated from A. ochraceus, P. verrucosum, and related species that contaminate grain. In 1951, yellowish colored rice imported from Thailand to Japan was found to be contaminated with P. citrinum, which contained citrinin. Synthesized citrinin is also used in biological research, as it induces mitochondrial permeability pore opening and inhibits respiration by interfering with complex I of the respiratory chain. The structure of citrinin is shown below. Both OTA and citrinin cause nephropathy in animals and they have also been implicated as the cause of Balkan endemic nephropathy in humans. The literature reveals that OTA has been studied to a greater extent than citrinin, partly because OTA is at least 10 times more toxic than citrinin. This chapter describes in detail the toxicity of ochratoxins and citrinin in animals.

Introduction. The fungi producing ochratoxins and citrinin are commonly encountered in animal feed and human food around the world. They are encountered with great frequency in the Balkan area. Both ochratoxins and citrinin are fungal metabolites. There are two major ochratoxins (A and B). OTAoccurs naturally with a greater frequency in a variety of cereal grains (barley, wheat, oats, corn, and beans), peanuts, dried fruits, grapes/raisins, cheese, and other food products.

OTAaccumulates in the food chain because of its long half-life. Citrinin usually co-occurs with OTA, and commonly contaminates cereal grains, including wheat, barley, oats, corn, and rice. Citrinin also contaminates peanuts and fruits. The levels of OTA and citrinin have been found far lower in human food than in raw animal feed, because during processing and baking of human food citrinin is almost eliminated and OTAis significantly reduced. Compared to OTA, ochratoxin B is rarely found and very less toxic. Both OTA and citrinin are well-known nephrotoxins. OTA is also carcinogenic to rodents and possesses teratogenic, immunotoxic, neurotoxic, mutagenic, and genotoxic properties. In humans, exposure to OTA and citrinin has been linked with Balkan endemic nephropathy, a chronic kidney disease associated with tumors of the renal system, which can be fatal. Cooccurrence with OTA of citrinin has been implicated in nephropathy of pigs in Denmark, Sweden, Norway, and Ireland. Citrinin and OTA are also involved in avian nephropathies. Residues of OTAhave been detected in the tissues of pigs in slaughterhouses, and it has been shown, under experimental conditions, that residues can still be detected in tissues 1 month after the end of exposure. Due to long half-life of OTA in the feed and biological system, serious concerns have been raised about the animal health, as well as for human consumption of meat. Thus, it appears that both OTA and citrinin seriously affect animal health and economic impact is enormous.

Toxicokinetics. In most animal species, OTAis absorbed from the stomach because of its lipid soluble, non-ionized, and acidic properties (pKa–7.1). Absorption of OTAalso takes place in the intestine, is involved in enterohepatic circulation, and its biliary excretion is very efficient. OTA is distributed to various organs, mainly to the kidneys. Liver, muscle, and fat contain lower concentrations. The overall percentage of OTA absorption is found to be 66% in pigs, 56% in rats, 56% in rabbits, and 40% in chickens. After a single oral dose, the maximum concentrations of OTA are found within 10–48 h in pigs and rats, 2–4 h in ruminant calves, after 1 h in rabbits, and after 0.33 h in chickens. Maximum tissue concentrations in rat tissues occur within 48 h. The serum half-life of OTA is long and varies widely among species, for example 24–39 h in mice, 55–120 h in rats, 6.7 h in quail, 510 h in Macaca mulata monkeys, 72–120 h in pigs, 4.1 h in chicken, and 840 h in a volunteer. In pigs, it has been observed that the kidney is generally the most heavily contaminated tissue with OTA, and that the levels in the blood are about 5-fold greater than in the kidney. Krogh et al. (1976) illustrated that if the level of OTA in swine kidney is 12.1 ng/g (resulting from about 1000 ng/g in the feed), its levels would be 7.8 ng/g in the liver, 4.2 ng/g in the muscle, and 2.8 ng/g in the adipose tissue. OTA in ruminants is usually hydrolyzed in the forestomach by protozoans and bacterial enzymes, and consequently little OTA is found in the tissues (Hult et al., 1976). Residue of OTA can be passed in the milk of rats, rabbits, and women, but very little is passed in the milk of ruminants because of its metabolism by the ruminal microflora. The major routes of excretion are urine and feces, and excretion is influenced by the extent of the enterohepatic circulation and binding to serum albumin and other macromolecules. The association constant for the binding of OTA to serum albumin is 7.1–104 per mol for pigs, 5.1–104 per mol for chickens, and 4.0 _ 104 per mol for rats. In various tissues of all species that are examined, OTAis hydrolyzed to ochratoxin alpha, which is the major metabolite.

This detoxication process takes place in cecum of rats and is facilitated by bacterial microflora. The enzymes responsible for hydrolysis to ochratoxin alpha in cows and rodents are carboxypeptidase A and chymotrypsin. Suzuki et al. (1977) demonstrated that the rat tissue homogenates of the duodenum, ileum, and pancreas also have a high activity to catalyze this reaction. Activity of these enzymes in liver and kidney is low. Studies in mice suggest that OTAcirculates from the liver into the bile and into the intestine, where it is hydrolyzed to ochratoxin alpha. About 25– 27% of OTA, given either i.p. or orally to rats, was found as ochratoxin alpha in the urine. Its presence in the urine can be explained by reabsorption from the intestine. A similar mechanism of intestinal reabsorption of ochratoxin alpha has been suggested to occur in ruminant calves. For details of biotransformation of OTA, refer to an extensive review by Benford et al. (2001). From animal studies, it is clear that OTA has a high degree of bioavailability, low plasma clearance rate, and long tissue half-life. All metabolites of OTA are less toxic than the parent compound. Mechanism of action. OTA produces a variety of toxic effects in several organs, and therefore multiple mechanisms are involved. In addition to nephrotoxicity, OTA disrupts blood coagulation and glucose metabolism. OTA is immunotoxic, teratogenic, and carcinogenic. A brief description of mechanisms involved in common toxic effects is given below. Protein synthesis. A dose-related inhibition of protein synthesis was found in mice given OTA intraperitoneally at a dose ~1mg/kg body weight. The degree of inhibition of protein synthesis 5 h after administration of OTAat 1 mg/kg dose was 26% in liver, 68% in kidney, and 75% in spleen as compared with controls.

Nephrotoxicity. Both OTA and citrinin are potent nephrotoxins. At high doses, OTA affects both renal function and morphology, as indicated by increased weight, urine volume, blood urea nitrogen, urinary glucose, and proteinuria. The last two findings indicate that the site of reabsorption (i.e. the proximal convoluted tubules) is damaged. OTA specifically causes defect of the organic anion transport mechanism located on the brush border of the proximal convoluted tubules and basolateral membranes. OTA also adversely affects the organic ion transport system by which OTA enters proximal tubular cells. The middle (S2) and terminal (S3) segments of the proximal tubule of the isolated nephron were found to be the most sensitive to the toxic effects of OTA(0.05 mmol/l), as shown by a significant decrease in cellular ATP and a dose-related decrease in mitochondrial ATP content (Jung and Endou, 1989). Mitochondrial dysfunction has been shown to be involved in the development of OTA-induced toxicity in proximal renal tubule cells. OTA toxicity is associated with inhibition of both protein and RNA synthesis.

OTA is known to interfere with the charging of tRNA with amino acids. OTA treatment can increase oxidative stress in peripheral organs. Administration of OTAto rats (1 mg/kg) resulted in a 22% decrease in á-tocopherol plasma levels and a 5-fold increase in the expression of the oxidative stress responsive protein heme oxygenase-1, specifically in the kidney. Neurotoxicity. Although toxic effects of OTA on the CNS have not yet been fully characterized, evidence strongly suggests that OTA has potential for neurotoxicity. Bruinink et al. (1998) demonstrated that OTA is neurotoxic and may affect selected structures of the brain. This mycotoxin has complex multiple mechanisms of action that include evocation of oxidative stress, bioenergetic compromise, inhibition of protein synthesis, production of DNA singlestranded breaks, and formation of OTA–DNA adducts. These authors found that administration of OTA in mice, at a single dose (3.5 mg/kg) that is approximately 10% of the reported LD50, caused widespread oxidative injury in six discrete brain regions. Immunotoxicity. There is ample evidence, resulting from studies conducted in several animal species, that under certain conditions of treatment, OTA can produce defects in the structure and/or function of elements comprising the immune system. The size of the mouse thymus was reduced to 33% that of controls after four i.p. injections of OTAat 20 mg/kg body weight on alternate days, a dose which caused minimal nephrotoxicity. Bone marrow depression was found to be dose related, significantly decreased marrow cellularity, including a reduction in bone marrow macrophage–granulocyte progenitors, a decreased number of hematopoietic stem cells, a significant decrease in erythropoiesis, and increased phagocytosis by macrophages (Boorman et al., 1984). The effects of OTA on the bone marrow and lymphatic cell population may reflect the sensitivity of these cells to the inhibition of protein synthesis induced by OTA. These effects on the structural components of the immune system indicated that OTAis likely to have an effect on immune function. In chickens fed diets containing OTAat a concentration of 2–4 mg/kg for 20 days, the lymphoid cell population of immune organs was decreased, and IgA and IgM in lymphoid tissues and serum were decreased. Complement activity was slightly affected in birds fed diets containing 2 mg/kg for 5–6 weeks. Immune suppression was observed in chickens fed diets containing OTA at 0.05 or 2 mg/kg for 21 days. Treated animals showed reduced total serum protein, lymphocyte counts, and weights of the thymus, bursa of fabricus, and spleen.

Carcinogenicity. The exact mechanism by which OTA induces carcinogenicity is unknown, although both genotoxic and nongenotoxic modes of actions have been proposed. Toxicity. The toxic effects of OTA have been studied extensively in a number of domestic, companion, and experimental animals. Toxic effects of OTA have also been studied in humans. Overall toxicity of OTA is greatly influenced by species, sex, and route of administration. Based on acute toxicity data, dogs and pigs are the most sensitive species and rats and mice the least sensitive. Oral LD50 values (expressed as mg/kg body weight) of OTAare reported to be 46–58 in mouse, 20–30 in rat, 3.9 in neonate rat, 0.2 in dog, 1 in pig, and 3.3 in chicken. LD50 values via i.p. route are reported to be 22–40 in mouse and 20–30 in rat; and with i.v. route, 26–34 in mouse and 13 mg/kg in rat. It causes renal toxicity, nephropathy, and immune suppression in several animal species. The acute LD50 (expressed as mg/kg body weight) of citrinin is reported to be 50 (oral) and 67 (s.c. or i.p.) in rat, 35–58 in mouse, and 19 (i.p. or i.v.) in rabbit. Citrinin causes kidney damage and mild liver damage in the form of fatty infiltration. Other toxic effects include vasodilatation, constriction of the bronchi, and increased muscular tone. All the animals studied so far have been found susceptible to orally administered OTAwith a varying degree of response. It is important to mention that at higher doses OTA causes alterations in kidneys and also in other organs and tissue, but renal lesions can be found at an exposure level that is identical to those occurring environmentally. Ochratoxin B is rarely found as a natural contaminant and is much less toxic than OTA. The other ochratoxins have never been encountered in natural products. Weanling Fischer 344/N rats of both sexes receiving OTA by gavage in maize oil at a dose of 0.06, 0.12, 0.25, 0.5, or 1 mg/kg body weight/day for 5 days/week for 91 days, showed growth retardation and a reduced relative kidney weight in males at the two higher doses. Karyomegaly of dose-related severity was observed in the proximal tubules at all doses. Milder renal changes consisting of tubular atrophy were seen at a dose of 1, 4, or 16 mg/kg body weight/day on 5 days/week for a total of 12 doses over 16 days. Rats receiving the highest dose had diarrhea and nasal discharge and died before the end of the study. Increased relative weights of kidneys, heart, and brain, thymus atrophy, forestomach necrosis and/or hyperplasia, and hemorrhage of adrenal glands were seen at the two higher doses.

Bone marrow hyperplasia and nephropathy were seen at all doses, involving renal tubular degenerative and regenerative changes. OTA has been shown to produce nephrotoxic effects in all animal species examined, with the exception of adult ruminants.

The nephrotoxic potential of OTA is well documented from all experimental studies, with a feed level of 200 ppb causing nephropathy in pigs and rats. Evidence strongly supports that OTA is involved in porcine nephropathy, which is characterized by degeneration of the proximal tubules, atrophy of the tubular epithelium, interstitial fibrosis in the renal cortex and hyalinized glomeruli. Field cases of OTA-induced nephropathy in farm animals have long been recognized. Benford et al. (2001) suggested that the adverse effect at the lowest effective dose in several mammalian species is nephrotoxicity, and this is likely also to be true in humans. Citrinin is also nephrotoxic, but it is 10 times less toxic than OTA. In a series of experiments, sows were given feed containing OTA at a concentration of 0.2, 1, or 5 mg/kg (equivalent to 0.008, 0.04, and 0.2 mg/kg body weight/day), for a period of 5 days, 8 or 12 weeks, or up to 2 years. Decreased renal function, nephropathy, and reduced renal enzyme activity were observed. Progressive nephropathy but no renal failure was seen in pigs given feed containing 1 mg/kg for 2 years. Beagle dogs receiving OTAin capsule form at a dose of 0.1 or 0.2 mg/kg body weight/day for 14 days showed tubular necrosis and ultrastructural changes in the proximal tubules at all doses. Necrosis of lymphoid tissues of the thymus and tonsils was also seen at all doses. In another set of experiments, young beagle dogs were given OTA and citrinin separately and combined for 14 days. OTA was administered by capsule at 0.1 and 0.2 mg/kg; and citrinin (5 and 10mg/kg) dissolved in ethanol was given by i.p. injection. Clinical signs of toxicosis with 10 mg/kg citrinin and the higher combined doses included anorexia, retching, tenesmus, weight loss, prostration, and death.

Severity of the clinical disease and mortality were increased when the mycotoxins were combined, which indicated synergism. The clinicopathological abnormalities reflected renal damage, cellular and granular casts, ketones, protein, and glucose were in the urine of dogs given large doses of citrinin alone or combined with OTA. In pathological studies, these authors found gross lesions, such as focal peritonitis and intestinal intussusceptions with citrinin. Changes in the kidneys of dogs given OTA were degeneration and necrosis with desquamation of tubular epithelial cells, primarily in the straight segment of the proximal tubules. Dogs given 10 mg/kg citrinin had similar changes in the distal dubules and collecting ducts. Dogs given combined doses of citrinin and OTA had degeneration and necrosis in proximal and distal tubules, and in thin segments and the collecting ducts, and ulceration of the mucosa of the intestine. In experimental studies, dogs given citrinin showed serous nasal discharge and lacrimation. It is important to mention that citrinin is a very strong emetic in dogs, which is a protective mechanism in this species. Therefore, it is very unlikely that dogs will be poisoned by citrinin alone because high amounts of this mycotoxin will induce emesis and feed refusal. Chicken, turkeys, and ducklings are all susceptible to OTA and it appears that OTA-contaminated feed has major economic impact on the poultry industry. Feed levels as low as 200 ppb produced renal changes in the course of 3 months in rats and pigs. Field cases of OTA induced nephropathy are regularly encountred in pigs and poultry. Clinical signs of ochratoxicosis include retarded growth rate, reduction in weight gains, poor feed conversion, reduced egg production, poor egg shell quality and nephrotoxicity/nephropathy, and mortality. Feed refusal has been observed in turkeys. In chickens, OTA at a dose rate of 3.6 mg/kg can cause 5% mortality. Ochratoxin B at the dose rate of 54mg/kg causes lowered growth rate, edema of visceral organs, and accumulation of uric acid in kidneys, liver, heart, and spleen. These mycotoxins induce suppression of blood formation in bone marrow, and lymph formation in spleen and bursa of fabricus. Highest toxicity of OTA is found to be in broiler chickens. Broiler chickens given OTA at a dietary concentration of 4 mg/kg for 2 months caused 42% mortality. This toxin is involved in reduced growth rate at 5 ppm, high mortality rate at 4–8 ppm, and cessation of egg production at 4 ppm. In chickens, nephrotoxicity and hepatotoxicity occurs at dietary levels of 250 μg/g of citrinin with liver and kidney enlargements of 11% and 22%, respectively.

Necropsy of affected birds revealed the presence of pale and swollen kidneys. Citrinin is at least 10 times less nephrotoxic than OTA. Griffiths and Done (1991) described an outbreak of citrinin toxicosis in a herd of cows, which ingested citrus pulp (visibly moldy) pellets that contained 30–40 ppb citrinin. Affected cows showed signs of pruritis, pyrexia, and hemorrhagic syndrome. Signs of the sindrome occurred within 3 days of ingesting the citrus pulp, which was fed for 21 days. Five calves whose dams had been fed citrus pulp were subsequently born with superior prognathism. Older animals were more susceptible to citrinin. The clinical signs, gross pathology, and histology were suggestive of citrinin involvement. OTA has been well tested for carcinogenicity by oral administration in mice and rats. When OTAwas administered in the diet, hepatocellular tumors (designated as welldifferentiated trabecular adenomas), renal-cell tumors (renal cystadenomas and solid renal- cell tumors), hepatomas (some exhibiting the trabecular structure), and hyperplastic hepatic nodules were observed in male mice. In another study, administration of OTAin the dietinduced hepatocellular carcinomas and adenomas in female mice. Gavage administration of OTA to male and female rats resulted in a dose-related increase in the incidence of renal-cell adenomas and adenocarcinomas. Furthermore, metastasis of the renal-cell tumors was also observed in male and female rats.

OTA also increased the incidence and multiplicity of fibroadenomas of the mammary glanda in female rats. In essence, these data suggest that OTAincreases the incidence of hepatocellular tumors in mice of each sex and produces renal-cell adenomas in male mice and in rats of each sex. Based on sufficient evidence of carcinogenicity in experimental animals, OTA is classified as a possible carcinogen in humans (Group 2B). OTA is known to induce teratogenicity in mice, rats, hamsters, and chicken.

Citrinin has been demonstrated to be mutagenic in hepatocytes. There is limited evidence for the carcinogenicity of citrinin to experimental animals.

Treatment. There is no specific antidote for ochratoxin(s) or citrinin toxicity. Recovery is usually slow. Immediate removal of the suspected feed and replacement with clean feed supplemented with increased vitamin levels can be rewarding. Growth of A. ochraceus in a common food, such as cereals, can be controlled or minimized by drying them rapidly and thoroughly. Effective approaches to grain storage include fumigation, aeration and cooling, sealed storage, and controlled atmosphere in tropical and subtropical regions where insect damage is a major problem. Citrinin is less of a problem because it is heat unstable. Citrinin is also likely to be destroyed during brewing. Presence of propionic acid destroys citrinin when added as a preservative to protect stored barley destined for animal feed from molding during storage.

Currently, highly sophisticated methods are available to detect these mycotoxins at ppb or below levels in food/feed or their products, so as to prevent animal health from toxicosis and economic loss.

Conclusions. OTAand citrinin both contaminate a wide range of animal feed and human food. Human risk is less because the levels of these mycotoxins are minimized during processing and baking, but the raw feed remains a potential source for animal poisoning. OTA accumulates and has a long halflife in feed. These mycotoxins have a serious impact on the health of animals, especially, pigs, dogs, and poultry. Pigs and dogs are most sensitive, while rats and mice are least sensitive. In general, females are more sensitive than males. The kidney is a major target organ (as evidenced by functional and morphological changes) for both mycotoxins, but a few other organs are affected as well. In animal studies, OTA has been found to be a nephrotoxin, neurotoxin, immune suppressant, mutagen, carcinogen, and teratogen.

ERGOTISM

This worldwide disease of farm animals results from continued ingestion of sclerotia of the parasitic fungus Claviceps purpurea , which replaces the grain or seed of rye and other small grains or forage plants, such as the bromes, bluegrasses, and ryegrasses. The hard, black, elongated sclerotia may contain varying quantities of ergot alkaloids, of which ergotamine and ergonovine (ergometrine) are pharmacologically most important. Cattle, pigs, sheep, and poultry are involved in sporadic outbreaks, and most species are susceptible.

Etiology. Ergot causes vasoconstriction by direct action on the muscles of the arterioles, and repeated dosages injure the vascular endothelium. These actions initially reduce blood flow and eventually lead to complete stasis with terminal necrosis of the extremities due to thrombosis. A cold environment predisposes the extremities to gangrene. In addition, ergot has a potent oxytocic action and also causes stimulation of the CNS, followed by depression. Ergot alkaloids inhibit pituitary release of prolactin in many mammalian species, with failure of both mammary development in late gestation and delayed initiation of milk secretion, resulting in agalactia at parturition.

Clinical Findings and Lesions. Cattle may be affected by eating ergotized hay or grain or occasionally by grazing seeded pastures that are infested with ergot. Lameness, the first sign, may appear 2-6 wk or more after initial ingestion, depending on the concentration of alkaloids in the ergot and the quantity of ergot in the feed. Hindlimbs are affected before forelimbs, but the extent of involvement of a limb and the number of limbs affected depends on the daily intake of ergot.

Body temperature and pulse and respiration rates are increased. Epidemic hyperthermia and hypersalivation may also occur in cattle poisoned with C purpurea (see also fescue poisoning,). Associated with the lameness are swelling and tenderness of the fetlock joint and pastern. Within ~1 wk, sensation is lost in the affected part, an indented line appears at the limit of normal tissue, and dry gangrene affects the distal part. Eventually, one or both claws or any part of the limbs up to the hock or knee may be sloughed. In a similar way, the tip of the tail or ears may become necrotic and slough. Exposed skin areas, such as teats and udder, appear unusually pale or anemic. Abortion is not seen. The most consistent lesions at necropsy are in the skin and subcutaneous parts of the extremities. The skin is normal to the indented line, but beyond, it is cyanotic and hardened in advanced cases. Subcutaneous hemorrhage and some edema occur proximal to the necrotic area. In pigs, ingestion of ergot-infested grains may result in reduced feed intake and reduced weight gain. If fed to pregnant sows, ergotized grains result in lack of udder development with agalactia at parturition, and the piglets born may be smaller than normal. Most of the litter die within a few days due to starvation. No other clinical signs or lesions are seen. Clinical signs in sheep are similar to those in cattle. Additionally, the mouth may be ulcerated, and marked intestinal inflammation may be seen at necropsy. A convulsive syndrome has been associated with ergotism in sheep.

Diagnosis. Diagnosis is based on finding the causative fungus (ergot sclerotia) in grains, hay, or pastures provided to livestock showing signs of ergotism. Ergot alkaloids may be extracted and detected in suspect ground grain meals. Identical signs and lesions of lameness, and sloughing of the hooves and tips of ears and tail, are seen in fescue foot in cattle grazing in winter on tall fescue grass infected with an endophyte fungus, in which the ergot alkaloid ergovaline is considered a major toxic principle. In gilts and sows, lactation failure not associated with ergot alkaloids is prevalent and must be differentiated from prolactin inhibition due to ergot.

Control. Ergotism can be controlled by an immediate change to an ergot-free diet. Under pasture feeding conditions, frequent grazing or topping of pastures prone to ergot infestation during the summer months reduces flower-head production and helps control the disease. Grain that contains even small amounts of ergot should not be fed to pregnant or lactating sows.

FESCUE LAMENESS, which resembles ergot poisoning, is believed to be caused by ergot alkaloids, especially ergovaline, in tall fescue (Festuca arundinacea). It begins with lameness in one or both hindfeet and may progress to necrosis of the distal part of the affected limb(s). The tail and ears also may be affected independently of the lameness. In addition to gangrene of these extremities, animals may show loss of body mass, an arched back, and a rough coat. Outbreaks have been confirmed in cattle and similar lesions have been reported in sheep.

Tall fescue is a cool-season perennial grass adapted to a wide range of soil and climatic conditions; it is used in Australia and New Zealand for stabilizing the banks of watercourses. It is the predominant pasture grass in the transition zone in the eastern and central USA. Fescue lameness has been reported in Kentucky, Tennessee, Florida, California, Colorado, and Missouri, as well as in New Zealand, Australia, and Italy.

The causative toxic substance has actions similar to those produced by sclerotia of Claviceps purpurea. However, ergot poisoning is not the cause of fescue lameness. Ergotism is most prevalent in late summer when the seed heads of grass mature. Fescue lameness is most common in late fall and winter and has been reproduced in cattle by feeding dried fescue free of seed heads and ergot.

Two fungi from toxic pastures have been implicated in fescue lameness. The clavicipitaceous endophyte fungus Acremonium coenophialum can synthesize ergot alkaloids in culture. The ergot alkaloid ergovaline has been detected in toxic fescue and is strongly implicated in some of the fescue toxicosis syndromes. However, the complete etiology of fescue foot remains unresolved.

Some reports indicate an increased incidence of fescue lameness as plants age and after severe droughts. Strains of tall fescue vary in their toxicity (eg, Kentucky-31 is more toxic than Fawn) due to variation in infection level with the fungus and to high variability within a strain. In some Kentucky-31 fescues, infection levels cannot be detected. High nitrogen applications appear to enhance the toxicity. Susceptibility of cattle is subject to individual variation.

Low environmental temperature is thought to exacerbate the lesions of fescue lameness; however, high temperatures increase the severity of a toxic problem known as epidemic hyperthermia or "summer syndrome," in which a high proportion of a herd of cattle exhibits hypersalivation and hyperthermia. It appears that the toxin is a vasoconstrictor that induces hyperthermia in hot weather and results in cold extremities during cold weather. Another cause of this is poisoning with C. purpurea (ergot alkaloids).

Erythema and swelling of the coronary region occur, and cattle are alert but lose weight and may be seen "paddling" or weight-shifting. The back is slightly arched, and knuckling of a hind pastern may be an initial sign. There is progressive lameness, anorexia, depression, and later, dry gangrene of the distal limbs (hindlimbs first). Signs usually develop within 10-21 days after turnout into a fescue-contaminated pasture in fall. A period of frost tends to increase the incidence. For control, all infected forage should be removed.

SUMMER FESCUE LAMENESS. This warm season condition is characterized by reduced feed intake and weight gains or milk production. The toxin(s) affects cattle, sheep, and horses during the summer when they are grazing or being fed tall fescue forage or seed contaminated with the endophytic fungus Acremonium coenophialum. The severity of the condition varies from field to field and year to year.

Signs other than reduced performance, which may appear within 1-2 wk after fescue feeding is started, include fever, tachypnea, rough coat, lower serum prolactin levels, and excessive salivation. The animals seek wet spots or shade. Lowered reproductive performance also has been reported. Agalactia has been reported for both horses and cattle. Thickened placentas, delayed parturition, birth of weak foals, and agalactia have been reported in horses. The severity increases when environmental temperatures are >24-27 °C and if high nitrogen fertilizer has been applied to the grass.

For control, toxic tall fescue pastures must be destroyed and reseeded with seed that does not contain endophytic fungus because transfer of the fungus from plant to plant is primarily, if not solely, through infected seed. Not using pastures during hot weather, diluting tall fescue pastures with interseeded legumes, clipping pastures to reduce seed formation, or offering other feedstuffs helps reduce severity.

Zearalenosis

Fusarium spp molds are extremely common and often contaminate growing plants and stored feeds. Corn (maize), wheat, and barley are commonly contaminated. In moderate climates under humid weather conditions, F graminearum may produce zearalenone, one of the resorcyclic acid lactones (RAL). Zearalenone (formerly called F2 toxin) is a potent nonsteroidal estrogen and is the only known mycotoxin with primarily estrogenic effects. Often, zearalenone is produced concurrently with deoxynivalenol. Depending on the ratio of these 2 mycotoxins, signs of reduced feed intake or reproductive dysfunction may predominate, but presence of deoxynivalenol may limit exposure to zearalenone, thus reducing its practical effect. Zearalenone binds to receptors for estradiol-17-b, and this complex binds to estradiol sites on DNA. Specific RNA synthesis leads to signs of estrogenism. Zearalenone is a weak estrogen with potency 2-4 times less than estradiol.

Under controlled administration, zearalanol, a closely related RAL, is widely used in cattle as an anabolic agent.

Estrogenism due to zearalenone was first clinically recognized as vulvovaginitis in prepubertal gilts fed moldy corn (maize), but zearalenone is occasionally reported as a disease-causing agent in sporadic outbreaks in dairy cattle, sheep, chickens, and turkeys. High dietary concentrations are required to produce disease in cattle and sheep, and extremely high dosages are required to affect poultry.

Etiology. Zearalenone has been detected in corn, oats, barley, wheat, and sorghum (both fresh and stored); in rations compounded for cattle and pigs; in corn ensiled at the green stage; and rarely in hay. It has been detected occasionally in samples from pastures in temperate climates at levels thought to be sufficient to cause reproductive failure of grazing herbivores.

Clinical Findings. The condition cannot be distinguished from excessive estrogen administration. Physical and behavioral signs of estrus are induced in young gilts by as little as 1 ppm dietary zearalenone. In pigs, zearalenone primarily affects weaned and prepubertal gilts, causing hyperemia and enlargement of the vulva. There is hypertrophy of the mammary glands and uterus, with occasional prolapse of the uterus in severe cases. In multiparous sows, signs include diminished fertility, anestrus, reduced litter size, smaller offspring, and probably fetal resorption. Constant estrus or pseudopregnancy may be seen.

Zearalenone causes reproductive toxicosis in sexually mature sows by inhibiting secretion and release of follicle-stimulating hormone (FSH) resulting in arrest of preovulatory ovarian follicle maturation. Reproductive effects in sexually mature sows depend on time of consumption.

Zearalenone fed at 3-10 ppm on days 12-14 of the estrous cycle in open gilts results in retention of the corpora lutea and prolonged anestrus (pseudopregnancy) for up to 40-60 days. Zearalenone fed at e"30 ppm in early gestation (7-10 days post-mating) may prevent implantation and cause early embryonic death.

In cattle, dietary concentrations >10 ppm may cause reproductive dysfunction in dairy heifers, while mature cows may tolerate up to 20 ppm. Clinical signs include weight loss, vaginal discharge, nymphomania, uterine hypertrophy, and in pregnant heifers, abortion 1-3 mo after conception—usually followed by multiple returns to service.

Young males, both swine and cattle, may become infertile, with atrophy of the testes.

Ewes may show reduced reproductive performance (reduced ovulation rates and numbers of fertilized ova, and markedly increased duration of estrus) and abortion or premature live births. Lesions. Lesions in pigs include ovarian atrophy and follicular atresia, uterine edema, cellular hypertrophy in all layers of the uterus, and a cystic appearance in degenerative endometrial glands. The mammary glands show ductal hyperplasia and epithelial proliferation. Squamous metaplasia is seen in the cervix and vagina.

Diagnosis. This is based on reproductive performance in the herd or flock, clinical signs, and history of diet-related occurrence. Chemical analysis of suspect feed for zearalenone and careful examination of reproductive organs at necropsy are required. As a bioassay, virgin prepubertal mice fed diets or extracts of zearalenone-contaminated feed demonstrate enlarged uteri and vaginal cornification typical of estrogens.

Differential diagnoses include reproductive tract infections and other causes of impaired fertility such as diethylstilbestrol in the diet of housed stock. In grazing herbivores, especially sheep, the plant estrogens (eg, isoflavones associated with some varieties of subterranean and red clovers, and coumestans in certain fodders [eg, alfalfa]) should be considered.

Control. Unless stock are severely or chronically affected, usually reproductive functions recover and signs regress 1-4 wk after intake of zearalenone stops. However, multiparous sows may remain anestrous up to 8-10 wk.

Management of swine with hyperestrogenism should include changing the grain immediately. Signs should stop within 1 wk. Animals should be treated symptomatically for vaginal or rectal prolapse and physical damage to external genitalia. For sexually mature sows with anestrus, one 10-mg dose of prostaglandin F2a, or two 5-mg doses on successive days, has corrected anestrus caused by retained corpora. Alfalfa and alfalfa meal fed to swine at 25% of the ration may reduce absorption and increase fecal excretion of zearalenone, but this is often not considered practical. Bentonite added to contaminated diets has been generally ineffective against zearalenone.

Fitomycotoxicosis

In this mycotoxic disease of grazing livestock, the toxic liver injury commonly results in photodynamic dermatitis. In sheep, the face is the only site of the body that is readily exposed to ultraviolet light, hence the common name. The disease is most common in New Zealand but also occurs in Australia, France, South Africa, several South American countries, and probably North America. Sheep, cattle, and farmed deer of all ages can contract the disease, but it is most severe in young animals.

Etiology and Pathogenesis. Sporidesmins are secondary metabolites of the saprophytic fungus Pithomyces chartarum, which grows on dead pasture litter. The warm ground temperatures and high humidity required for rapid growth of this fungus restrict disease occurrence to hot summer and autumn periods shortly after warm rains. By observing weather conditions and estimating toxic spore numbers on pastures, danger periods can be predicted and farmers alerted.

The sporidesmins are excreted via the biliary system, in which they produce severe cholangitis and pericholangitis as a result of tissue necrosis. Biliary obstruction may be seen, which restricts excretion of bile pigments and results in jaundice. Similarly, failure to excrete phylloerythrin in bile leads to photosensitization. Previous ingestion of toxic spores causes potentiation, thus a succession of small intakes of the spores can lead to subsequent severe outbreaks.

Clinical Findings, Lesions, and Diagnosis. Few signs are apparent until photosensitization and jaundice appear ~10-14 days after intake of the toxins. Animals frantically seek shade. Even short exposure to the sun rapidly produces the typical erythema and edema of photodermatitis in unpigmented skin. The animals suffer considerably, and deaths occur from one to several weeks after photodermatitis appears.

Characteristic liver and bile duct lesions are seen in all affected animals whether photosensitized or not. In acute cases showing photodermatitis, livers are initially enlarged, icteric, and have a marked lobular pattern. Later, there is atrophy and marked fibrosis. The shape is distorted, and large nodules of regenerated tissue appear on the surface. In subclinical cases, livers often develop extensive areas in which the tissue is depressed and shrunken below the normal contour, which distorts and roughens the capsule. Generally, these areas are associated with fibrosis and thickening of corresponding bile ducts. The bladder mucosa commonly shows hemorrhagic or bile-pigment-stained ulcerative erosions with circumscribed edema. The clinical signs together with characteristic liver lesions are pathognomonic. In live animals, high levels of hepatic enzymes may reflect the extensive injury to the liver.

Control. To minimize intake of pasture litter and toxic spores, short grazing should be avoided. Other feedstuffs should be fed during danger periods; encouraging clover dominance in pastures helps to provide a milieu unsuited to growth and sporulation of P. chartarum on litter.

The application of benzimidazole fungicides to pastures considerably restricts the buildup of P chartarum spores and reduces pasture toxicity. A pasture area calculated at 1 acre (0.45 hectare)/15 cows or 100 sheep should be sprayed in midsummer with a suspension of thiabendazole. When danger periods of fungal activity are predicted, animals should be allowed only on the sprayed areas. The fungicide is effective within 4 days after spraying, provided that no more than 1 in. (2.5 cm) of rain falls within 24 hr during the 4-day period. After this time, heavy rainfall does little to reduce the effectiveness of spraying because the thiabendazole becomes incorporated within the plants. Pastures will then remain safe for ~6 wk, after which spraying should be repeated to ensure protection over the entire dangerous season. Sheep and cattle can be protected from the effects of sporidesmin if given adequate amounts of zinc. Zinc may be administered by drenching with zinc oxide slurry, by spraying pastures with zinc oxide, or by adding zinc sulfate to drinking water. Sheep may be selectively bred for natural resistance to the toxic effects of sporidesmin. The heritable trait for resistance is high. Ram sires are now being selected in stud and commercial flocks for resistance either by natural field challenge or by low- level, controlled dosage of ram lambs with sporidesmin.

FUMONISINS TOXICOSIS

Equine leukoencephalomalacia is a mycotoxic disease of the CNS that affects horses, mules, and donkeys. It occurs sporadically in North and South America, South Africa, Europe, and China. It is associated with the feeding of moldy corn (maize), usually over a period of several weeks.

Fumonisins are produced worldwide primarily by Fusarium moniliforme Sheldon and F. proliferatum. Conditions favoring fumonisin production appear to include a period of drought during the growing season with subsequent cool, moist conditions during pollination and kernel formation. Three toxins produced by the fungi have been classified as fumonisin B1 (FB1), B2 (FB2), and B3 (FB3). Current evidence suggests that FB1 and FB2 are of similar toxicity, whereas FB3 is relatively nontoxic. Major health effects are observed in Equidae and swine.

Signs in Equidae include apathy, drowsiness, pharyngeal paralysis, blindness, circling, staggering, and recumbency. The clinical course is usually 1-2 days but may be as short as several hours or as long as several weeks. Icterus may be present if the liver is involved. The characteristic lesion is liquefactive necrosis of the white matter of the cerebrum. The necrosis is usually unilateral but may be asymmetrically bilateral. Some horses may have hepatic necrosis similar to that seen in aflatoxicosis. Horses may develop leukoencephalomalacia from prolonged exposure to as little as 8-10 ppm fumonisins in the diet.

Fumonisins have also been reported to cause acute epidemics of disease in weanling or adult pigs, characterized by pulmonary edema and hydrothorax. Porcine pulmonary edema (PPE) is usually an acute, fatal disease and appears to be caused by pulmonary hypertension with transudation of fluids in the thorax resulting in interstitial pulmonary edema and hydrothorax. Acute PPE results after consumption of fumonisins for 3-6 days at dietary concentrations >100 ppm. Morbidity within a herd may be >50%, and mortality among affected pigs ranges from 50 to 100%. Signs include acute onset of dyspnea, cyanosis of mucous membranes, weakness, recumbency, and death, often within 24 hr after the first clinical signs. Affected sows in late gestation that survive acute PPE may abort within 2-3 days, presumably as a result of fetal anoxia. Prolonged exposure of pigs to sublethal concentrations of fumonisins results in hepatotoxicosis characterized by reduced growth; icterus; and increased serum levels of cholesterol, bilirubin, AST, lactate dehydrogenase, and ã-glutamyltransferase.

The biochemical mechanism of action for PPE or liver toxicosis is believed to be due to the ability of fumonisins to interrupt sphingolipid synthesis in many animal species.

Cattle, sheep, and poultry are considerably less susceptible to fumonisins than are horses or swine. Cattle and sheep tolerate fumonisin concentrations of 100 ppm with little effect. Dietary concentrations of 200 ppm cause inappetence, weight loss, and mild liver damage. Poultry are affected by concentrations >200-400 ppm and may develop inappetence, weight loss, and skeletal abnormalities.

No treatment is available. Avoidance of moldy corn is the only prevention, although this is difficult because it may not be grossly moldy or it may be contained in a mixed feed. However, most of the toxin is present in broken kernels or small, poorly formed kernels. Therefore, cleaning grain to remove the screenings markedly reduces fumonisin concentration. Corn suspected of containing fumonisins should not be given to horses.

Lupinosis

Lupines (Lupinus spp.) cause 2 distinct forms of poisoning in livestock—lupine poisoning and lupinosis. The former is a nervous syndrome caused by alkaloids present in bitter lupines; the latter is a mycotoxic disease characterized by liver injury and jaundice, which results mainly from the feeding of sweet lupines. Lupinosis is important in Australia and South Africa and also has been reported from New Zealand and Europe. There is increasing use of sweet lupines, either as forage crops or through feeding of their residues after grain harvest, as strategic feed for sheep in Mediterranean climate zones. Sheep, and occasionally cattle and horses, are affected, and pigs are also susceptible.

Etiology and Pathogenesis. The causal fungus is Phomopsis leptostromiformis, which causes Phomopsis stem-blight, especially in white and yellow lupines; blue varieties are resistant. It produces sunken, linear stem lesions that contain black, stromatic masses, and it also affects the pods and seeds. The fungus is also a saprophyte and grows well on dead lupine material (eg, haulm, pods, stubble) under favorable conditions. It produces phomopsins as secondary metabolites on infected lupine material, especially after rain. Clinical changes are mainly attributable to toxic hepatocyte injury, which causes mitotic arrest in metaphase, isolated cell necrosis, and hepatic enzyme leakage, with loss of metabolic and excretory function.

Clinical Findings, Lesions, and Diagnosis. Early signs in sheep and cattle are inappetence and listlessness. Complete anorexia and jaundice follow, and ketosis is common. Cattle may show lacrimation and salivation. Sheep may become photosensitive. In acute outbreaks, deaths occur in 2-14 days. In acute disease, icterus is marked. Livers are enlarged, orange-yellow, and fatty. More chronic cases show bronze- or tan-colored livers that are firm, contracted in size, and fibrotic.

Copious amounts of transudates may be found in the abdominal and thoracic cavities and in the pericardial sac. Feeding of moldy lupine material, together with clinical signs and increased levels of serum liver enzymes, strongly indicate lupinosis.

Control. Frequent surveillance of sheep and of lupine fodder material for characteristic black spot fungal infestation, especially after rains, is advised. The utilization of lupine cultivars, bred and developed for resistance to P leptostromiformis is advocated. Oral doses of zinc (e”0.5 g/day) have protected sheep against liver injury induced by phomopsins.

PASPALUM STAGERS

This incoordination results from eating paspalum grasses infested by Claviceps paspali. The life cycle of this fungus is similar to that of C. purpurea (see ergotism). The yellow-gray sclerotia, which mature in the seed heads in autumn, are round, roughened, and 2-4 mm in diameter.

Ingestion of sclerotia causes nervous signs in cattle most commonly, but horses and sheep also are susceptible. Guinea pigs can be affected experimentally. The toxicity is not ascribed to ergot alkaloids; the toxic principles are thought to be paspalinine and paspalitrem A and B, tremorgenic compounds from the sclerotia. A sufficiently large single dose causes signs that persist for several days. Animals display continuous trembling of the large muscle groups; movements are jerky and incoordinated. If they attempt to run, the animals fall over in awkward positions. Affected animals may be belligerent and dangerous to approach or handle. After prolonged exposure, condition is lost and complete paralysis can occur. The time of onset of signs depends on the degree of the infestation of seed heads and the grazing habits of the animals. Experimentally, early signs appear in cattle after ~100 g/day of sclerotia has been administered for >2 days. Although the mature ergots are toxic, they are most dangerous just when they are maturing to the hard, black (sclerotic) stage.

Recovery follows removal of the animals to feed not contaminated with sclerotia of C. paspali. Animals are less affected if left alone and provided readily available nutritious forages. Care should be taken to prevent accidental access to ponds or rough terrain where accidental trauma or drowning could occur. Topping of the pasture to remove affected seed heads has been effective in control.

Slaframine toxicosis

Trifolium pratense (red clover) may become infected with the fungus Rhizoctonia leguminocola (black patch disease), especially in wet, cool years. Rarely, other legumes (white clover, alsike, alfalfa) may be infected. Slaframine is an indolizidine alkaloid recognized as the toxic principle, and it is stable in dried hay and probably in silage. Horses are highly sensitive to slaframine, but clinical cases occur in cattle as well. Profuse salivation (salivary syndrome) develops within hours after first consumption of contaminated hay; signs also include mild lacrimation, diarrhea, mild bloat, and frequent urination. Morbidity can be high, but death is not expected, and removal of contaminated hay allows recovery and return of appetite within 24-48 hr. A related alkaloid, swainsonine, produced by R. leguminicola, has caused a lysosomal storage disease from prolonged exposure, but its importance in the salivary syndrome is not confirmed. Diagnosis is tentatively based on recognition of the characteristic clinical signs and the presence of “black patch” on the forages. Chemical detection of slaframine or swainsonine in forages helps to confirm the diagnosis. There is no specific antidote to slaframine toxicosis, although atropine may control at least some of the prominent salivary and GI signs. Removal of animals from the contaminated hay is essential. Prevention of Rhizoctonia infection of clovers has been difficult.

Some clover varieties may be relatively resistant to black patch disease. Reduced usage of red clover for forages or dilution with other feeds is helpful.

Trichotecenoses

The trichothecene mycotoxins are a group of closely related secondary metabolic products of several families of imperfect, saprophytic, or plant pathogenic fungi such as Fusarium, Trichothecium, Myrothecium, Cephalosporium, Stachybotrys, Trichodesma, Cylindrocarpon and Verticimonosporium spp. On the basis of molecular structure, the trichothecenes are classed as nonmacrocyclic (eg, deoxynivalenol [DON] or vomitoxin, T-2 toxin, diacetoxyscirpenol, and others) or macrocyclic (satratoxin, roridin, verrucarin).

The trichothecene mycotoxins are highly toxic at the subcellular, cellular, and organic system level. They swiftly penetrate cell lipid bilayers, thus allowing access to DNA, RNA, and cellular organelles. Trichothecenes inhibit protein synthesis by affecting polyribosomes to interfere with the initiation phase of protein synthesis. At the subcellular level, these toxins inhibit protein synthesis and covalently bond to sulfhydryl groups.

Trichothecene mycotoxins are generally cytotoxic to most cells, including neoplastic cells; they are not mutagenic. Toxicity of the trichothecenes is based on direct cytotoxicity and is often referred to as a radiomimetic effect (eg, bone marrow hypoplasia, gastroenteritis, diarrhea, hemorrhages). The cutaneous cytotoxicity that follows administration of these compounds is a nonspecific, acute, necrotizing process with minimal inflammation of both the epidermis and dermis. Stomatitis, hyperkeratosis with ulceration of the esophageal portion of the gastric mucosa, and necrosis of the GI tract have been seen after ingestion of trichothecenes.

Given in sublethal toxic doses via any route, the trichothecenes are highly immunosuppressive in mammals; however, longterm feeding of high levels of T-2 toxin does not seem to activate latent viral or bacterial infections. The main immunosuppressive effect of the trichothecenes is at the level of the T-suppressor cell, but the toxins may affect function of helper T cells, B cells, or macrophages, or the interaction among these cells.

Hemorrhagic diathesis may occur after thrombocytopenia or defective intrinsic or extrinsic coagulation pathways. It appears that hemorrhage results from depression of clotting factors, thrombocytopenia, inhibition of platelet function, or possibly a combination of these.

Refusal to consume contaminated feedstuff is the typical sign, which limits development of other signs. If no other food is offered, animals may eat reluctantly, but in some instances, excessive salivation and vomiting may occur. In the past, the ability to cause vomiting had been ascribed to DON only, hence the common name, vomitoxin. However, other members of the trichothecene family also can induce vomiting.

Feed refusal caused by DON is a learned response known as taste aversion. It may be related to neurochemical changes in serotonin, dopamine, and 5-hydroxyindoleacetic acid. Feed refusal response to vomitoxin varies widely among species. DON in swine causes conditioned taste aversion, and swine would be expected to recognize new flavors (eg, flavoring agents) added to DON-containing feed and thus develop aversion to the new taste as well. Provision of uncontaminated feed usually leads to resumption of eating within 1-2 days.

In swine, reduced feed intake may occur at dietary concentrations as low as 1 ppm, and refusal may be complete at 10 ppm. Ruminants generally will readily consume up to 10 ppm dietary vomitoxin, and poultry may tolerate as much as 100 ppm. Horses may accept as much as 35-45 ppm dietary DON without feed refusal or adverse clinical effects. Related effects of weight loss, hypoproteinemia, and weakness may follow prolonged feed refusal. There is little credible evidence that vomitoxin causes reproductive dysfunction in domestic animals.

Irritation of the skin and mucous membranes and gastroenteritis are another set of signs typical of trichothecene toxicosis. Hemorrhagic diathesis can occur, and the radiomimetic injury (damage to dividing cells) is expressed as lymphopenia or pancytopenia. Paresis, seizures, and paralysis occur in almost all species. Eventually, hypotension may lead to death. Many of the severe effects described for experimental trichothecene toxicosis are due to dosing by gavage. From a practical perspective, high concentrations of trichothecenes often cause feed refusal and therefore are self- limiting as a toxic problem.

Due to the immunosuppressive action of trichothecenes, secondary bacterial, viral, or parasitic infections may mask the primary injury. The lymphatic organs are smaller than normal and may be difficult to find on necropsy.

Although no specific name has been given to most nonmacrocyclic trichothecene-related diseases, the term fusariotoxicosis is often used. Some other names used are moldy corn poisoning in cattle, bean hull poisoning of horses, and feed refusal and emetic syndrome in pigs. A condition in chickens, referred to as "rickets in broilers," is also thought to be caused by trichothecenes.

Macrocyclic trichothecene-related diseases have received a number of specific names. The best known is stachybotryotoxicosis of horses, cattle, sheep, pigs, and poultry, first diagnosed in the former USSR but occurring also in Europe and South Africa. Cutaneous and mucocutaneous lesions, panleukopenia, nervous signs, and abortions have been seen. Death may occur in 2-12 days.

Myrotheciotoxicosis and dendrodochiotoxicosis have been reported from the former USSR and New Zealand. The signs resemble those of stachybotryotoxicosis, but death may occur in 1-5 days.

Diagnosis. Because the clinical signs are nonspecific, or masked by secondary infections and disease, diagnosis is difficult. Analysis of feed is often costly and time consuming but ideally should be attempted. Interim measures are carefully examining feedstuff for signs of mold growth or caking of feed particles and switching to an alternative feed supply. Change of feed supply often results in immediate improvement and thus may provide one more clue that the original feed was contaminated.

Control. Symptomatic treatment and feeding of uncontaminated feed are recommended. Steroidal anti-shock and anti-inflammatory agents, such as methylprednisolone, prednisolone, and dexamethasone, have been used successfully in experimental trials. Poultry and cattle are more tolerant of trichothecenes than are pigs.

10

POISONING BY ANIMAL BAITING POISONING BY SNAKES

Venomous snakes fall into 2 classes: 1) the elapines, which include the cobra, mamba, and coral snakes; and 2) the 2 families of viperines, the true vipers (eg, puff adder, Russell's viper, and common European adder) and the pit vipers (eg, rattlesnakes, cottonmouth moccasin, copperhead, and fer-de-lance). Poisonous North American snakes include pit vipers and coral snakes. Elapine snakes have short fangs and tend to hang on and "chew" venom into their victims. Their venom is neurotoxic and paralyzes the respiratory center. Animals that survive these bites seldom have any sequelae. Viperine snakes have long, hinged, hollow fangs; they strike, inject venom (a voluntary action), and withdraw. Many bites by vipers reportedly do not result in injection of substantial quantities of venom. Viperine venom is typically hemotoxic, necrotizing, and anticoagulant, although a neurotoxic component is present in the venom of some species, eg, the Mojave rattlesnake (Crotalus scutulatus scutulatus).

Fatal snakebites are more common in dogs than in any other domestic animal. Due to the relatively small size of some dogs in proportion to the amount of venom injected, the bite of even a small snake may be fatal. Because of their size, horses and cattle seldom die as a direct result of snakebite, but deaths may follow bites on the muzzle, head, or neck when dyspnea results from excessive swelling. Serious secondary damage sometimes occurs; livestock bitten near the coronary band may slough a hoof. Snakebite, with envenomation, is a true emergency. Rapid examination and appropriate treatment are paramount. Owners should not spend time on first aid other than to keep the animal quiet and limit its activity.

Diagnosis. In many instances, the bite has been witnessed, and diagnosis is not a problem. However, many conditions thought by the owner to be snakebites are actually fractures, abscesses, spider envenomations, or allergic reactions to insect bites or stings.

When possible, owners should be instructed to bring the dead snake along with the bitten animal; they should be warned not to mutilate the snake's head because identification may depend on the morphology of the head.

Many bites do not result in envenomation, or are made by nonpoisonous snakes.

Typical pit viper bites are characterized by severe local tissue damage that spreads from the bite site. The tissue becomes markedly discolored within a few minutes, and dark, bloody fluid may ooze from the fang wounds if not prevented by swelling. Frequently, the epidermis sloughs when the overlying hair is clipped or merely parted. Hair may hide the typical fang marks. Sometimes, only one fang mark or multiple punctures are present. In elapine snakebites, pain and swelling are minimal; systemic neurologic signs predominate.

Treatment. Intensive therapy should be instituted as soon as possible because irreversible effects of venom begin immediately after envenomation. Animals bitten by an elapine may be treated with antivenin (which may be available on an as-needed basis through larger human hospital emergency rooms) and supportive care, including anticonvulsants if necessary. A polyvalent antivenin (horse-serum origin) against North American pit vipers is readily available and should be used in all cases of substantial pit viper envenomation.

The progression of events after pit viper envenomation can be divided into 3 phases: the first 2 hr, the ensuing 24 hr, and a variable period (usually ~10 days) afterward. The first 2 hr is the acute stage in which untreated, severely envenomized animals usually die. If death does not occur during this period, and the untreated animal is not in shock or depressed, the prognosis usually is favorable. The acute phase can be prolonged for several hours by use of corticosteroids and, if they are administered, prognostication should be withheld. If the animal is active and alert after 24 hr, death due to the direct effects of the venom is unlikely. The third phase is a convalescent period in which infection (possibly anaerobic) may be of concern. If necrosis has been extensive, sloughing occurs and may be so severe as to involve an entire limb.

An attempt to estimate the severity of envenomation should be made. Although not infallible, it is prudent to consider the size of the snake both as an indicator of the quantity of venom injected, and as it relates to the size of the victim. In dogs and cats, mortality is generally higher from bites to the thorax or abdomen than from bites to the head or extremities. However, this may relate to the size and vulnerability of the victim because smaller animals are more likely to be bitten on the body. Sensitivity to the venom of pit vipers varies among domestic animals. In decreasing order, sensitivity is reportedly horse, sheep, goat, dog, rabbit, pig, and cat. If there has been a previous bite, the victim may have developed some degree of active humoral immunity and be less vulnerable to the toxic effects of the venom.

Treatment for pit viper envenomation should be directed toward preventing or controlling shock, neutralizing venom, preventing or controlling disseminated intravascular coagulation, minimizing necrosis, and preventing secondary infection. Any dog or cat presented within 24 hr of a snakebite showing signs of pit viper envenomation requires intensive treatment, starting with IV fluids to combat hypotension. The use of corticosteroids has been questioned, principally because they alone do not alter the ultimate outcome. They do, however, prolong the clinical course and therefore allow more time in which to institute curative measures. Rapid-acting corticosteroids may help to control shock, protect against tissue damage, and minimize the likelihood of allergic reactions to antivenin. Antivenin is highly beneficial because its action is the only direct and specific mechanism for neutralizing snake venom. Smaller animals probably receive a larger dose (per unit body wt) of venom than more massive animals and, accordingly, require proportionally larger doses of antivenin. Up to 100 ml of antivenin may be necessary for small dogs bitten by a large snake; 5-10 mL may be injected into the tissues around the bite, and the remainder given IV. The efficacy of antivenin is diminished if the bite occurred >24 hr previously. In the event of an anaphylactoid reaction to the heterologous (horse) serum components in antivenin, 0.5-1 mL of 1:1,000 epinephrine should be administered SC. If disseminated intravascular coagulation occurs, appropriate treatment, including blood products and heparin sodium (in mini dose at 5-10 U/kg/hr or low dose at 50-100 U/kg, tid), should be administered SC.

Broad-spectrum antibiotics should be given to prevent wound infection and other secondary infections. Several potential pathogens, including Pseudomonas aeruginosa, Clostridium spp, Corynebacterium spp, and staphylococci have been isolated from rattlesnakes' mouths.

Antibiotics should be continued until all superficial lesions have healed.

Tetanus antitoxin also should be administered; other supportive treatment (eg, blood transfusion in the case of hemolytic or anticoagulant venoms) is administered as needed. In most cases, surgical excision is impractical or unwarranted. Antihistamines have been reported to be contraindicated, but diphenhydramine hydrochloride is frequently given along with antivenin to treat snakebite in humans. Other procedures to neutralize venom (high-voltage, low-amperage electric shock and trypsin) have not proved effective in controlled studies.

POISONING BY BEES

Bees: (Apis spp.); Wasp (Polistes SpD.) and Hornet-Yellow jacket (VesDula SDp.). Envenomation by bees and wasps-potentially, all animals. Effects vary depending on species, season nutritional status, and age of bee. Dog's frequently snaps at insects, therefore swelling of the tongue or in the oral cavity is frequently observed with insect stings. Envenomation by Yellow jacket-Dogs and people. Yellow jackets are attracted to food.

- Bee Venom contains allegens (Phospholipase A2, acid phosphatase, hyaluronidase, melittin, and an unidentified allergen Ag-l) and non-allergens (histamine, dopamine, nor adrenaline, amino acids, and volatile substances).
- Clinical Signs. Syndromes-Immediate hypersensitivity; Local inflammatory response at sting site; Systemic toxicity from non-allergen (Human). Local swelling, edema, and erythematous plaque. Classical anaphylactic reaction reported in human is not documented in livestock. Deaths have been reported in pigs envenomated by yellow jacket, and numerous other livestock by swarms of
- Africanized honey bees. Multiple stings-pain, excitement, followed by tachycardia, diarrhea, hemoglobinuria, icterus, and prostration. Head sting near nostrils or mouth-dyspnea from swelling which accludes air passage (livestock). Animal presented with facial, aural, or periorbital edema. Severe dyspnea when animal is stung in oral cavity and the swelling acc1udes the air passage.
- Multiple stings-Prostration, convulsion, bloody diarrhea, vomiting of bloody fluid, leukocytosis with left shift, and elevated serum urea nitrogen, and alanine transaminase indicating renal and hepatic involvement.
- Treatment. I. V. electrolyte fluid (100 ml/kg/day-dog; 50-75 ml/kg/day-horse); Corticosteroids, antihistamines, diazepam to control convulsions; careful monitoring hepatic and renal functions; oxygen supplementation may be necessary for some animals.

Poisoning by arthropods baiting

Brown Spiders. Envenomation of animals by spiders is relatively uncommon and difficult to recognize. It may be suspected on clinical signs, but confirmatory evidence is rare. Spiders of medical importance in the USA do not inflict particularly painful bites, so it is unusual for a spider bite to be suspected until clinical signs appear. It is also unlikely that the offending spider will remain in close proximity to the victim for the time (30 min to 6 hr) required for signs to develop. Almost all spiders are venomous, but few possess the attributes necessary to cause clinical envenomation in mammals-=mouth parts of sufficient size to allow penetration of the skin and toxin of sufficient quantity or pptency to result in morbidity. The spiders in the USA that are capable of causing clinical envenomation belong to 2 groups-widow spiders (Latrodectus spp.) and brown spiders (mostly Loxosceles spp).

Widow Spiders. Widow spiders usually bite only when accidental skin contact occurs. The most common species is the black widow, Latrodectus mactans, characterized by a red hourglass shape on the ventral abdomen. In the western states, the western black widow,

L.hesperus, predominates, while the brown widow, L.bishopi, is found in the south, and the red widow, L.geometricus, is found in Florida. Latrodectus venom is one of the most potent biologic toxins.

The most important of its 5 or 6 components is a neurotoxin that causes release of the neurotransmitters norepinephrine and acetylcholine at synaptic junctions, which continues until the neurotransmitters are depleted. The resulting severe,· painful cramping of all large muscle groups accounts for most of the clinical signs. Unless there is a history of a widow spider bite, diagnosis must be based on clinical signs, which include restlessness with apparent anxiety or apprehension; rapid, shallow, irregular respiration; shock; abdominal rigidity or tenderness; and painful muscle rigidity, sometimes accompanied by intermittent relaxation (which may progress to clonus and eventually to respiratory paralysis). Partial paresis also has been described. An antivenin (equine origin) is commercially available but is usually reserved for confirmed bites of high-risk individuals. Symptomatic treatment is usually sufficient but may require a combination of therapeutic agents. Calcium gluconate IV is reportedly helpful. Meperidine hydrochloride or morphine, also given IV, provides relief from pain and produces muscle relaxation. Muscle relaxants and diazepam are also beneficial. Tetanus antitoxin also should be administered.

Recovery may be prolonged; weakness and even partial paralysis may persist for several days. Brown Spiders. There are at least 10 species of Loxosceles spiders in the USA, but the brown recluse spider, L reclusa, is the most common, and envenomation by it is typical. These spiders have a violin-shaped marking on the cephalothorax, although it may be indistinct or absent in some species. In the northwestern USA, the unrelated spider Tegenaria agrestis reportedly causes a clinically indistinguishable dermonecrosis in humans and presumably in other animals. Brown recluse spider venom has vasoconstrictive, thrombotic, hemolytic, and necrotizing properties. It contains several enzymes, including a phospholipase (sphingomylinase D) that attacks cell membranes. Pathogenetic mechanisms of the characteristic dermal necrosis are poorly understood, but activation of complement, chemotaxis, and accumulations of neu~rophils affect (or amplify) the process. A history of a bite by a "fiddleback" brown spider is useful but rare. A presumptive diagnosis may be based on the presence of a discrete, erythematous, intensely pruritic skin lesion that may have irregular ecchymoses. Wjthin 4-8 hr, a vesicle develops at the bite wound, and sometimes a blanched zone circumscribes the erythematous area, imparting a "bull's-eye" appearance to the lesion. The central area sometimes appears pale or cyanotic. The vesicle may degenerate to an ulcer that, unless treated in a timely manner, may enlarge and extend to underlying tissues, including muscle. Sometimes, a pustule follows the vesicle and, on its breakdown, a black eschar remains. The final tissue defect may be extensive and indolent and require months to heal. However, medical authorities claim that not all brown recluse spider bites result in severe, localized dermal necrosis. Systemic signs sometimes accompany brown recluse spider envenomation and may not appear for 3-4 days after the bite. Hemolysis, thrombocytopenia, and disseminated intravascular coagulation are more likely to occur in cases with severe dermal necrosis. Fever, vomiting, edema, hemoglobinuria, hemolytic anemia, renal failure, and shock may result from systemic loxoscelism.

Scorpions. Centruroides spp. They may inhabit arid deserts. The bark scorpion C.sculpturatus is the most dangerou~ U.S. species (Arizona, Texas, New Mexico, and some areas in California). Cases are possible outside these areas (Tourism). They are nocturnal, and lives where moisture is available. Pet exposure is purely accidental. Movement may initiate a stinging response.

Venom. Complex mixture of neurotoxic proteins (enzymes and others). Phospholipase A2 is common, hyaluronidase, phosphomonoesterase, acetylcholinesterase, and others are only found in venoms of some species. Other compounds are-amino acids, histamine, 5 hydroxytryptamine, and serotonin. Scorpion venoms are antigenic.

Clinical signs. Local swelling with ecchymosis but no tissue necrosis. Systemic signs-Salivation, muscle fasciculation, generalized weakness, and paralysis including respiratory paralysis (parasympathomimetic). Hypertension, respiratory failure, and skeletal muscle stimulation have been reported in dogs and cats experimentally administered.

Treatment. Antivenin though available, its usefulness is questionable due to delay before treatment. Atropine (0. 04 mg/kg) or to effect to counter parasympathomimetic effects. Corticosteroids for shock and edema. Parenteral fluids must be carefully monitored pulmonary edema. Positive pressure respiration assistance-Respiratory failure. Meperidine hydrochloride (Demerol) and other narcotics are contraindicated –acts synergistically to increase toxicity of venom.

Poisoning by Marine toxins

Natural toxins represent an increasing hazard to the seafood consuming public. Toxins are produced by marine algae and are accumulated through the food chain and are ultimately deposited in higher predator fish or filter-feeding bivalves. Toxins may also affect people through the air and drinking water. Marine and freshwater toxin diseases in human populations are due to the contamination of seafood with a myriad of natural toxins created by minute marine and freshwater organisms such as dinoflagellates found throughout the marine and freshwater world, especially in coral reefs and their surroundings. In general, the natural marine and freshwater toxins are tasteless, odorless, and heat and acid stable. Therefore, normal screening and food preparation procedures do' not prevent intoxication if the fish or shellfish is contaminated. The marine and freshwater toxin diseases are divided into two groups by their primary transvectors, i.e. those associated with the ingestion of shellfish, and those with fish consumption. In addition, there is a group of marine and freshwater diseases associated with water exposure. The shellfish- associated diseases and those associated with water exposure generally occur with algal blooms or "red tides" while fishassociated diseases are more localized to specific reef areas and/or types of fish.

Paralytic Shellfish Poisoning (PSP), Red Tide/Neurotoxic Shellfish Poisoning (NSP), Diarrheic Shellfish Poisoning (DSP), and Amnesic Shellfish Poisoning (ASP) are primarily associated with the ingestion of contaminated shellfish.

Ciguatera Poisoning and Tetrodotoxin Poisoning (Fugu or Pufferfish Poisoning) occur with the ingestion of contaminated fish.

Aerosolized Florida Red Tide/Brevetoxins, the Blue Green Algae (Cyanobacteria), and Pfiesteria and the Pfiesteria-like Organisms (PLOs) can affect humans through exposure to water droplets or even drinking water contaminated with the organisms and/or their toxins.

The primary toxic effect of the marine and freshwater toxins is to the neurologic system, however, affected individuals usually present a wide range of symptoms resulting in a confusing clinical picture. The acute onset of severe gastro-intestinal distress after eating the contaminated fish occurs within minutes to hours. In the case of PSP, Fugu, and Ciguatera, this gastro-intestinal picture can be accompanied by acute respiratory distress which may be fatal within hours (up to 50% of cases of Fugu and 10% of PSP).

In the case of Ciguatera and ASP, debilitating chronic neurologic symptoms lasting months to years have been reported. The majority of people with Ciguatera, especially in the Caribbean, suffer for weeks to months with debilitating neurologic symptoms, including profound weakness, temperature sensation changes, pain, and numbness in the extremities. ASP has left people with permanent and severe memory loss, even years after their initial illness. In addition to increasing worldwide seafood consumption, anthropogenic causes may have furthered the spread of the dinoflagellates and their toxins. There is a body of evidence to indicate that man-induced transportation of the cysts or "seeds" of the toxic marine and freshwater organisms such as dinoflagellates, or of the dinoflagellates themselves located inside the 'spat' (young bivalve shellfish sold commercially to global markets for aquaculture) and ship ballast water (international regulations are now changing to require ship ballast water purging in the open ocean prior to docking).

Paralytic Shellfish Poisoning. A seasonal (summer) allergic, digestive and/or paralytic syndrome observed in man and other animals ingesting toxic shellfish (clams, oysters, mussels). Shellfish become toxic following the filtration of toxic dinoflagellates through their digestive and respiratory systems. Species of dinoflagellates of the genus Gonyaulaux produces the water soluble, heat stable toxin saxitoxin, which becomes localized in the digestive organs, gills, and siphon of the shellfish.

Clinical signs. Allergic-previous sensitization necessary (anaphylactic reaction). Digestive-Seen in 10-12 hours post ingestion of toxic shellfish (vomiting, abdominal pain, nausea, recovery).

Paralytic-Tingling and burning sensation (lips, gums, and tongue) rapidly spreading to all parts of the body. Numbness, difficulty moving, joint pain, weakness, increased salivation, intense thirst. Generalized paralysis with death (20%). Syndromes observed commonly together in varying degrees.

Mechanism of action. Membrane permeability to sodium ion is altered; action potential is blocked thus preventing nerve conductance. Negligible effects on potassium and chloride ions to permeability. Minimal effects on neuromuscular snapsis and cardiovascular system.

- Diagnosis. Difficult without a history of shellfish consumption and the observed clinical signs. No postmortem lesion. Digestive disturbances could be attributed to endotoxins, histadine and/or thiaminase exposure.
- Treatment. With no specific antidote available supportive care becomes essential.
- Ciguatera Poisoning A sporadic, complex syndrome (digestive, cardiovascular, and nervous systems) in man and other animals within 30 hours of ingesting certain food fish species (snapper, barracuda, grouper, king fish, herring). Three water soluble, heat stable toxins (Ciguatoxin, Scaritoxin, and Maitotoxin), produced by the dinoflagillate Gambierdiscus toxicus have been implicated in this condition.
- Clinical signs. Digestive-Nausea, vomiting, diarrhea, and abdominal pain. Cardiovascular Bradycardia and hypotension, followed by tachycardia and hypertension. Nervous severe pruritis, temperature reversal, paresthesia, convulsions, muscle paralysis, and loss of equilibrium.
- Neurologic signs can persist for months, and recur when stressed, consumed alcohol and/or eating
- non-toxic fish. Most commonly, varying degrees of each syndrome is expressed. Death, though rare (7 %), is generally due to respiratory paralysis. Cats are very sensitive to ciguatoxin.

Mechanism of action. Inhibits acetylcholiesterase, cholinergicomimetic effects. Competitively inhibits calcium regulates sodium ion channel.

Diagnosis. History of fish consumption and clinical signs. Differentiate between other conditions Paralytic shellfish poisoning, Type E botulism, Organophosphate/Carbamate poisoning.

Treatment. No antidote available. Symptomatic and supportive therapy essential. Acetaminophen and indomethacin are recommended for chronic ciguatoxin poisoning. As a precaution, viscera of tropical marine should not be eaten. Since the toxin is water soluble, it would be advisable to soak the fish meat for days, and discard the water before cooking.

Tetrodotoxin Poisoning. Marine puffer fish (toad fish, globe fish, swell fish, porcupine fish, and balloon fish) have in their ovaries (roe), liver, intestines, and skin, the neurotoxin tetrodotoxin (basic, water soluble). Ingestion of improperly prepared fish results in toxicosis. Tetrodotoxin originated from either of four possible sources-Endogenous, Exogenous, Symbiotic microorganisms, or multiple origins. Clinical signs Within minutes (45 minutes) of ingesting toxic fish, there is a tingling (lips and tongue) followed by motor incoordination; excessive salivation, nausea, leading to vomiting and diarrhea. Generalized paralysis with convulsions and death (60%) from skeletal muscle paralysis (human). Dogs and cats may become poisoned from scavenging.

Cats experimentally administered tetrodotoxin showed-gross mydriasis, flaccid paralysis, tachycardia, and depressed respiration.

Mechanism of action. Tetrodotoxin selectively blocks' sodium ion channel along the axon. Diagnosis. History of ingestion of the puffer fish and the rapid onset of neurologic signs are suggestive. Agonal hypoxic ecchymoses on serosal surface and pulmonary edema may be seen but are not pathognomonic. Differentiate from tick paralysis and ciguataria poisoning. Treatment. No known treatment. Ventilatory support, atropine, dopamine, crystalloid infusion are primary modalities currently used.

Toxicity of Clostridium botulinum toxin

Botulism, or "sausage poisoning", was reportedly first recognized in Germany around the late 1700s. However, it was not until the 1820s that Justinus Kerner systematically studied and described the fatal paralytic disease (aka, "Kerner's disease") associated with the ingestion of spoiled sausage. Kerner recognized that a poisonous substance isolated from the spoiled sausage was responsible for the clinical signs associated with botulism; however, he was unable to identify the origin of the deadly poison. With the subsequent introduction of anaerobic microbiological techniques in the late 1800s, the source of the poison was finally determined. In 1897, van Ermengem (1979) was able to identify the offending etiological agent as a bacterium, now known as Clostridium botulinum (C. botulinum), in spoiled ham. The deadly poison produced by the bacteria is now known as botulinum neurotoxin (BoNT), the most potent biological toxin ever encountered, with lethal doses as low as 0.03 ng/kg body weight depending on both toxin type and animal species. Here we describe the toxicology of BoNT in relation to various animal species.

Background. BoNT is produced under anaerobic conditions by primarily, C. botulinum, a rod- shaped, spore-forming (subterminal) bacterium; however, other clostridial species such as C. barati and C. butyricum are also capable of producing the neurotoxin. Although often referred to collectively as a single toxin, there are actually seven immunologically distinct serotypes of BoNT, and they are designated alphabetically, A–G. Serotypes A, B, C1, and D have been associated with outbreaks of botulism in our domestic animals, livestock, poultry, and wildlife; while serotypes A, B, E, and, rarely F, are known to cause disease in humans. When isolated from the bacterium, native BoNTs are found in complex with hemagglutinins and other non- hemagglutinin, non-toxic proteins. These accessory proteins are thought to protect the toxin from harsh environmental conditions such as those found in the gut following ingestion of toxincontaminated food products. The molecular masses of these toxin complexes range between 300 and 500 kDa, depending on the toxin serotype. Although serologically distinct, the seven toxin serotypes share both structural and functional similarities. When not complexed with other proteins, the active neurotoxin has a molecular mass of 150 kDa and exists as a polypeptide di- chain molecule. The di-chain consists of a heavy chain linked by a single disulfide bond to a light chain. The 100 kDa heavy chain is responsible for membrane targeting and cellular uptake, while the 50 kDa light chain mediates its intracellular action. The neurotoxin molecule can be further divided both structurally and functionally into three domains. The heavy chain contains both a binding domain and a translocation domain, while the smaller light chain contains a catalytically active domain.

Mechanism of action. BoNTs are potent zinc-dependent metalloproteases that exquisitely target acetylcholine (ACh) containing nerve terminals. Since BoNTs are too large to cross the intact blood brain barrier, their selective action on cholinergic terminals is generally limited to peripheral cholinergic systems. The primary target site is the neuromuscular junction (NMJ), where they act within the motor nerve terminal to prevent the release of ACh, the primary neurotransmitter at this major synapse. Intoxication by BoNT has been described as a multistage process. In the first stage of intoxication, the toxins must bind selectively, via their heavy chains, to protein receptors (and gangliosides) located on the plasma membrane of the motor nerve terminal. The binding of the toxins to the nerve terminal membrane initiates the second stage of the intoxication process which is characterized by receptormediated uptake of the toxins into endosomes (e.g. endocytosis). The third stage of intoxication occurs within the endosome, where the disulfide linkage between the toxin heavy and light chains is reduced, and a subsequent drop in endosomal pH promotes a conformational change in the toxin molecule, allowing the light chain to escape across the endosomal membrane into the cytosol, possibly mediated by the translocation domain of the toxin (e.g. translocation). Following this translocation to the cytosol, the toxin light chain is free to act upon its intracellular target during the final stage of intoxication. During this last stage of the process, the neurotoxins enzymatically cleave one of three specific proteins found within the presynaptic terminal. These three proteins, SNAP-25, synaptobrevin, and syntaxin, known collectively as SNARE (soluble n-ethylmaleimide sensitive factor attachment protein receptor) proteins, are necessary for neurotransmitter release. Synaptobrevin, an integral membrane protein, is found on the synaptic vesicle, while SNAP-25, a membrane associated protein and syntaxin, another integral membrane protein, are both localized to the presynaptic terminal membrane. Under normal circumstances, these SNARE proteins interact to form a four helical "fusion" or "SNARE" complex that brings the ACh containing synaptic vesicles into close apposition with the terminal membrane.

Fusion of the vesicular and terminal membranes and the subsequent release of ACh into the synaptic cleft are triggered by fast Ca2_ signaling. The enzymatic cleavage of any one of the SNARE proteins by BoNT either destabilizes or prevents the formation of functional SNARE complexes, inhibiting vesicular fusion and neurotransmitter release. The catalytic active sites of the different BoNT serotypes vary slightly, giving each serotype both substrate and cleavage site specificity. Toxin serotypes A, C, and E cleave SNAP-25; serotypes B, D, F, and G cleave synaptobrevin, and serotype C cleaves syntaxin. Clinical diagnosis and treatment. Clinically, botulism is recognized as a lower motor neuron disease resulting in progressive flaccid paralysis. Although deficits in somatic neuromuscular transmission are the most prominent effects, motor deficits in cranial nerve function, as well as the autonomic nervous system have also been reported. With the exception of impaired vision (most likely related to disruption in autonomic function), neither altered sensation nor mentation has been specifically reported in botulism. In animals, paresis begins in the hind limbs and progresses cranially, often resulting in quadriplegia and recumbency. As in humans, death may result from respiratory muscle paralysis. The onset of clinical signs may be within hours or days of exposure, and is dependent on exposure conditions (dose and duration), as well as individual sensitivity. Susceptibility to the different serotypes varies across species. Botulism in large animals (herbivores) is commonly, but not exclusively, due to serotype B; while botulism in small animals (carnivores) and avian species is most commonly due to serotype C1. With the exception of serotype C1, the different toxin serotypes produce similar clinical disease. Serotype C1 toxin producing strains of C. botulinum bacteria may produce multiple exotoxins, including the C1 neurotoxin, and the C2 and C3 cytotoxins. The concurrent presence of these cytotoxins with C1 neurotoxin may account for the different clinical scenarios that have been reported in botulism associated with serotype C1 intoxication, including dysautonomias in horses and cats.

Botulism is most often acquired from the ingestion of preformed toxin in either spoiled vegetation or carrioncontaminated foodstuffs, although this may vary among species. In addition, two other forms of botulism are well recognized in veterinary medicine and include wound contamination with C. botulinum, or gastrointestinal (GI) colonization by C. botulinum (toxico-infection). The mainstay of therapy for all animal species with botulism is supportive care, although the specifications of this therapy may differ slightly between species. Antitoxin (equine origin) may be administered in certain instances; however, since the antitoxin is only protective against toxin in the general circulation, it must be given early in the disease course to be effective. Once the patient displays symptoms, antitoxin administration may be ineffective. Further, antitoxin administration can cause adverse effects. Antibiotic administration may be indicated to reduce the risk of secondary infections; however, aminoglycosides, tetracycline, and procaine penicillin should be avoided as these drugs potentiate neuromuscular weakness. C. Botulinum is not sensitive to metronidazole, an antimicrobial used to treat anaerobic infections. Further, metronidazole has been associated with an increased risk of botulism in laboratory animals and workers exposed to the bacterium, possibly due to alterations in GI flora that permit clostridial growth.

Laboratory diagnosis. The diagnostic gold standard for botulism is the mouse bioassay (MBA). Although the MBA may have limited sensitivity, especially in horses, the test is highly specific. Samples of serum, vomitus, gastric contents, feces, wound tissues, and suspect food can be submitted to specific diagnostic laboratories for MBA analysis. All samples, with the exception of wound samples, should be kept at 4°C immediately following collection. Blood samples should be collected in red top tubes or serum separating tubes. Serum volumes between 10 and 15 ml should be collected as soon as clinical signs are detected; volumes less than 3 ml may yield inconclusive results. Whole blood should not be submitted as hemolysis may occur during shipment. For fecal samples, 20–50 g should be submitted. For constipated patients, 20 ml of rectal contents post- enema can be submitted; however, it is important to use as little fluid as possible in the enema to avoid over-dilution of samples. Suspect food should be left in its original container, and placed inside a sterile, unbreakable and well-labeled container. Empty food containers can also be submitted for testing. Environmental samples can also be tested with the MBA; submission of 50– 100 g of soil and approximately 100 ml of water is recommended. All specimens should be placed in leak-proof containers marked "Medical Emergency, Biological Hazard, Refrigerate on Arrival." Specimens should be shipped with ice packs in order to maintain a temperature of 4°C. In cases of wound botulism, debrided tissues, exudates, and/or tissue swabs may be submitted.

Wound specimens should be placed in anaerobic transport devices and shipped without refrigeration. For specific details on packaging and shipping, refer to the American Society of Microbiology Procedures for Transportation and Transfer of Biological Agents (www.asm.org). If samples cannot be shipped for several days, specimens should be frozen and then subsequently shipped on dry ice. Freezing samples may compromise the detection of C. botulinum bacteria; however, freezing will not affect toxin detection. Only qualified diagnostic laboratories can perform both the MBA and neutralization tests to determine the presence, as well as the serotype of BoNT in submitted samples. During these tests, mice (20–30 g IRC strain) are injected IP with prepared samples of serum and/or other sample extracts. For the MBA, a group of mice is injected with sample only to determine the presence of the toxin. For the neutralization assay, a separate group of mice is injected with the sample combined with a serotypespecific antitoxin. In both cases, mice are observed for clinical signs of botulism including ruffled coats, abdominal breathing patterns, dyspnea, weakness, ataxia, respiratory failure, and ultimately death. C. Botulinum bacteria can be isolated from anaerobic enrichment cultures of submitted samples.

Following enrichment, culture supernatants are collected and tested similarly for the presence and serotype of toxin using the MBA and neutralization tests. In the former case, a positive MBA test is suggestive of the presence of C. botulinum bacteria in the sample. It should be emphasized that humans are extremely sensitive to certain serotypes of BoNT; therefore, both toxin and clostridial bacterial isolation should only be attempted by qualified immunized personnel in diagnostic laboratories approved by the Centers for Disease Control and Prevention (CDC). Although the MBA is still the most widely utilized test for botulism, alternative diagnostic avenues are being explored in an effort to minimize the use of animals in toxicity testing. For example, enzyme- linked immunoabsorption (ELISA) techniques have been used to identify BoNT in both avian and bovine specimens. In one study comparing the capabilities of the ELISA and the MBA to detect BoNT/C and D in bovine serum and tissue samples, Thomas (1991) determined that the ELISA was not sensitive enough to replace the MBA. Further, a cross-reaction with C. novyi serotype A was reported. Similarly, Rocke et al. (1998) found that the MBA was more sensitive than the ELISA (immunostick method) in detecting BoNT/C in small serum volumes from wild birds with botulism. However, when larger sample volumes were tested, the sensitivity of the ELISA improved, and actually surpassed that of the MBA. When comparing these two methods, it should be noted that only the MBA detects biologically active toxin, while the ELISA detects both active and inactive forms as well as subunits of the toxin. The use of polyvalent antibodies to capture toxin antigen and the slight antigenic heterogenicity within serotypes may further complicate ELISA results. More recently, polymerase chain reaction (PCR) methods have also been used to detect the genes encoding BoNT. Trampel et al. (2005) used PCR to identify BoNT/C genes from cecal samples in caponized chickens. Further, Szabo et al. (1994) determined that PCR detection of the genes encoding BoNT/B following culture enrichment of equine serum, tissues, and fecal samples were comparable to results obtained by the MBA.

Species-specific disease. Equine botulism. Horses are thought to be among the most susceptible of species to intoxication by BoNT. The serotype most commonly reported to cause equine botulism in North America (greater than 80% of equine cases) is serotype B;

however, serotypes Aand C1 have also been associated with clinical disease in the horse. Serotype B producing strains of C. botulinum are found ubiquitously in the soils of the northeastern and central United States, particularly in regions extending from Kentucky to the mid-Atlantic states. Conversely, serotype A producing strains are more prominent in the western states (California, Utah, Idaho, Oregon) and have also been reported in Ohio. Although intoxication with BoNT/C1 occurs with less frequency, cases have been reported in California, Florida, the New England States, and Canada. Horses acquire botulism in one of three ways: (1) from the ingestion of preformed toxin in contaminated foodstuffs, (2) from the contamination of wounds with C. botulinum, and (3) from the colonization of the intestinal tract with C. botulinum bacteria (toxico-infection). The ingestion of preformed toxin, in either spoiled vegetation or carrioncontaminated foodstuffs, is the most common scenario for equine botulism. Contaminated feed sources such as alfalfa cubes, alfalfa hay, baled hay, wheat, oats, potatoes, bale silage, rye silage, grass clippings, oat chaff, and brewer's grains have all been purported sources of botulism in large animals. Serotype B is more commonly associated with contamination of spoiled foodstuffs or moldy hay. Interestingly, the feeding of silage or hay stored in large plastic bags has been implicated as an increased risk factor for botulism.

The damp, alkaline conditions of spoiled vegetation provide an optimal environment for clostridial growth, sporulation, and toxin production. For these reasons it is recommended that silage with a pH _ 4.5 should not be fed to horses. Cases of equine botulism associated with the ingestion of carrion-laden foodstuffs are more often the result of intoxication with BoNT serotype C1. A herd outbreak of BoNT/C1 intoxication in California was determined to be the result of the ingestion of preformed toxin in alfalfa cubes contaminated with rodent carcasses. Interestingly, another report of an outbreak of BoNT/C1 intoxication in horses revealed that birds may act as mechanical vectors, transporting toxin or bacterial spores from a rotting carcass to nearby horse farms. Wound botulism occurs from the contamination of a wound with the Clostridium bacteria. The anaerobic environment of the wound provides optimal conditions for C. botulinum growth and toxin production. The neurotoxin (most commonly serotype B) is produced within the wound and enters the peripheral circulation where it is distributed to the toxin's site of action, the NMJ. Distal limb wounds, castration sites, umbilical hernias, and injection site abscesses have all been associated with wound botulism. "Shaker foal syndrome" is a form of toxico-infectious botulism that occurs in foals usually between 2 and 5 weeks of age.

The infection occurs most commonly in fast growing foals on high planes of nutrition. GI ulcerations and liver abscesses have been documented postmortem in foals that succumbed to botulism. Thus, exposure to stress, high nutrient diets, or corticosteriod use may all play a role in the foal's susceptibility to toxico- infection. Such underlying conditions may lead to gastric ulcers, which then serve as a nidus for C. botulinum colonization.

Further, as in human neonatal toxico-infection, the immature GI tract of foals may be more permissible to overgrowth by C. botulinum. Shaker foal syndrome is commonly associated with the production of toxin serotype B; however, a case of serotype C1 intoxication was documented in a foal in Florida. Serotype C1 producing clostridial strains have been identified in Florida soils, and in this particular case, it was concluded that the foal developed a toxico-infection after ingesting bacterial spores from the soil. Additionally, this foal was also diagnosed with sand colitis; an irritated GI mucosa likely increased this foal's susceptibility to toxico-infection. Toxico-infection with BoNT serotype C1 has also been implicated as a causative agent of equine dysautonomia, also known as equine grass sickness (EGS). EGS is frequently diagnosed in Great Britain, and its occurrence correlates with the grazing season. EGS presents as an acute or subacute illness resulting in death (or euthanasia) within days of onset; however, classic neuromuscular symptoms of botulism are not observed in EGS. Rather, the primary symptoms associated with this syndrome include dysphagia, ileus, and weight loss. If the horse survives the subacute phase, a chronic form may persist.

In addition, histological examination revealed neuronal degeneration in autonomic ganglia, the myenteric plexus, and the submucosal plexus. Such histological findings are not reported with classic botulism. Increased BoNT/C1 has been detected in ileal and fecal contents of horses diagnosed with EGS; however, it should be noted that in this same study, BoNT/C1 was also isolated from a small number of control animals. Conversely, Garrett et al. (2002) demonstrated an increased number of bacterial colonies with a prominent number of clostridial species in the GI tract of horses with EGS, compared with healthy control horses. The possibility, however, that the increased colonization of the gut by clostridia was secondary to the ileus produced in EGS cannot be ruled out. McGorum et al. (2003) recently documented the occurrence of clinical and pathological signs associated with both EGS and toxico-infectious botulism in the same foal. One theory proposes that in addition to the blockade of acetylcholine release at cholinergic nerve terminals, the C1 neurotoxin also causes nerve cell degeneration. However, it is worth noting that serotype C1 producing clostridial strains may also produce the ADPribosylating cytotoxic toxins, C2 and C3. Interestingly, the C2 toxin has been associated with an increase in vascular permeability leading to the development of edema, congestion, and hemorrhage in in vivo models. It is plausible then these exotoxins may play a role in producing the clinical symptoms of EGS that deviate from those of classic botulism. To date, however, the exact cause of EGS and the potential role of a C. botulinum toxico-infection in its manifestation remain elusive.

Clinical signs. The onset of clinical signs associated with equine botulism is variable and can occur anywhere between 12 h and 10 days post-exposure. The clinical presentation of poisoned horses may be gradual, acute, or peracute depending on exposure dose and duration, as well as on individual sensitivity to the toxin. Adult horses that ingest low doses of toxin may show only mild dysphagia, and recover with minimal treatment; whereas, ingestion of large doses is associated with peracute illness and a grave prognosis. In peracute illness, muscle paralysis progresses rapidly and the animal is recumbent within 8–12 h; ultimately, paralysis of the respiratory muscles results in death. With the exception of serotype C1 and EGS, clinical symptoms of equine botulism vary little between the different botulinum toxin serotypes. Myasthenia and dysphagia are usually the first symptoms observed. Astute horse owners may first notice mild signs such as depression, exercise intolerance, and difficulty with grain consumption.

Ataxia, gait stiffness, and muscle tremors (particularly in the triceps muscles) may be noted early in the course of disease. In addition, mydriasis, ptosis, and decreased pupillary light responses, as well as decreased palpebral reflexes are characteristic of early botulism. As the disease progresses, pupillary light responses diminish further. Periods of exercise may worsen paresis due to the reduction of acetylcholine release at the NMJ. As dysphagia and pharyngeal weakness progress, the swallowing of food becomes more difficult and secondary aspiration pneumonia may ensue. Horses may also have difficulty drinking as they tend to stand with their muzzles submerged in water troughs without swallowing. In horses infected experimentally with serotype C, there was a more pronounced mydriasis and an inability to lift the head; facial edema and inspiratory stridor resulted from low head carriage. Initially, vital signs such as heart rate, respiration rate, body temperature, and capillary refill time are within normal limits. However, decreased borborygmi, ileus, colic, and constipation develop as botulism progresses. Diarrhea is often associated with serotype C, possibly in association with C2 toxin. Urine retention with resultant bladder distention often occurs, thereby increasing the risk of urinary tract infection. As the disease progresses, horses spend more time in sternal recumbency, and ultimately become laterally recumbent. Heart and respiratory rates may increase as recumbent horses struggle to stand. In late stages, dyspnea and other signs of respiratory distress may be observed. With serotype C intoxication, an exaggerated expiration and “prolonged abdominal lift” may be noted. In the final stages of botulism, horses are laterally recumbent, demonstrante significant respiratory difficulty and develop anoxia. As the anoxia progresses, horses may exhibit agonal paddling. At this point, the patient either dies due to respiratory failure or is euthanized. In the foal, botulism most commonly occurs between 2 and 5 weeks of age. The first clinical signs usually observed are increased periods of recumbency and muscle tremors. Shortly after the foal rises, muscle tremors are evident and after brief periods of standing, the foal collapses from weakness. Recumbent foals appear to be bright and alert. Foals may dribble milk from their muzzles shortly following nursing due to dysphagia and pharyngeal muscle paresis. Thus, aspiration pneumonia is a common sequela in the foal. Constipation and ileus are also frequently observed. Other symptoms are similar to those observed in the adult horse.

Diagnosis. Atentative diagnosis of botulism can be made following a comprehensive neurological assessment. Typically, abnormalities in palpebral reflexes, pupillary light responses, tail tone, tongue tone, prehension of food, dysphagia, and gate are detected with botulism. Both the tongue stress test and grain test are sensitive measures for botulism. In the tongue test, the examiner pulls the tongue laterally through the interdental space. As the horse attempts to retract the tongue, muscular tone is assessed. Although this test is both subjective and variable, a flaccid tongue or weak retraction effort is suggestive of botulism. In the grain test, the consumption of 8 oz of grain is timed; typically, a healthy horse consumes 8 oz of grain in 2min. Difficulty in prehension, slow consumption, and a characteristic grain/saliva mixture hanging from the lips are indicative of botulism. These examination findings may also be useful in the assessment of disease progression and treatment efficacy. Abnormalities in routine diagnostic indicators (complete blood cell (CBC) count, blood chemistry, cerebral spinal fluid (CSF) analysis, and urinalysis) are not typically detected in early botulism, but do usually accompany other neurological diseases. Therefore, normal laboratory values in light of neurological deficits support the diagnosis of botulism.

Differential diagnoses for botulism include infectious diseases such as equine protozoal myeloencephalitis, equine herpes virus-1, eastern and western equine encephalitis, rabies, guttural pouch mycosis, listeriosis; other toxicoses such as leukoencephalomalacia (moldy corn poisoning), ionophore poisoning (monesin, salinomycin, and narasin), yellow star thistle poisoning, yew poisoning, white snake root poisoning, and organochlorine poisoning; metabolic disorders such as equine motor neuron disease, azoturia, eclampsia, hypocalcemia, hyperkalemic periodic paralysis, white muscle disease; and pharyngeal ulceration. A tentative diagnosis of equine botulism may be confirmed by: (1) MBA detection of formed toxin in horse sera, GI contents, viscera, or wounds, (2) detection of C. botulinum spores or toxin in suspect foodstuffs in association with clinical signs, and (3) ELISA detection of serum antitoxin antibodies in unvaccinated horses with clinical signs (Whitlock, 1996). However, a definitive diagnosis of botulism is often difficult to achieve in the horse. There are no gross or pathognomonic histological lesions associated with botulism, and serum toxin levels in the horse are often too low to be detected by the MBA. Because the horse is more sensitive to BoNT than the mouse, the MBA is most valuable in early, peracute equine botulism, when higher concentrations of toxin may be present in the bloodstream. In addition to serum, GI contents and liver samples can also be submitted for the MBA, although greater diagnostic success may be achieved through detection of BoNT in the foodstuff rather than within the patient. Fecal or tissue culture enrichment can be used to enhance bacterial spore numbers and toxin levels for greater detection. However, since spores may be present in the feces of healthy horses, direct detection of BoNT within the animal is a more reliable finding. Following a positive result from the MBA, the serotype can be identified using the mouse neutralization test.

Treatment. Once botulism is suspected, the patient should be confined to the stall to prevent exertion. Polyvalent antiserum (antitoxin) should be given as soon as possible; the recommended dose for an adult horse is 70 000 IU (500 ml) and 30 000 IU (200 ml) for the foal. One dose usually provides passive immunity for approximately 60 days. The use of parasympathomimetics should be avoided as these agents deplete acetylcholine stores and exacerbate paresis/ paralysis. Antibiotic therapy is indicated in cases of wound botulism or secondary infections; however, as previously stated aminoglycosides, tetracycline, procaine penicillin, and metronidazole are contraindicated. Aminoglycosides block neurotransmission at the NMJ and will exacerbate muscle weakness and paralysis. Although Gram-positive anaerobes are sensitive to penicillin and metronidazole, administration of these drugs is controversial.

These antimicrobials may cause more bacterial lysis, thus increasing the release of toxin (in the case of a toxico-infection) or they may promote C. botulinum colonization by altering the normal intestinal flora. Drugs such as the aminopyridines and guanidines should also be avoided as they will further deplete acetylcholine stores. Second only to antitoxin administration, supportive care is the mainstay of therapy for botulism. H2 blockers and proton pump inhibitors may be indicated, especially for foals. Topical ophthalmic ointments should be used to prevent corneal abrasions and ulceration. Adult horses may need to be sedated with xylazine or diazepam to reduce anxiety and exertion. Patients should be muzzled between feedings to prevent aspiration pneumonia. Nutritional support should be provided to dysphagic patients. Alfalfa slurries with adequate amounts of water may be administered through a nasogastric tube to adult horses. Foals should receive milk replacer through a nasogastric tube, or parenteral nutrition if ileus is present. Patients should be maintained in sternal recumbency to prevent aspiration pneumonia and checked periodically for gastric reflux as ileus may lead to the accumulation of ingesta/ fluid in the stomach. If gastric reflux is not present, some authors recommend that mineral oil be administered via a nasogastric tube to alleviate ileus and constipation; however, this should be done under close supervision due to the increased risk of aspiration in these patients. Recumbent patients should be turned frequently or suspended periodically by full body slings to prevent decubital ulcer formation, myopathies, and other complications of prolonged recumbency. Recumbent stallions and geldings should be catheterized twice daily to empty the bladder and prevent pressure necrosis or cystitis.

A tracheostomy should be performed in cases of botulism where horses show signs of upper airway obstruction as a result of paralysis of the nares or larynx. In more complicated cases, patients may require intravenous (IV) fluids to correct respiratory acidosis resulting from decreased ventilation. For foals in particular, arterial blood gases should be monitored frequently to determine the need for artificial ventilation. Intranasal oxygen insufflation and mechanical ventilation can be instituted in foals with poor arterial blood gas values and/

or metabolic acidosis. Unfortunately, mechanical ventilation is not practical in the adult horse. The overall prognosis is favorable for horses that are exposed to low doses of the toxin, exhibit a slow disease progression (3–7 days), or display mild symptoms. Likewise, a grave prognosis is given to patients exposed to high doses of toxin, manifest a rapid onset of clinical disease, or become recumbent within 8–12 h. Patients responsive to antitoxin therapy should be able to eat within 7–10 days post-treatment and regain full strength within a month. Recumbent foals are often able to stand within 7–10 days post- antitoxin administration. Although the prognosis for recumbent adult horses is poor, if the patient does not become distressed or show severe respiratory compromise, recovery may be achieved with extensive supportive care. The most common complications associated with botulism are decubital ulcers and aspiration pneumonia; these problems can be resolved with supportive care and antimicrobial therapy.

Prevention. Following recommended vaccination protocols, along with sound husbandry methods, reduces the occurrence of equine botulism. Forages should be examined for carrion, while pastures should be cleared of decaying vegetation and rotting animal carcasses. Appropriate wound management is also an important preventative measure. To date, only serotype B toxoid vaccine is marketed for horses in the United States. The American Association for Equine Practitioners (2005) recommends vaccination only for horses in endemic areas. Adult horses in endemic areas should be vaccinated annually. Mares should be boostered 4–6 weeks prior to parturition to achieve adequate antitoxin immunoglobulin (Ig) levels in colostrum. Foals born to vaccinated mares should receive a series of three vaccinations, each 1 month apart, starting at 2–3 months of age. Foals born to unvaccinated mares should be vaccinated at 2, 4, and 8 weeks of age. Avian botulism. Background. Avian botulism, otherwise known as "limberneck" or "western bird disease", has been a significant problem worldwide in both domestic and wild fowl. The occurrence of avian botulism has been globally widespread, having been documented in as many as 17 countries and on every continent except Antarctica.

The majority of the natural outbreaks of avian botulism have occurred in fowl. Carnivorous, omnivorous, carrion-scavengers, and insectivorous birds, as well as aquatic bottom-feeding birds are all susceptible to botulism. In 1984, 117 avian species were determined to have been affected by the disease. Specifically, botulism has been reported in chickens, ducks, turkeys, pheasants, and ostriches. Although broiler outbreaks are not uncommon, botulism is a more significant problem for waterfowl, resulting in millions of deaths worldwide. Avian species are sensitive to serotypes A, B, C1, and E, although serotype C1 is most commonly associated with outbreaks (Smith, 1975; Gross, 1984; Dohms, 2003). Outbreaks of serotype C1 intoxication have been reported worldwide, while outbreaks of serotype A botulism have only been reported in western regions of North and South America; serotype B in the eastern United States, England, Europe, and China;

and serotype E in the Great Lakes and North Sea. Interestingly, serotype Awas found to be more toxic than serotype C1 when administered IV to chickens; however, when given orally, serotype C1 demonstrated greater toxicity.

The etiology of botulism among wild avian species and waterfowl differs from that observed in other animals. The process is a complex cycle involving environmental contamination, toxico-infection, bird die offs, bacterial proliferation in bird carcasses, and invertebrate vectors. C. botulinum often colonizes the intestinal tract and cecum of clinically normal birds, increasing the potential for toxico-infection in avian species. Since these birds are already seeded with the bacteria, upon death, avian carcasses provide an excellent substrate for C. botulinum growth. The proliferating bacteria spread from the GI tract to other tissues, the carcass becomes flyblown and toxin accumulates in the fly larvae. Invertebrates concentrate the bacterium or toxin after feeding on contaminated carcasses; however, due to their neurophysiological differences, BoNT does not affect insects and aquatic invertebrates. Subsequently, birds ingest these animals and accumulate lethal amounts of BoNT. One gram of fly larvae may contain 1.8 _ 105 mouse LD50s, and ingestion of as little as eight fly larvae was sufficient to kill a pheasant.

Bird and invertebrate die offs perpetuate botulism outbreaks by increasing the levels of C. botulinum in soils, lakes, rivers, and estuaries. Environmental factors such as shallow alkaline waters, warm seasons/summer months, and flooding of mudflats or dried out lakes may promote invertebrate die offs, further enhancing environmental levels of C. botulinum. As C. botulinum levels increase in the environment, the intestinal tracts of wild birds and waterfowl become seeded with the bacteria, and any cause of bird deaths can trigger an outbreak of botulism. Contaminated feed, water, litter, carcasses, and insects may be associated with botulism in broilers. Often the source of BoNT cannot be identified and toxico-infection has been hypothesized to be the

perpetuating factor. C. botulinum has been isolated from the intestinal tract and cecum of healthy birds; further, the chicken body temperature (41°C) and cecal pH (7.4) are optimum for C. botulinum growth. Most broiler outbreaks have occurred in chickens between 2 and 3 weeks of age; however, an outbreak in post-caponized chickens was documented in birds as old as 14 weeks. Coprophagy has also been implicated as a causative factor in poultry outbreaks since both BoNT/C1 and C. botulinum are secreted in cecal droppings. Broiler outbreaks are also more likely to occur in hot weather. Morbidity and mortality of avian botulism increase with the dose of BoNT ingested. The onset of clinical symptoms may be anywhere from a few hours to 2 days post-exposure. The mortality rate in broilers has been reported to be as high as 27%, whereas thousands to millions of birds may have been lost as a result of outbreaks in waterfowl. In fact, it had been suggested that botulism may have been the limiting factor of waterfowl population growth in predisposed areas of the United States.

Clinical signs. As in other species, avian botulism is characterized by lower motor neuron deficits resulting in flaccid muscle paralysis. Paresis begins in the legs and progresses cranially to involve the wings, neck, and eyelids. Mildly affected birds may appear ataxic, reluctant to move, have a ruffled coat, and easily epilated feathers. The wings may droop and the neck become flaccid, hence the name "limberneck". Diarrhea is often noted in broilers. As the disease progresses, birds become recumbent. Neck muscles become paralyzed and birds eventually lie down with necks extended out, resting on the ground. Birds may appear comatose due to eyelid paralysis. Dyspnea may develop as paralysis progresses. Birds usually die from respiratory failure and dehydration. Broilers may succumb to hyperthermia as sick birds are smothered by others and the respiratory mucosal cooling mechanism is compromised.

Diagnosis. The diagnosis of avian botulism is based on clinical signs, a lack of specific pathological changes, and the isolation of toxin from serum/tissues of clinically ill birds. Although no pathognomonic changes have been described, post-mortem hepatic and renal congestion along with signs of dehydration may be found. The most definitive diagnosis of botulism is the isolation of BoNT from the sick bird. Ten milliliters of blood is the suggested minimum amount for the MBA; however, if necessary, equal aliquots of blood from individual sick birds may be pooled to accommodate volume requirements of the assay. Following a positive result from the MBA, the serotype can be identified using the mouse neutralization test. Most outbreaks of avian botulism are due to BoNT/C1; therefore, antiserum for serotype C1 is usually tested first. One IU of antiserum/mouse typically neutralizes BoNT/C1 levels found in chickens suffering from botulism. BoNT/C1 and other serotypes can also be detected in bird serum using ELISA technology. For small sample volumes, the MBA appears to be more sensitive; however, for larger serum samples, the ELISA sensitivity may be comparable to the MBA. Isolation of BoNT or C. botulinum from the bird intestines, cecum, or other tissues may aid in a diagnosis; however, these tests are less valuable as the bacterium can be isolated from the intestinal tract of healthy birds. Further, isolation of BoNT or the bacterium from carcass tissues is not definitive since C. botulinum may proliferate and spread from the intestinal tract to surrounding tissues of the carcass. The MBA can be performed on intestinal, cecal, and crop flushes, or samples can be assayed for toxin or bacterium after culture enrichment. PCR methods have been used to identify genes encoding BoNT/C light chain in cecal contents. In order to identify the source of contamination, feed, water, litter, carcasses, and insects should be assayed for toxin, or cultured to isolate the bacterium. Both ELISA and the passive hemagglutination test can be performed to identify serum antibodies to BoNT. However, the levels of toxin that produce illness are usually insufficient to stimulate an immune response in chickens and ducks. Interestingly, antibody titers to several BoNT serotypes have been identified in healthy carrioneating birds such as vultures and crows.

Differential diagnoses for avian botulism in poultry include transient brain paralysis, coccidiostat toxicity, pesticide or other chemical toxicity, New Castle disease, Marek's disease, avian encephalomyelitis, avian reovirus, and musculoskeletal problems. Fowl cholera and chemical toxicity, lead poisoning in particular, are the common differentials for botulism in waterfowl.

However, eyelid paresis and the lack of postmortem lesions are supportive of botulism as the diagnosis.

Treatment. When possible, clinically ill birds should be isolated and provided fresh water; once these measures are taken, birds often recover fully within a few days. Waterfowl should be herded to uncontaminated shores, and carcasses should be removed daily in poultry operations. Antitoxin therapy may be administered for valuable birds or zoo animals, but it is impractical for most production operations or wildlife. Furthermore, antitoxin protection is transient and birds may again become susceptible to BoNT. In broiler outbreaks, antimicrobial therapy may be instituted through watering systems or feed. Administration of bacterin (100 g/ton of feed) or streptomycin (500–1000 g/ton of feed or 1 g/l of water for 3 days) was shown to decrease mortality rates in chickens. Penicillin may also be administered, but a mixed efficacy has been reported with this treatment. Periodic use of chlortetracycline was reported to reduce botulism outbreaks on one poultry farm. Additives such as sodium selenite (6 g/1000 l of water for 5 days) and vitamins A, D3, and E may also reduce mortality. Conversely, elevated iron levels in water or feed may promote the intestinal proliferation of C. botulinum; therefore, citric acid, an iron chelator, may be added to water as a preventative. Further, citric acid may lower the pH of the GI tract, inhibiting the growth of C. botulinum and promoting the growth of normal flora.

Prevention. Immunization with the toxoid vaccine has been explored in broilers, pheasants, and ducks with mixed results. Protection in broilers between 3 and 8 weeks of age was variable after vaccination at 1 and 14 days of age. Chickens are most susceptible to botulism between 2 and 8 weeks of age, and vaccinations to protect this group may be less efficacious due to interference from maternal antibody and immaturity of the immune system. Routine vaccination further increases production costs, and the toxoid may not provide adequate protection against the high doses of toxin obtained from maggot ingestion. Toxoid immunizations are also impractical for waterfowl. Therefore, preventative measures to minimize outbreaks of avian botulism should be aimed at flock and environmental management in both production birds and waterfowl. In broiler outbreaks, the goals are to limit further exposure and eliminate C. botulinum or BoNT from the environment. Unaffected birds should be moved to uncontaminated houses. Carcasses should either be incinerated or buried in a deep hole. Rodents should be eliminated from broiler houses as rodent carcasses may harbor C. botulinum. Chicken houses associated with outbreaks should be emptied and cleaned. All litter should be removed. Houses should be washed with high-pressure steam and cleaned with a detergent agent. A surface-active solution should be sprayed on the interior walls. The walls should then be disinfected with an organic iodine solution or an organic iodine and calcium hypochlorite solution. Twenty-four hours later, the interior walls should be sprayed with 10% formalin. Soil in contaminated areas may also be treated with calcium hypochlorite. Houses should also be sprayed with pesticides to limit flies. Iron levels in feed and water sources should be monitored. Prevention of waterfowl outbreaks is best achieved by reducing the potential for environmental contamination associated with the proliferation of C. botulinum in the carcasses of dead vertebrate and invertebrate animals. Carcasses should be collected and flocks should be herded away from shores associated with outbreaks. Pond management should maintain deep waters, steep banks, and smooth bottoms to prevent deaths of invertebrates and vertebrates. Routine flooding, which may lead to the death of terrestrial invertebrates, should be avoided in areas utilized by waterfowl. Water in wetland areas should be maintained as fresh as possible as oxygen depletion in shallow, stagnant waters leads to aquatic animal die offs. Any factors that may increase deaths in susceptible wetlands, such as overhead power lines, should be removed or avoided. The possibility for transmission of botulism from birds to their predators may exist. Coincidence of avian outbreaks with botulism in omnivorous animals have been documented. For instance, Weiss et al. (1982) reported botulism in a fox and a weasel in association with a waterfowl outbreak. In addition, there have been several reports of canine botulism in hunting breeds. Farrow et al. (1983) reported the occurrence of botulism in three dogs after the consumption of a rotten duck carcass. Outbreaks of botulism (BoNT/C and D) in cattle and sheep have been associated with the feeding of contaminated poultry litter in silage.

No cases of human botulism resulting from the consumption or handling of contaminated birds have been reported although both scenarios have likely occurred. The risk for the human acquisition of botulism from avian species appears to be limited. Although Smart et al. (1980) reported an outbreak of BoNT/C in non-human primates, humans do not appear susceptible to BoNT/C or D following oral exposure. Further, proper cooking of poultry should denature any toxin protein and eliminate the possibility of transmission through consumption.

Bovine botulism. Cattle are susceptible to BoNT serotypes B, C1, and D, and acquire botulism most commonly from the ingestion of preformed toxin in spoiled silage, carrion-laden silage, or silage contaminated with poultry litter. As with horses, toxicoinfectious and wound botulism are also potential routes of intoxication in cattle. All three toxin serotypes (B, C1, D) have been associated with clinical disease caused by the ingestion of spoiled or carrion-laden feedstuffs.

Intoxication with BoNT/B is associated with the ingestion of poorly ensiled or spoiled silage, while BoNT/C1 is associated with the ingestion of carrion- or poultry-litter-laden feedstuffs. Although less frequent, BoNT/D has also been implicated with the ingestion of contaminated silage. Interestingly, intoxication with BoNT/D has also been associated with the ingestion of bones by phosphorus-deficient cattle or cattle with pica. There are numerous studies documenting the association between outbreaks of bovine botulism and the ingestion of improperly ensiled silage or spoiled haylage contaminated with BoNT/B. Wet hay or soil-contaminated hay, wrapped in plastic bags for storage, can provide the ideal moist anaerobic environment for C. botulinum growth. Contamination of a total mixed ration (TMR) with a cat carcass was determined to be the source of a BoNT/C1 outbreak in a herd of adult Holstein dairy cattle in California. The practice of feeding ensiled poultry litter to cattle has also been associated with outbreaks of serotype C1 botulism as documented in an Irish beef herd by McLoughin et al. (1988). Although serotype D is less commonly associated with food-borne botulism, an outbreak occurred on a Canadian feedlot following the feeding of a TMR containing spoiled bakery waste. A separate outbreak of BoNT/D in a Holstein dairy herd occurred where the source was suspected, but not proven, to be contaminated haylage.

Clinical signs. Bovine botulism usually occurs in the context of a herd outbreak. The classical signs of bovine botulism are similar to those observed in horses; however, cattle exhibit a more gradual progression of clinical signs, improving the prognosis and probability of recovery in cattle. Further, at least one study has reported that ruminal microbes degrade BoNT, decreasing the absorption of active toxin in cattle compared to horses. The clinical course ranges anywhere from 2 to 30 days, depending on exposure dose and duration, and the administration of treatment. Early botulism may be confused with milk fever as generalized muscle weakness, increased ataxia, and muscle tremors may occur in both conditions. Cattle with botulism also exhibit depression, dysphagia, decreased tongue and jaw tone, hypersalivation, dehydration, decreased tail tone, decreased pupillary light responses, and mydriasis. Rumen contractions decrease and constipation may develop. Diarrhea and/or putrid smelling feces may also be noted. Cattle with botulism tend to spend significant amounts of time in sternal recumbency. At terminal stages of botulism, cattle are laterally recumbent, exhibit abdominal breathing patterns, and finally succumb from respiratory failure. Vital signs are often normal in early stages of botulism; however, as the disease progresses, increased heart and respiratory rates may be noted, while body temperature may decrease. Recently, a syndrome resembling equine dysautonomia has been described in German cattle and a link to BoNT has been proposed. These cattle may present with a subclinical to chronic "visceral" disease. Non-specific symptoms such as weight loss, decreased milk production, depression, alternating constipation and diarrhea, edema, laminitis, ataxia, retracted abdomen, emaciation, tachypnea, and unexpected death are associated with this syndrome. In cattle exhibiting these symptoms, Böhnel and associates demonstrated the presence of both C. botulinum and BoNT in lower GI tract contents. Further, neither BoNT nor C. botulinum was isolated from asymptomatic herds. This study hypothesized that small levels of C. botulinum colonized the lower intestinal tract and created a low level, chronic exposure of BoNT.

This low level of toxin may not reach the systemic circulation, and thus toxin may only disrupt nearby parasympathetic ganglionic innervation of the GI tract, altering intestinal function. In cases of bovine botulism, clinical pathology may reveal signs of dehydration such as increases in packed- cell volume and total plasma protein. Bicarbonate loss from excessive ptyalism may lead to a metabolic acidosis. Increases in muscle enzymes such as aspartate transaminase (AST) and creatinine kinase may be present due to muscle atrophy or trauma resulting from prolonged lateral recumbency. Electrolyte abnormalities and hyperglycemia may also be detected. One study documented indicators of renal failure in a herd poisoned by BoNT/B; increased ã-glutamyl transpeptidase (GGT), urea, creatinine, and phosphorus were also detected. As with other species, there are no definitive gross pathological or pathognomonic histological signs of botulism.

Aspiration pneumonia and pulmonary emphysema are the most frequent sequelae of botulism in cattle. Other lesions such as gastric ulcerations, thickened intestinal mucosa, hepatic lipidosis, suppurative rumenitis, and renal failure have been documented in concurrence with botulism; however, these findings are not consistent in all cases of bovine botulism.

Diagnosis. Botulism in cattle is usually a presumptive field diagnosis made on the basis of clinical signs and the ruling out of other diseases. Differential diagnoses include hypocalcemia, hypomagnesia, hypokalemia, hypophosphotemia, listeriosis, lead poisoning, polio- encephalomalacia, ionophore toxicity, nutritional or plant toxin induced myopathies, molds, organophosphate poisoning, and tick paralysis. Clinical diagnosis is usually made through the detection of neurological deficiencies in light of relatively unremarkable laboratory diagnostic findings. The neurological examination should assess cranial nerve responses, gait, posture, and attitude. Specifically, a tongue tone test, tongue stress test, and a jaw tone test should be performed. The tongue stress test is performed by placing a hand at the base of the cow's tongue and putting pressure on the tongue followed by an assessment of muscular tone. The tongue tone test is performed as in the horse. Cattle with botulism will exhibit weak tongue strength. The jaw test is performed by grasping the mandible near the symphysis and attempting to move the mandible laterally. This test assesses the strength of the masseter muscles. A"loose" jaw is suggestive of botulism.

Adefinitive diagnosis is made by identifying toxin in the patient's serum, ruminal fluid, or tissues. Identification of BoNT or C. botulinum in suspect feedstuffs previously consumed by clinically ill animals may further support a diagnosis. Isolation of BoNT from the rumen may prove difficult because the toxin is often diluted by rumen contents and/or degraded by ruminal microbes. Similar to botulism in other species, the MBA is the gold standard for a definitive diagnosis in cattle; however, as in horses, the MBA is often not sensitive enough to detect the low levels of toxin in the general circulation. The MBA may also be used to detect toxin in rumen contents, the liver and other organ tissues, milk, or feedstuffs. Due to the relatively slower progression of clinical signs in cattle, diagnostic samples are often obtained long after ingestion of toxin. Thus, the level of toxin in these samples may have fallen below the level of detection. Specimens may also be cultured to isolate C. botulinum. An ELISA test has been developed to detect BoNT/C and D in cattle; however, this test is considered less sensitive than the MBA. ELISA tests for the detection of antibodies to BoNT/C and D in cattle have also been developed.

Treatment. Supportive care is the core of therapy for bovine botulism, and treatment should only be pursued in standing cattle. Affected cattle should be kept in confinement to minimize movement and exertion. Dehydration, electrolyte deficiencies, acid/base abnormalities, and glucose deficiencies should be managed with fluid therapy. Fluids can be administered orally (via an orogastric tube) or IV. Mineral oil or sodium sulfate can be administered with care as cathartics to treat ileus; however, magnesium sulphate should be avoided as it may potentiate muscle weakness. Rumen transfaunation may also be performed. Alfalfa gruels may be administered via an orogastric tube to maintain caloric intake. Equine origin polyvalent antiserum may be administered to cattle. However, anti-toxin therapy may be less efficacious in cattle since most of the toxin will have been internalized into the neuron or degraded by the time the diagnosis is made. Antibiotics may be administered for secondary complications such as aspiration pneumonia; as in horses, those that produce muscle weakness should be avoided. Although toxoid vaccinations for serotypes B, C, and D are administered to cattle in other countries, there are no FDA approved vaccinations for cattle in the United States (www.vmtrc.ucdavis.edu).

Public health. The Food Safety Act of the United States (1990) requires that meat or milk products be withheld from market for a minimum of 14 days after the onset of the last clinical case of botulism cin an affected herd. However, the public health concern for transmission of BoNT through milk appears to be minimal. No cases of human botulism acquired from the consumption of meat or milk from botulism affected cattle have been reported; further, it does not appear that calves acquire botulism through nursing from affected cows. Only a single report to date has been able to detect BoNT in milk from a dairy cow affected with botulism. In this report, BoNT/B was isolated in milk collected from one udder quarter that was simultaneously affected with mastitis. The toxin concentration in the milk was determined to be approximately 104 mouse LD50s. However, the milk did not test positive for C. botulinum bacteria. It is likely that the concurrent mastitis infection enhanced the passage of the rather large toxin protein

(150 kDa) across the normally protective blood:milk barrier by altering its permeability. This is supported by a much earlier report from Moberg and Sugiyama (1980), who isolated BoNT in milk using an infected rat model. Other studies have not been able to detect BoNT in milk from affected cows using either ELISA or MBA techniques. Regardless, the pasteurization process would likely denature any toxin protein that was able to pass into milk, reducing the risk to the consumer. It should be noted, however, that the potential for BoNT contamination of milk may be more relevant in regions where unpasteurized milk is available for public consumption.

Canine and feline botulism. Background. Although carnivores are thought to be more resistant to the development of botulism, cases of canine botulism have been documented in the United States, Great Britain, continental Europe, and Australia. Most reported cases of botulism in dogs result from the ingestion of BoNT/C1-contaminated carrion; however, a few cases of serotype D have been documented in Senegal. Barsanti et al. (1978) described an outbreak of type C1 botulism in a hunting colony of American foxhounds; however, the source of the toxin was not identified. Farrow et al. (1983) described type C1 botulism in three young Australian Cattle Dogs following the ingestion of rotting duck carcasses found around a local Sydney park. Canine botulism has also been associated with the ingestion of contaminated raw meat. Until recently, the only documented cases of feline botulism were experimentally induced; however, Elad et al. (2004) have described a natural outbreak of botulism in eight cats who ingested parts of a pelican carcass contaminated with BoNT/C1. Interestingly, BoNT/C1 botulism has also been reported in lions.

Clinical signs. The onset of canine botulism can occur within hours or as late as 6 days post- exposure. Severe cases are associated with an earlier onset of clinical signs. The course of the disease usually ranges from 12 to 24 days. In the clinical report of an outbreak of feline botulism, clinical symptoms were first noted 3 days post-ingestion of contaminated pelican muscle.

Although 50% of the exposed cats died, those that survived recovered significantly by 6 days post-intoxication. Lower motor neuron dysfunction as well as to a lesser extent, cranial nerve, and autonomic nervous system deficits are observed in canine botulism. Paresis begins in the hind limbs and progresses cranially, ultimately resulting in flaccid muscle paralysis and quadriplegia. Interestingly, dogs with botulism maintain the ability to wag their tail. Tremors of the masseter and temporal muscles may be noticed. Muscle atrophy is variable throughout the course of the disease.

Mydriasis, decreased pupillary light response, decreased palpebral reflexes, and decreased or weak vocalizations may occur. Hyperemic conjunctiva and decreased Shirmer tear tests may be noted. Heart rates and respiratory patterns are variable; however, as abdominal muscle tone diminishes, diaphragmatic breathing may be noted.

Regurgitation, megaesophagus, urinary retention, and constipation are also observed. Secondary complications include aspiration pneumonia, bilateral keratoconjunctivitis sicca, and urinary tract infections. If paralysis progresses to the respiratory muscles, death may occur from respiratory failure; however, death may also result from progressive secondary pneumonia or urinary tract infections. If secondary complications do not arise, the prognosis for canine botulism is good. Recovery occurs in the reverse order from that of the onset of paralysis; cranial nerve function and motor function of the neck and limbs return first. In the one case study of natural botulism in cats, clinical signs were similar to those of dogs. Motor deficits and paresis were noted; however, cranial nerve reflexes were normal. Depression, anorexia, mild dehydration, tachycardia, and urinary retention were also noted. As with EGS in the horse, there has also been speculation of an association between feline dysautonomia ("Key– Gaskell's disease") and BoNT/C. Clinical signs for feline dysautonomia include depression, anorexia, vomiting, regurgitation, mydriasis, constipation, and urinary retention; however, the somatic lower motor neuron paralysis characteristic of classical botulism is not observed. Histological evidence of neuronal degeneration in autonomic ganglia confirms the diagnosis of dysautonomia. Interestingly, BoNT/C was detected in feces, ileal contents, and foodstuffs of cats displaying symptoms of dysautonomia. Further, affected cats had higher levels of anti-BoNT/C and C. botulinum surface antigen IgA in their feces when compared to control cats. Additional studies are warranted to determine the potential role of BoNT/C in feline dysautonomia.

Diagnosis. With the exception of dehydration or secondary infection, the CBC count, blood chemistry, urinalysis and CSF analysis are usually within normal limits in canine and feline botulism. Thoracic radiographs may reveal a megaesophagus and aspiration pneumonia.

Electromyographic (EMG) findings may indicate lower motor neuron disease in clinically ill animals. Decreases in the amplitudes of compound muscle action potentials and motor unit potentials are often detected. Furthermore, fibrillation potentials and decreases in nerve conduction velocity may also be detected. In order to make a definitive diagnosis, toxin must be identified in serum, vomitus or gastric contents, feces, or food samples from animals showing clinical signs. The gold standard MBA appears to have adequate sensitivity for the detection of toxin in canine and feline biological samples or in carrion. It should be noted that the isolation of C. botulinum bacteria through cultures of feces, GI contents, or viscera is not a definitive diagnosis, as this bacterium can be isolated from the GI tract and viscera of healthy dogs. Differential diagnoses for canine botulism should include tick paralysis, polyradiculoneuritis (coonhound paralysis), myasthenia gravis, coral snake envenomization, and the dumb form of rabies. Both the lower motor neuron deficits and EMG findings are similar to those of tick paralysis and polyradiculoneuritis; however, due to its action on cholinergic terminals, botulism also causes cranial nerve and autonomic deficits. The nature of botulism outbreaks to affect multiple animals further differentiates the disease from other causes of lower motor neuron dysfunction.

Treatment. Treatment of canine botulism mainly consists of supportive care. If the ingestion of toxin-contaminated food has been recent, gastric lavage, cathartics, and enemas may be used to decrease toxin absorption from the GI tract. However, as in other species, magnesium sulfate should be avoided. Supplemental fluids should be administered as needed to maintain hydration. Nutritional support via orogastric or parenteral administration may also be needed. Animals should be monitored for aspiration pneumonia due to megaesophagus and decreased gag reflexes. If constipation develops, enemas and stool softeners may be administered. Manual expression of the bladder may be required to decrease the occurrence of urinary tract infections. Topical ophthalmic ointments should be used to prevent corneal ulcers, which may result from diminished palpebral tone and tear production. Adequate bedding and frequent repositioning are necessary to prevent the development of decubital ulcers. In cases where respiration is compromised, mechanical ventilation may be necessary. Antimicrobial therapy may be needed for secondary infections; however, as in other species aminoglycosides, tetracycline, procaine penicillin, metronidazole, aminopyridines, and guanidines should be avoided. Administration of the equine antitoxin in small animals is controversial. By the time clinical signs are noted, antitoxin is likely to be ineffective as most of the toxin is already bound to the nerve cell or has translocated into the neuron. Only the trivalent antitoxin vaccine for serotypes A, B, and E is available in the United States; this antitoxin is less useful for dogs and cats, which are usually affected by serotype C. However, the heptavalent antitoxin is available in other countries. Since adverse reactions to antitoxin may occur, and patients with mild disease often recover with supportive care alone, antitoxin administration is usually reserved for severe cases.

Prevention. Limiting exposure of small animals to carrion, and fully cooking all meat products to be fed to companion animals are simple measures to prevent botulism in these species. Heating foodstuffs to 80°C for 30 min or 100°C for 10 min denatures the toxin protein. Although a toxoid vaccine is available for other species, no such vaccine exists for dogs or cats.

11

ECOTOXICOLOGY

Because of the breadth of this subject, readers are encouraged to capitalize on other sources as well. There are numerous scientific journals that publish papers on a broad array of topics in ecotoxicology, including Ecotoxicology, Environmental Toxicology and Chemistry, Environmental Pollution, Journal of Environmental Quality, Environmental Health Perspectives, Archives of Environmental Contamination and Toxicology, and others. In addition, there are numerous texts available on this subject, including Fundamentals of Ecotoxicology (Newman, 1998), Handbook of Ecotoxicology (Hoffman et al., 2003), and Principles of Ecotoxicology (Walker et al., 2001), to name just a few. There is also a growing number of books that are focused on specific issues or taxa, e.g. Ecotoxicology of Amphibians and Reptiles (Sparling et al., 2000), Radiotelemetry Applications for Wildlife Toxicology Field Studies (Brewer and Fagerstone, 1998), Principles and Practices for Evaluating Endocrine Disruption in Wildlife (Kendall et al., 1998), Wildlife Toxicology and Population Modeling: Integrated Studies of Agroecosystems (Kendall and Lacher, 1994), and many others.

Defining ecotoxicology. Ecology is defined as the science of all living organisms, and all their interactions among one another and with the environment. We define toxicology as the science of all the adverse biochemically mediated effects of all chemicals on all life forms. Newman (1998:13) defined ecotoxicology as the "science of contaminants in the biosphere and their effects on constituents of the biosphere." More recently, Hoffman et al. (2003:1) defined ecotoxicology as "the science of predicting effects of potentially toxic agents on natural ecosystems and on nontarget species." For the purposes of this chapter, we define ecotoxicology as the science of all the adverse biochemically mediated effects of all chemicals on all living organisms including all their interactions within organisms and among species in the environment (Beasley, 1993). As such, ecotoxicology is the most encompassing specialty within the discipline of toxicology. The breadth of ecotoxicological studies is therefore immense. It ranges from: (a) tightly controlled laboratory studies of the pathophysiological effects of single chemicals on one strain of a native or surrogate microbe, plant, or animal species, to (b) studies of the transport and fate over time of one or more contaminants in the environment, (c) studies of the pathophysiological effects of single chemicals on a suite of organisms with various inanimate factors in microcosms or mesocosms, (d) studies of toxic interactions among groups of contaminants on one or more organisms in concert with inanimate factors in microcosms or mesocosms, and (e) small- and large-scale field research on impacts of one or more chemicals on suites of abiotic and biotic components in the environment.

Successes and challenges. The necessity of an interdisciplinary approach and a combination of field and laboratory studies was illustrated in early ecotoxicological studies of spent lead shot pellets. Although lead poisoning resulting from lead shot ingestion was known to be a mortality factor in waterfowl in the late 1800s, the scope of the problem and the nature of the syndrome were not understood until Dr Frank Bellrose began comprehensively investigating this issue in the 1940s and 1950s. Bellrose (1959) and colleagues from the Illinois Natural History Survey addressed: the geographic distribution, frequency, and magnitude of die-offs; the accessibility of spent shot; ecological determinants of differential susceptibility; dose response; clinical signs in moribund individuals; lesions in exposed specimens; and the influence of diet on toxicity of lead shot. Others began examining lesions of lead poisoning in waterfowl and other species, and developing diagnostic criteria to support diagnoses of lead poisoning in birds. Their findings and others that followed eventually prompted restrictions and later a complete ban on lead shot for hunting waterfowl in the United States, Canada, and several other nations.

Government regulations in these countries required the testing of potential non-toxic substitutes for lead shot. Some, such as iron (steel), bismuth, and tungsten, were deemed functionally non-toxic after ingestion by waterfowl and were approved for use in waterfowl hunting, whereas other candidates such as zinc were found to be toxic. Such restrictions soon led to decreased lead exposure and reduced mortality due to lead shot ingestion in some duck species. The agricultural application of dichlorodiphenyltrichloroethane (DDT) began shortly after World War II. At that time, large amounts were sprayed on farm fields in the United States, Great Britain, and elsewhere. After noting increased incidence of raptor nests in the United Kingdom with broken eggs, Ratcliffe (1967) examined the thickness of eggshells, using the ratio of shell weight/size of eggs of three species of raptorial birds collected from 1900 through 1967. He found a precipitous decline in shell weight/size ratio from 1946 to 1950, after which shell thinning persisted. Shell thickness of eggs collected during 1900–1946 was significantly greater than of those collected during 1947–1967. Ratcliffe (1967) noted that the sharp decline in eggshell thickness coincided with beginning of widespread agricultural use of DDT (1945–1946), and Jefferies 1967 linked DDT and delayed ovulation in birds (suggesting a hormonal mechanism, i.e. endocrine disruption). This work was followed by many corroborative field and laboratory studies as discussed below, which, taken together provided an overwhelming weight of evidence as to the insidious and devastating impacts of DDT on avian reproduction. Acombination of pest resistance (resulting in development of alternatives), decades of accumulated evidence on environmental effects, and publication of Silent Spring by Rachel Carson (1962) were necessary to drive a US ban on DDT use in 1972.

- Many other developed nations banned these compounds also, and pressured other, less-affluent nations to discontinue or curtail its use. Recently, the World Health Organization and Stockholm Convention on Persistent Organic Pollutants condoned the use of DDT for indoor mosquito control in tropical malaria-endemic (World Health Organization, 2005).
- Dichlorodiphenyldichloroethylene (DDE), the most persistent, bioactive environmental product of DDT, though declining in regions where its use was discontinued, is still widely disseminated in the environment.

The field of ecotoxicology has: helped terminate or reduce the manufacture and dissemination of some major environmental pollutants [e.g. lead, DDT, polychlorinated biphenyls (PCBs)]; prompted development of new, manmade chemicals of lower environmental risk (e.g. insect growth hormone mimics such as methoprene); helped applied ecology become problem driven; and is an important component of conservation medicine (Munson and Karesh, 2002; Tabor, 2002). In recent decades, ecotoxicology has begun to be applied in conjunction with other components of ecological restoration science to help re-establish ecological health in previously contaminated systems (Cairns, 2003; Linder et al., 2003). Despite benefits from ecotoxicology and other sciences that support rational actions to enhance ecosystem health, in many ways we have witnessed nearly 45 additional years of global mismanagement of ecological resources since the publication of Silent Spring in 1962. Although the general public in developed and developing nations has become more environmentally aware in recent decades, powerful economic interests e.g. mining, energy generation, chemical, construction, and agricultural and forestry industries have often hampered the application of sound ecological and ecotoxicological science to resource extraction and development. Decisions sometimes fail to address long-term environmental needs because of misguided political pressure. Unlike progress seen recently in Europe, the United States has failed to maintain leadership in environmental stewardship with regard to contaminants. In fact, no new, major legislation in support of ecosystem health has been passed by the US Congress since 1990.

Sources of pollution. Myriad toxic substances are released into our environment daily. Toxic agents are either deliberately manufactured (pesticides, drugs, construction chemicals, household chemicals) or accidentally produced (by-products in final formulations or in gaseous, particulate, liquid, and solid waste streams). Inappropriate siting and operation of chemical manufacturing and waste storage facilities can lead to "accidental" releases of chemicals into groundwater, surface water, soil, and air. Although it is increasingly recognized as ill-conceived, chemical and textile

manufacturing plants, refineries, smelters, chemical storage facilities, and electricity generating plants were often located near waterways for ease of transportation of raw materials and finished products, because they use large quantities of water in manufacturing, and/or to allow for dissemination of wastes. Such siting predisposes to deliberate (sometimes permitted), accidental, and even malicious releases of toxic chemicals into aquatic ecosystems. Environmental contamination can emanate from a definable location (point-source pollution), such as atmospheric emissions from the stacks of a coal-burning power plant or liquid effluent from the discharge of a sewage treatment plant. Contamination can also be more widespread, emanating from a larger surface area (non-point source pollution), such as pesticides washed off the landscape by precipitation or airborne effluents from a host of automobiles.

Chemical disasters. The stewardship of chemically mediated disasters ranges from responsible and efficient to neglect and indifference. Localized disasters involving a single, acutely toxic, shortlived chemical (e.g. a volatile solvent released from damaged rail tanker) can be relatively easy to address. Ecological recovery is also more likely when ecological resources (locally adapted microbes, plants, and animals) from the periphery are intact; local emergency agencies are well equipped, staffed, and responsive; funding is available from responsible parties or government to support containment, cleanup, and ecological rehabilitation; and the public is strongly engaged. Conversely, recovery is less likely when releases involve multiple long-lived chemicals (e.g. complex, halogenated, higher molecular-weight wastes) spread over a wide area when ecological resources from periphery are insufficient to support recovery, when responsible parties deny involvement, when neither the responsible parties nor the governments involved put forth needed funds, and the public is largely disengaged. On March 24, 1989, the oil-tanker ship the Exxon Valdez hit a reef, spewing nearly 11,000,000 gallons of crude oil into the pristine Prince William Sound in Alaska. Exxon Corporation (now Exxon Mobil), the ship's owner, the Alyeska Pipeline Service Company, a consortium of oil companies with interests in Alaska, as well as state and federal disaster response planning, were criticized (Exxon and Alyeska were eventually sued) as being wholly unprepared to provide a quick and effective response for such a disaster, as was required by law. Containment and cleanup were to begin within 5 h, but crews and equipment did not begin arriving for 10–12 h after the incident, while oil continued to gush from the leaking tanker. Confusion, poor communications and coordination, lack of sufficient personnel, equipment and supplies for a spill of that magnitude, and the remoteness of the location hampered efforts to contain and begin effective cleanup during the first 2 days after the ship ran aground, after which rough seas created further difficulties. Although the ecological and socioeconomic costs were tremendous, and could have been lessened, the response among the public, conservation organizations and agencies, academic institutions and media was unprecedented for an environmental disaster. Frustrated with a lack of action on the part of industry and government, local citizens began to clean up what oil they could with resources at hand. Within hours of the spill, scientists arrived to begin documenting the environmental impacts of the spill. Particularly sensitive environments were quickly identified and prioritized for protection and cleanup. Crews began to arrive from throughout the world to help clean oil-covered beaches, seabirds, and sea otters. An estimated 36,000 birds and 1000 sea otters eventually were treated for oil exposure; however, untold numbers of fish and wildlife died from direct impacts of oiling.

According to the Exxon Valdez Oil Spill Trustee Council, the populations of only 4 of 11 bird species impacted were recovering or had recovered as of 2002. Impacts to sensitive intertidal communities may not be fully realized for many years to come. The global news media followed many phases of the cleanup and recovery operations, focusing attention on the environmental and economic costs of this disaster, as well as on Exxon's initial response and failure to accept responsibility for the accident. The Oil Pollution Act of 1990, in response to the Exxon Valdez disaster, required increased spill preparedness and the phase-in of double-walled tanker ships, which greatly reduce the chances of a spill should a tanker run aground or hit an iceberg or reef. Unfortunately, the petroleum industry has found loopholes in the law and effectively delayed the deadline for new orretrofitted tankers until 2015. In the meantime, sensitive marine environments remain at high risk from oil spills.

Ecotoxicological stress in concert with other mechanisms of ecosystem disease. Aquatic systems throughout the world are increasingly under stress from human activities. Wetlands, lakes, streams, rivers, and estuaries are being degraded due to loading with nutrients, metals, salts, synthetic organic chemicals, and combustion products, as well as a wide array of other anthropogenic stressors. For example, many of the world's coral reefs are being degraded through the use of destructive fishing and collecting methods, eutrophication, increasing ocean temperatures, and coral diseases produced by bacteria, fungi, and cyanobacteria. Although the role of environmental degradation in producing coral disease remains unclear, the vast majority of impacted reefs in the Caribbean were near areas of human activity, and in the Netherlands Antilles, changes in the structure of coral reef bacterial, including cyanobacterial (blue-green algal), communities were observed in proximity to large inputs of contaminants. Surface runoff, coupled with stream channelization and subsurface tiling of fields, decreases groundwater recharge and increases the "flashiness" of streams, sometimes resulting in prolonged, catastrophic flooding. Rapid flow of water due to runoff from cropland, cleared forests, lawns, roofs, stockyards, roads, parking lots, demolition and construction sites, and poorly managed industrial and dump sites also carries pulses of complex pollutant mixtures to water bodies. For example, precipitation events can mobilize pesticides from soils into runoff and increase concentrations in surface waters. Lack of water retention, due to a diminished wetland base, results in abnormally low flows during dry periods, resulting in decreased dilution and thus increased concentrations of contaminants. Reduced flows can also result in increased water temperatures in summer, thereby decreasing the oxygen-holding capacity of the water. Removal of tree canopies through logging, development, or other activities also increases stream temperatures to the detriment of aquatic species requiring cool, highly oxygenated water, such as trout and salmon. In addition to being sources of toxicants, road and bridge construction, deforestation, agriculture and mining erode soils and increase sedimentation of waterways.

High suspended- sediment loads increase turbidity, shading plants so that primary productivity from photosynthesis, biomass, structure, and dissolved oxygen are reduced. Excess suspended sediments in waterways can also degrade spawning habitat, produce lesions in gills, and modify animal behavior, resulting in reduced fish productivity and altered fish community assemblages. Prior to the advent of modern intensive agriculture, farms consisted of a mosaic of small fields of vegetables, grains, and forage crops, interspersed with fencerows, woodlots, pastures, and wetlands. Crop rotations offset the need for fertilization and pest control. By contrast, modern agricultural and forestry practices have produced vast monocultures of genetically similar plants that are susceptible to attack by plant pathogens such as fungi and herbivorous insects, triggering the use of toxic pesticides. Such monocultures deplete nutrients, prompting fertilization which increases the availability of free nutrients. High levels of nutrients such as nitrogen can be directly toxic (ionic forms, e.g. nitrite or nitrate) or have indirect impacts, e.g. cause harmful algal blooms that may produce toxins and usurp dissolved oxygen.

Modern agriculture and forestry, along with urbanization and suburban sprawl, have created a patchwork of habitat islands in a largely inhospitable landscape. Biota inhabiting such islands are prone to extirpation through environmental perturbations such as catastrophic weather events; competition with invasive, exotic species, habitat degradation that prompts starvation, desiccation, or predation; and mortality or reduced productivity via exposure to environmental contaminants. This loss of species from habitat patches, combined with reduced recolonization rates, can result in local or regional reductions in populations, leading to extirpation, and increasing the risks of extinction events. Anthropogenic air pollution has been reported since 13th century London, when smog from coal fires blanketed the city; and the situation worsened with the Industrial Revolution of the 1700s. An estimated 12,000 people died in London during the Great Smog of the winter of 1952. Pollutants from tall stacks of power plants may travel long distances, harming the environment far from the source. In addition to power generation, other major anthropogenic sources of atmospheric pollutants include waste incineration, oil refinement, industrial emissions, fires from slash and burn agriculture; large wildfires that follow misguided forest-fire suppression, and transportation sources. Widespread deforestation, especially in the tropics (the "lungs of the earth"), and coastal pollution that limits phytoplankton photosynthesis reduce the ability of the environment to remove CO2 and clean the air.

Excessive reliance on fossil fuels (natural gas, petroleum, coal) and clearing of forests are increasing CO2 and methane levels, which trigger global climate change. Small changes in average temperatures appear to be increasing the rate of glacial retreat and causing reductions in ice/snow packs, and increasing the frequency of extreme weather (e.g. drought, flooding, and hurricanes). Rising global temperatures are producing shifts in species tolerating warmer temperatures, with cold-dependent species moving nearer to the poles and into higher elevations. The United States is the leader in greenhouse emissions, according to the Oak Ridge National Laboratory. North American releases amounted to 1.73 billion tons in 2002 (approximately 92% of this generated in the United States), which was 26% of the world total. Half of the estimated 290 billion tons released due to human activities since 1751 have been since mid-1970s. Carbon dioxide emissions in the United States increased an estimated 9-fold during 20th century. Of even greater concern at present is China, which has a much higher human population and has been rapidly industrializing, culminating in increased carbon dioxide emissions in the same time frame by 6000-fold.

Environmental and ecotoxicological problems related to fossil fuels go well beyond the aforementioned oil spills and global warming from carbon dioxide. Placing oilfields in sensitive arctic and marine environments, especially offshore in the Arctic, may have immense adverse impacts on marine mammal populations and indigenous human communities. Emissions during oil refining, and polycyclic aromatics produced through combustion from petroleum and coal have caused widespread pollution. Many of the polycyclic aromatic hydrocarbons are oxidized by P450 enzymes in lung, liver, and other tissues to form reactive epoxide derivatives that form adducts to DNA, resulting in mutations, with the potential to cause cancers as well as heritable defects. Among the best known polycyclic aromatic hydrocarbons is benzo(a)pyrene. Also of concern are volatile organic compounds (VOCs) such as benzene. Both benzo(a)pyrene and benzene known human carcinogens. Aluminum smelters using obsolete technologies have also polluted the environment with polycyclic aromatic hydrocarbons, and were incriminated in digestive tract cancers of beluga whales Martineau et al. (1991, 2002). Through emissions of sulfur dioxide, coal-fired power plants in the United Kingdom have affected the environments of Sweden and Norway; and likewise those in the Midwestern United States have impacted lakes and forests in the northeast part of that country as well as Canada. Oxidized nitrogen (NOx) from burning gasoline, diesel fuel, heating oil, natural gas, and coal and oxidized sulfur from refining oil and burning coal form strong mineral acids when combined with precipitation. This can lead to changes in soil and water chemistry, including solubilization of aluminum, with adverse impacts on soil and aquatic flora and fauna.

Aluminum is a ubiquitous constituent of soils, and excess concentrations in solution can precipitate out on gills of fishes and aquatic invertebrates to impair oxygen utilization. Highly acidic waters with excess aluminum can also reduce calcium uptake via gills, impacting growth and development in aquatic and invertebrate species (Malley and Chang, 1985). Semi-volatile compounds such as PCBs are released into the atmosphere from soils at contaminated locations and are transported great distances before being deposited, often over cool surfaces such as large bodies of water. Thus, large lakes, seas, and oceans have become repositories for such chemicals. This process largely accounts for the comparatively high concentration of persistent organochlorines in polar ocean foodwebs. Other semi-volatile chemicals, such as halogenated dibenzodioxins and halogenated dibenzofurans, are toxic by- products of the incineration of solid wastes and manufacturing processes such as the Kraft paper bleaching process. Such compounds have multiple toxic effects including cognitive impairment, liver damage, and endocrine disruption.

Considering that milliseconds can greatly influence the likelihood of successful predation as well as avoiding becoming prey, that overall health is essential for competition to be among the breeding population in the wild, that high functioning is needed by many parents to feed and protect the next generation until it can survive on its own, it should be no surprise that biodiversity is being harmed not only by habitat loss, climate change, exotic species introduction, pathogen pollution, and noise pollution, but also by a host of chemical contaminants – both manmade and produced by anthropogenically altered populations of microbes, plankton, plants, and animals.

Environmental toxicology and contaminants

Nutrients. Nutrients, primarily nitrogen and phosphorus, are released into the environment via chemical fertilizers, manure/sewage effluent, burning of fossil fuels, fires, and industrial processes such as pulp/paper milling and nitric acid production. Globally, the rate of transfer of atmospheric nitrogen (N) to the available nitrogen pool has doubled due to anthropogenic activities (Vitousek et al., 1997). Agriculture is the primary source of excess nutrients in the environment, both through soil management and livestock operations. World consumption of fertilizer in 2005/2006 totaled 154 million tons [International

Fertilizer Industry Association, Statistics: Total fertilizer nutrient consumption _ N + P2O3 + K2O (million tons nutrients)]. Fertilizer use in developed nations has plateaued, reaching a point where further inputs do not increase yield. The United States alone used 21 million tons of plant fertilizers in 2003, primarily as N, P, and K, which is about the same level as in 1976 (USDA Economic Research Service - US consumption of nitrogen, phosphate, and potash for 1960–2003). However, as populations and associated demand for cereal grains in developing nations grow, fertilizer use will likely increase. Inorganic nitrogen, including ammonia, ammonium, nitrate, and nitrite, is toxic to aquatic life including fish, invertebrates, and amphibians. Ammonium nitrate fertilizer produced toxic effects in American toad (Bufo americanus), western chorus frog (Pseudacris triseriata), northern leopard frog (Rana pipiens), and green frog (Rana clamitans) tadpoles at concentrations found in the agricultural environment (Hecnar, 1995). High nitrogen concentrations contributed to lower amphibian reproductive success (Knutson et al., 2004) and species richness in wetlands in agriculturalized areas. Rouse et al. (1999) reported that 20% of over 8000 water samples from the Great Lakes Watershed exceeded concentrations considered toxic to amphibians. Nutrients may also interact with pesticides to produce toxic effects different than those of either alone. For example, a mixture of atrazine and nitrate had greater than expected impacts (synergistic effect) on snout- vent length of Xenopus laevis tadpoles (Sullivan and Spence, 2003). Conversely, the addition of higher nitrate concentrations to the same atrazine concentrations resulted in greater snout-vent lengths, an antagonistic effect. Much work remains to be done in examining the effects of complex mixtures of chemicals on amphibians and other biota.

In addition to direct toxic effects on amphibians, large inputs of nutrients into water bodies increase algal growth, which, at high levels, can result in decreased dissolved O2 concentrations through algal respiration and decay, and shading, reducing light needed by submerged aquatic vegetation. Metabolism of dead vegetation further reduces dissolved oxygen concentrations, and a reduction in aquatic plants results in reduced nitrogen storage. Excess nutrient input promotes increased snail populations (Chase, 1998), due to increased algal and periphyton resources that comprise the main food source for snails. Johnson et al. (2002) found a positive relationship between the frequency of limb malformations and infection with Ribeiroia trematodes. Trematode infections may produce limb deformities or potentially debilitating kidney infections in frogs. Infections by Ribeiroia and Echinostoma at the early stage in development can vastly reduce survival of amphibian tadpoles. Moreover, sublethal trematode infections may also make tadpoles less tolerant of other stressors. Higher snail densities coupled with increased waterbird (definitive hosts) use of a diminished wetland base and the reduced structural complexity of habits through use of herbicides and fertilizers appear to result in higher trematode infection rates and intensities. The presence and abundance of Ribeiroia infection were associated with the presence and abundance of snails that function as intermediate hosts for larval trematodes. These parasites undergo extremely high levels of asexual reproduction in snails.

Furthermore, the presence and abundance of planorbid snails was associated with manmade wetlands (e.g. ponds) and higher orthophosphate concentrations. Eutrophication can also promote growth of toxigenic algae such as certain blue- green algae (cyanobacteria), which produce neurotoxins, such as anatoxin-A, anatoxin- A(s), saxitoxin, and neosaxitoxin, and the yet-to-beidentified toxin believed to account for avian vacuolar myelopathy (AVM). A current theory is that acyanobacterium of the order stigonematales, a cyanobacteria that grows on the exotic invasive Hydrilla, which is consumed by avian species such as coots underlies AVM. Eagles are exposed via the food web. In addition to producing neurotoxins, cyanobacteria may produce hepatotoxins, such as microcystins, nodularin, and cylindrospermopsin. Both neurotoxic and hepatotoxic cyanobacteria have been implicated in die-offs of lesser flamingos, an obligate algal feeder. However, additional research will be needed before the importance of cyanobacterial toxins in population trends of lesser flamingos can be deduced. The causes of harmful marine algal blooms, which appear to be an emerging problem, increasing in frequency and distribution worldwide, are less well understood. For example, diatoms have shown to produce neurotoxic domoic acid. Other phytoplankton such as dinoflagellates have been shown to produce saxitoxins, ciguatoxins, and brevetoxins. Shellfish and finfish may accumulate these toxins and cause secondary poisoning in humans, and have been implicated in mass mortalities of bottlenosed dolphins (Tursiops truncates) on the Atlantic coast of the United States. Brevetoxin exposures via aerosols and ingestion of sea grasses contaminated with toxins from dinoflagellates seem to account for a number of major death losses in manatees in Florida over the past decade.

Organochlorine compounds. World pesticide use in 2001 totaled 2.3 billion kg. In that year, the United States alone used about 546 million kg, including herbicides (44%), insecticides (10%), fungicides (6%), and others (40%, including nematicides, fumigants, rodenticides, avicides, molluscicides, and piscicides) (US Environmental Protection Agency, 2000-2001 Pesticide Market Estimates). Agriculture accounts for the largest proportion of total pesticides applied (75- 80% in United States, primarily glyphosate, atrazine, and metam sodium). The widespread use of organochlorine insecticides (OCIs), DDT, aldrin, chlordane, dieldrin, heptachlor, and others began after WWII. These compounds were initially a boon to agriculture and public health, as they were both effective and persistent. However, pests soon began to develop resistance, requiring higher application rates and the continual development of new compounds. DDT increases sodium conductance across nerve cell membranes, increasing excitability and resulting in tremors and the potential for seizures.

The cyclodiene pesticides such as aldrin, chlordane, dieldrin, endrin, and heptachlor, the structures of which differ markedly from DDT, exert their influence on ã- aminobutyric acid (GABA) receptors in the brain and are more acutely toxic than DDT. The slow biodegradation rate and high lipid solubility of many organochlorine compounds, which prompts accumulation in adipose tissue, causes vast increases in the net uptake as the chemical is passed from lower to higher trophic levels, resulting in biomagnification, or successively greater residue concentrations in consuming organisms. Predatory animals thus often accumulate very high concentrations, and the nursing young of top predators feeding in contaminated aquatic food webs are often massively exposed. Among the species most at risk are marine mammals, the young of which may consume milk with extremely high fat content (e.g. up to 60% fat). 4,4'-(2,2,2- trichloroethane-1,1-diyl)bis(chlorobenzene), best known as DDT, was used extensively in agriculture and public health, at peak production in 1962, 85,000 tons were produced in the United States alone, and an estimated 675,000 tons were applied in that country during 1945– 1972 [USEPA-DDT Regulatory History: A Brief Survey (to 1975)]. This pesticide was considered safe, as it has a low acute toxicity to most bird and mammal species. However, over time, it became apparent that DDT had insidious effects; its metabolic by-product p,p'-DDE inhibits prostaglandin synthetase, resulting in reduced calcium uptake by the shell gland mucosa (Lundholm, 1993, 1997) during eggshell formation. This results in thinning, so that adults crush eggs during incubation (e.g. herons and egrets; sparrowhawks). Eggshell thinning resulting from DDE exposure, and subsequent impacts on productivity, has been observed in a variety of species of carnivorous birds, such as: herons and other wading birds; eagles; falcons (Newton et al., 1989); hawks; gulls; and others. Some of the other OCIs, such as the cyclodienes aldrin and dieldrin, resulted in direct mortality in predatory birds, such as sparrowhawks and kestrels. Such mortality, when combined with sublethal effects such as starvation and accidents due to neurological deficits and reduced productivity due to eggshell thinning caused by DDE exposure, soon led to population declines in such species. DDE also may co-occur in egg contents with PCBs, which are embryotoxic and therefore contributed to decreased productivity in many waterbird populations. Exposure of Great Lakes fish-eating birds to organochlorine compounds including dioxins and dioxinlike compounds such as PCBs may result in embryo and chick mortality, growth retardation, and deformities associated with edema, hepatomegaly, gastroschisis, and other lesions, otherwise known as the Great Lakes Embryo Mortality, Edema, and Deformity Syndrome, or GLEMEDS. Populations of other vertebrates apart from birds may also have been negatively impacted by DDT. For example, Reeder et al. (2005) examined spatial and temporal patterns of intersexuality in cricket frogs collected in Illinois during 1852–2001.

Compared with the preorganochlorine era (1852–1929), the percentage of intersex cricket frogs increased during the period of industrialization and initial uses of PCBs

(1930–1945), was highest during the peak of manufacture and use of DDT and PCBs (1946–1959), began declining with increased public concern and environmental regulation that led restrictions and eventual ban on the use of DDT in the United States (1960–1979), and continued to decline through the period of gradual reductions in environmental residues of organochlorine pesticides and PCBs in the midwestern United States (1980–2001).

Cholinesterase-inhibiting pesticides. Organophosphorous (OP) and carbamate insecticides such as diazinon, chlorpyrifos, malathion, and carbofuran are cholinesterase-inhibiting compounds. As such, they prevent breakdown of acetylcholine after transmission of a nerve impulse. The first OP cholinesterase inhibitors were developed as chemical weapons during WWII. After WWII, the use of OP compounds increased as pests developed resistance to OCIs and their use was restricted.

The OP compounds are not persistent in the environment in comparison to OCIs, but they are highly acutely toxic and broad spectrum; human deaths from mishandling occur annually.

Secondary poisoning can occur when predators consume invertebrates or vertebrate wildlife that have been poisoned by these chemicals. In one of the best-documented secondary-poisoning events, approximately 6000 wintering Swainson's hawks (Buteo swainsoni) were poisoned in Argentina during 1995–1996 after feeding on grasshoppers sprayed with the OP insecticide monocrotophos. In another incident, hundreds of laughing gulls (Larus atricilla) were poisoned after consuming insects killed or debilitated by application of parathion to cottonfields; this is especially noteworthy in that the mortality included nestlings which died after being fed contaminated prey by adult gulls. It is likely that large numbers of individual poisonings of birds and other species by OP insecticides go unnoticed, following both the intended use and misuse of licensed products. Organophosphorous insecticides have also been used illegally as avicides to protect grain crops from bird depredations, and in baits to kill livestock predators such as coyotes. Such misuse has resulted in additional mortalities from secondary poisoning of predatory birds and mammals eating carcasses of poisoned animals. Two of the three previously most widely used OP insecticides, diazinon and chlorpyrifos (the third is malathion), are now severely restricted in the US. The removal of diazinon and chlorpyrifos from household products early in the 21st century led to almost immediate reduction in concentrations in umbilical cord blood and was associated with increased birth weights of babies of low-income families. Granular carbofuran pellets have been consumed by birds that apparently mistook them for seeds; and one pellet of this carbamate insecticide is enough to kill a bird. Carbofuran was the insecticide most commonly associated with wildlife pesticide poisoning events (e.g. red-winged blackbirds (Agelaius phoeniceus), Augspurger et al., 1996; bald (Haliaeetus leucocephalus) and golden (Aquila chrysaetos) eagles, coyotes

(Canis latrans), and red foxes (Vulpes vulpes), Wobeser et al., 2004; bald eagles and red-tailed hawks (Buteo jamaicensis), Elliot et al., 1996; buzzards (Buteo buteo), Dietrich et al., 1995; herons, Hunt et al., 1995), and was even implicated in declines of some species, e.g. burrowing owls (Athene cunicularia). The phase-out of the granular formulation began in 1991 and registrations for most solid forms have been cancelled. In addition to direct toxic effects to avian wildlife, insecticides may reduce populations of non-target invertebrates used as food by birds, which may require increased foraging effort or prey switching by adults with young to feed. Additional research in this topic area is needed.

Other insecticides. Pyrethrins from chrysanthemums and structurally related synthetic insecticides (pyrethroids) bind to sodium channels preventing closure, or to GABA-mediated chloride channels inhibiting chloride influx at sodium channels resulting in tremors or seizures. These chemicals are generally of low toxicity to mammals and birds; however, they are highly acutely toxic to arthropods, including non-target taxa such as spiders, parasitoids, and bees. Also, they are often highly acutely toxic to tadpoles and fish. Although early formulations had a short half-life – measured in hours in direct sunlight – newer compounds are designed to be more photostable, extending their effectiveness for up to 10 days out-of-doors. Unfortunately this greater stability increases environmental risks associated with their use. There are a variety of other classes of pesticides used for control of insects, arachnids (e.g. mites), and other "pests." Some are nerve poisons, others inhibit energy production, chitin synthesis, water balance, or growth. A number of lower-risk pesticides include microbes (e.g. the soil bacterium Bacillus thuringiensis (Bt) for mosquito control), microbial products (e.g. spinosyn to control caterpillars is a fermentation product of a soil actinomycete), a range of other "natural chemicals" (e.g. pheromones, florals, sulfur), and "plant-incorporated protectants" (e.g. transgenic crops that may produce toxins originally synthesized in bacteria, such as Bt). Many of these new methods of pest control are more environmentally benign alternatives to traditional insecticides. However, they are intended to disrupt normal physiological processes, and thus, each carries potential environmental risks, which are only partially understood.

Herbicides. During 2000, herbicides represented 36% of the world pesticide market; the agricultural sector accounted for 78% of the herbicide use in the United States in 2001. Atrazine is perhaps the world's most used agricultural herbicide, with an estimated 74–80 million pounds applied in the United States alone in 2001. Atrazine or its degradation products have been found in air, rainwater, surface water, and groundwater. Herbicides are often considered benign with regard to impacts on animals; however, these compounds can have toxic effects at concentrations found in the environment. Atrazine has come under increased scrutiny due to evidence that it may produce endocrine-disrupting effects in amphibians, and has been banned by several European Union countries. Reeder et al.

(1998) associated detection of atrazine with the finding of intersex gonads in cricket frogs (Acris crepitans). Hayes et al. (2002a, b, 2003) found that concentrations of atrazine as low as 0.1 parts per billion (ppb) affected gonadal development, produced hermaphroditism, and reduced the laryngeal muscle (needed for calling to attract mates and ward off potential competing males) in larval male frogs. At a range of test concentrations that can be found in the environment, atrazine was associated with reduced survival of tadpoles, increased length of larval period, reduced size at metamorphosis, gonadal dysgenesis, and reduction of spermatogenesis. However, there have been contrasting results from others studies, and authors have suggested that intersex may be found in unexposed frogs. However, in most such reports to date, the animals in control groups or at reference sites were also exposed to low concentrations of atrazine. Additional, carefully designed studies are needed. Glyphosate is a broad-spectrum herbicide that has become the most used herbicide in the United States. This chemical has a low toxicity to mammals; however, surfactants used in some formulations to cause glyphosate to adhere to plant surfaces greatly increases the chemical's toxicity. Thus, some formulations (those not intended for aquatic vegetation control) can result in direct mortality of larval and juvenile amphibians and subsequent loss of biodiversity. Glyphosate readily adsorbs to soil or is rapidly degraded by bacteria, and therefore is not highly bioavailable to animals and has low potential for runoff. Nevertheless due to its high water solubility, glyphosate is sometimes detected in surface waters due to runoff. Moreover, some formulations are intended for control of aquatic vegetation.

Due to its effectiveness and non-specific herbicidal properties, use of glyphosate to control undesirable plant species put, non-target plants at risk. Studies conducted by our laboratory have revealed a complex set of interactions involving herbicides that affect the health of amphibians inhabiting agriculturalized regions. Beasley et al. (2005) studied impacts of herbicides on trematode infections on frogs. They found that, in herbicide-impacted farm ponds, recruitment of juvenile cricket frogs was reduced and trematode infections in the frogs were greatly increased.

The reduced recruitment might have resulted from deaths due to the infection with trematode cercariae, because other species of frogs died when early stages were exposed to the parasite. Sousa and Grosholz (1991) had previously suggested that a more complex habitat structure impedes parasite transmission, and Beasley et al. (2005) hypothesized that severe trematode infections noted in their studies were due to interacting factors related to fewer plants in the water. In such simplified ecosystems, predation would likely be facilitated so that the survivors would receive a greater infective load from trematodes in the water. Similarly, the motile cercariae would likely have less difficulty finding the tadpole intermediate host thereby facilitating infection. Herbicide contamination of water bodies may also reduce dissolved oxygen and the algal food needed by tadpoles, which may slow the growth of tadpoles, possibly delaying metamorphosis in some species or reducing size at metamorphosis, which may reduce fitness.

Both hypoxic water and reduced food may stress the tadpoles, so that they are less able to avoid cercarial infection and encystment. After a reduction in algal and macrophyte communities due to herbicide contamination, there is likely to be a subsequent rebound to produce higher than normal algal concentrations. This is because the competition for nutrients among surviving plant species is likely to be greatly reduced. Although algal food during a rebound might help later season tadpoles, it might also provide food for snails. Under conditions where snail numbers increase relative to tadpole numbers – and considering the immense amplification of infective load related to asexual reproduction of trematodes within snail intermediate hosts – it seems plausible that some wetlands impacted by herbicides may serve as sources of super-infection of tadpoles. This combination of events and nutrient loading as previously discussed may help explain a number of the outbreaks of supernumerary limbs in frogs at levels noted first in the 1990s that were previously seen only infrequently. In addition, pesticide exposure can lead to immunosuppression and increased susceptibility to trematode infections in tadpoles and frogs. Such stressors may result in prolongation of tadpole stage; longer time in the earlier tadpole stages may result in greater lethality due to trematodes, and higher body burdens of pesticides to pass to avian predators. In addition to herbicidal impacts on vegetation and periphyton, exposure to insecticides might influence trematode infections by impacting populations of benthic and plankton predators of cercariae. Members of our team (Labak and Schotthoefer, unpublished data) have demonstrated that invertebrates such as hydra, copepods, daphnids, and especially damselfly larvae and dragonfly larvae attack and consume cercariae in a controlled setting.

Insecticidemediated changes in invertebrate community structure might therefore release cercariae or other trematode life stages from this predation pressure. Whether these "micropredators" are important in determining infective loads of cercariae in natural and human-altered aquatic environments remains to be thoroughly investigated. Herbicides typically occur in the environment along with other anthropogenic chemicals. Studies examining mixtures of herbicides and insecticides have found additive, synergistic, or even antagonistic effects on organisms, depending on the chemicals used, test organisms, and conditions. Other stress-ors may enhance the toxicity of pesticides. For example, Relyea (2005c) found that predatory stress increased the lethality of Roundup® (a popular glyphosate product) to larval amphibians. As with many chemicals, there is a need for further examination of the environmental distribution, fate and toxicity of herbicide and their transformation products, considering both direct and indirect adverse effects.

Rodenticides. Anticoagulant rodenticides such as brodifacoum inhibit blood clotting through inhibition of vitamin K reductase, an enzyme essential to the reuse of the vitamin in producing clotting proteins. Many such compounds are highly and acutely toxic. These rodenticides may directly impact non-target small mammal populations. Also, secondary poisonings (relay toxicoses) have had impacts on birds of prey (e.g. buzzards, Berny et al., 1997; barn owls (Tyto alba), Hegdal and Bloskiewicz, 1984; red kites (Milvus milvus),

Ntampakis and Carter, 2005)), as well as carnivorous mammals such as mink (Mustela vison) (Fournier-Chambrillon et al., 2004), polecats (Mustela putorius) (Shore et al., 1996), and red foxes.

Endocrine-disrupting compounds in sewage receiving waters. A variety of endocrine-disrupting compounds have been detected in sewage effluent, including steroidal hormones, pesticides, breakdown products of surfactants and plasticizers, pharmaceuticals, and others (e.g. PCBs and dioxins). Kolpin et al. (2002) detected 82 compounds in 139 US streams, 34 of which are known or suspected to have estrogenic activity. In that study, 75% of the streams sampled had more than one of the organic compounds present in detectable concentrations. Non-hormonal compounds that influence estrogen receptors include those that mimic the effects of hormones (mimics) and those that interfere with normal hormonal activity (antagonists). To our knowledge, the first study to establish a link between disruption of the endocrine system and exposure to both sewage effluent and synthetic estrogens was that of Purdom et al. (1994). Following reports from fishermen of hermaphroditic fish in sewage lagoons, caged rainbow trout (Salmo gairdnari) and carp (Cyprinus carpio) were placed in sewage effluent in the field, and others were exposed to an oral contraceptive formulation in the laboratory. Marked increases in plasma vitellogenin (VTG) concentrations were observed, particularly in the trout. VTG is an egg-yolk precursor protein produced in the liver under control of estrogen, and in adult fish, it is normally present in measurable amounts only in the blood or tissues of mature females. Thus, plasma VTG concentrations in males have been widely used as a marker of exposure to environmental estrogens (Hansen et al., 1998). Induction of VTG and increased incidence of intersexuality (presence of ova in testicular tissue) have been observed in male fish collected below sewage outfalls in numerous studies. Young male roach (Rutilus rutilus) exposed to graded dilutions of sewage effluent experienced VTG induction and feminization of reproductive ducts, but no effects on germ cell development. Rodgers-Gray et al. (2000) found that VTG response was dose and time dependent; fish chronically exposed to low concentrations of effluent in river water had a lower threshold for VTG induction than did fish that were exposed for a shorter time period. Fish may show a sustained VTG response to estrogen exposure. High levels of plasma VTG in male fish may lead to reduced testicular growth and be accompanied by kidney and liver damage.

Plasma VTG concentrations in male fathead minnows remained elevated 21 days after exposure to estradiol. Also, reduced gonad mass relative to body mass, increased relative liver mass, testicular damage, reduced gonopodium length (the modified male anal fin, critical for sperm transfer in some species), and reduced serum testosterone concentrations in males have been observed in fish collected downstream of wastewater treatment plants. In addition, gamete production and quality were reduced in intersex male roach collected from sites that received sewage effluent. Natural (e.g. 17â-estradiol, estrone) and synthetic (e.g. 17á-ethynylestradiol) steroid hormones are among the most commonly occurring and potent endocrine disrupting compounds in effluents from wastewater treatment plants handling domestic wastewater.

However, sensitivity to estrogen exposure is speciesspecific, highlighting the need for proper selection of species for monitoring purposes.

Although mature female fish are not as responsive to estrogen exposure as are males, normal patterns of vitellogenesis in adult females may be altered by exposure to high concentrations of ethynylestradiol. Also, increased VTG production has also been documented in mussels exposed to sewage effluent, either directly or in receiving streams. Research studies have documented a variety of endocrine-disrupting effects of steroidal hormones on fish, including reductions in gonadosomatic indices (organ weights as a proportion of body weight) and the number of males with milt and spawning tubercles, reduced gonopodium (used as a phallus for sperm transfer) length, and reduced fecundity (number of eggs) without a reduction in fertility. Others have found changes in breeding behavior of male fish exposed to estradiol, including fewer courtship displays, chasing/following behaviors, and copulation attempts, and reduced aggressiveness toward other males. Male fathead minnows exposed to estradiol were able to acquire nests and spawn successfully in the absence of competing males. However, estradiol-treated males were less aggressive in encounters with control males, and subsequently acquired and held fewer nests and produced 5 times fewer offspring than controls. This decreased productivity of individual males could reduce their reproductive fitness (genes less likely to be passed to the next generation) and reduce the growth rates of exposed populations.

The long-term impacts of exposure to complex mixtures of endocrine disrupting compounds on populations of fish, frogs, and other aquatic life are not well understood. Grist et al. (2003) modeled the effects of ethynylestradiol exposure on populations of fathead minnows (Pimephales promelas) and determined that reductions in population growth rates due to reduced fertility could be anticipated. Questions such as whether males with subtle intersex characters breed successfully in the wild (impacting the reproductive fitness of affected individuals) and whether a population with fewer mature males can be sustainable long-term need to be addressed.

Metals, metalloid and a non-metal. Metals are electropositive elements with metallic bonds that readily form cations. The heavy metals such as cadmium, lead, and mercury have high atomic weights, have no known physiological role, and can produce toxic effects at low concentrations. Others, such as the essential elements copper and zinc, are necessary for life, but they can be toxic at elevated concentrations. Metalloids are elements with properties intermediate between metals and non-metals. Some metalloids such as arsenic and non-metals such as selenium can be toxic at relatively low doses. Of course, selenium is an essential element needed in minute amounts for optimal health. Metals are usually complexed with other elements in the environment, and the form, or species, of a metal, as well as its valence state, are important in determining its toxicity to organisms. Trivalent chromium occurs naturally, and small amounts are necessary for optimal health. In contrast, hexavalent chromium is produced and used in industry and is highly toxic. Use of fertilizers, irrigation, and pesticides may increase the soil's toxic load of elements and salts.

Early pesticides were metal salts such as lead arsenate; in some areas soils remain contaminated from early use of these compounds. Other anthropogenic activities such as processing and burning of fossil fuels, mining, smelting, and steel making have increased concentrations of metals in the biosphere. Mercury is a highly toxic metal that is released naturally into the biosphere from volcanic eruptions, exposed bedrock, as well as through anthropogenic activities including chloralkali plants used to produce chlorine (now being phased out in much of the world), mining, reservoir development, burning fossil fuels, medical procedures and waste, metal processing, smelting, waste and incineration. Mercury is used in thermometers and barometers, electrical switches, fluorescent lights, dental amalgams and in the past was used in the making of felt hats.

Mercury is also released during the burning of coal. Another important source has been use of liquid mercury metal to capture gold from stream bed deposits. Because human and wildlife populations are already experiencing the impacts of toxic exposures to methylmercury, the continued atmospheric transport and deposition of mercury are major concerns. When mercuric salts and metallic mercury are metabolized by organisms in anaerobic sediments, they form the more toxic methylmercury. The rate of conversion of other forms of mercury to methylmercury is much faster in certain warm climates that have favorable geochemistry (e.g. parts of the Amazon and of Florida) than in areas more distant from the equator Veiga et al. Methylmercury bioaccumulates via the aquatic food chain, readily crosses the placenta and blood–brain barrier, and is highly toxic to the developing nervous system. Cadmium is a toxic element used in Ni–Cd batteries and in certain pigments, plastics stabilizers, coatings, alloys, and electronics. It also is an impurity found in, or released during the processing of, other metals such as copper, iron, and zinc, manufacturing of steel, cement, fertilizers, and the combustion of fossil fuels. Cadmium is readily taken up by some plants (e.g. leafy vegetables, rice) and is then ingested by humans and other animals in their food. Among the principal target organs of cadmium are bone and kidneys.

Lead has been used in shotgun pellets, bullets, fishing sinkers, paint, batteries, gasoline, solder, water pipes, and other products. One of the largest sources of lead in the atmosphere is leaded gasoline; alkylated lead is added to reduce engine noises or "knocking." To help control harmful emissions, beginning in 1975 passenger cars and light trucks manufactured in the United States included catalytic converters that required lead-free gas. The US Environmental Protection Agency mandated a phase-down of lead concentrations from 2–3 to 1/10th g/

gallon during 1977– 1986; prohibitions on lead in paint, water pipes, and tin cans were also enacted during this period. Leaded gasoline was still available for use in passenger automobiles in some parts of the United States until 1996, and is still used in aviation, motorboat, farming, and racing vehicle gasoline. In the United States, reductions in lead use were reflected in a 37% decrease in average blood lead concentrations of human beings during 1976–1980, and 41% reduction between the periods 1988/ 1994 and 1999/2002. However, lead exposures among children of some socioeconomic groups in the United States remain high due in part to contaminated dust, lead paint chips, lead dissolved from plumbing, and other sources.

Worldwide, internal combustion engines are still a major source of environmental lead, although an increasing number of nations have banned or are beginning to restrict its usage in gasoline. Mining of metals and minerals often carries large environmental impacts. Not only do these activities leave the land degraded and scarred, but processes to remove the product of interest often result in the release of toxic elements or other chemicals into the environment. For example, sulphides released from newly exposed rock may combine with water and oxygen to form deadly sulfuric acid, which often makes its way into soils, groundwater and streams. This acid also has the added effect of leaching or dissolution of toxic metals such as arsenic, cadmium, lead, iron, and mercury from mine tailings or "waste" rock.

The legacy of the mining boom of the mid- to late-1800s in the American west are many miles of streams and rivers still contaminated with metals and sulfuric acid. In addition to devastating environmental impacts from effluents of mining in developing countries that are due to the lack, or poor enforcement, of environmental protection laws, the pursuit of the metals produced may also lead to armed conflict and human rights abuses. The use of chromated copper arsenic (CCA) in treating wood to prevent insect damage and rot has received much recent attention with regard to risks to children using treated playground equipment, as well as playing in areas where such equipment exists, due to leaching of the metals from wood into soil. The US Environmental Protection Agency has banned sales of CCA-treated wood for most residential uses.

Nevertheless, much of the wood remains in use today, and CCA-treated wood produced before 2003 can still be sold. Of importance, animals attracted to the salty taste of the ash left from burning CCAtreated wood the pursuit of the metals produced may readily ingest a lethal dose of arsenic. The process of removing metals from ore, or smelting, results in the deposition of aluminum, cadmium, fluoride, lead and zinc, and/or other elements that can be highly toxic to plants, soil-dwelling invertebrates, and vertebrates. Such toxicity can reduce the diversity of flora and soil fauna, leaving only more tolerate species, so that ecological function in contaminated landscapes is altered for centuries and likely millennia.

An example is provided by two zinc smelters that operated near Palmerton in the mountains of eastern Pennsylvania for much of the 20th century. Approximately 485 ha of vegetation were severely impacted, and damage extended 10 km downwind of the smelters. Drought and fire followed by erosion of as much as 30–60 cm of topsoil Oyler, 1988 exacerbated the damage to the mountainside.

Areas that remained forested became dominated by species that re-sprouted from roots or stumps instead of seeds. There was also a loss of diversity in moss and lichen communities. Retardation of decomposition due to greatly reduced populations of soil and litter macro- and microorganisms resulted in a build-up of plant litter on top of mineral soil. These habitat changes, along with accumulations of toxic levels of cadmium, lead, and zinc, caused reductions in populations of forest birds, small mammals, and salamanders (forest floor salamanders were extirpated), especially insectivorous species. These impacts extended many kilometers downwind and beyond the area of obvious environmental impacts. The sheer volume of contaminated soil, along with a cinder bank or "dross" pile containing an estimated 33 million tons of slag (waste ore), makes removal of contaminated material impractical. Efforts to re-vegetate the site began in the 1990s with the application of a mixture of fly ash, sewage sludge, and limestone to the barren mountainside, in an effort to control pH, prevent further erosion, and provide a medium for plant growth. However, efforts to restore the steep, rocky, highly contaminated slopes have been largely unsuccessful to date.

Most natural soils contain low concentrations of toxic elements, although there are some areas where soils are naturally high in certain elements, e.g. seleniferous soils of the western United States. Arsenic and selenium can be present in levels of concern in ancient groundwater, and pumping from wells for drinking and irrigation makes them available at the surface. Due to the marine origins of the adjacent mountains, the soils of the San Joaquin Valley in California are high in salts and certain trace elements including selenium. Irrigation is necessary to leach salts from the soil to allow the production of vegetables and other crops away agricultural wastewater is then carried by a network of canals. In the past, the water was directed to a basin that formed 12 constructed wetlands on what became the Kesterson National Wildlife Refuge. These wetlands, located in an arid region with a diminished wetland base, were an attractive habitat to a variety of breeding waterbirds. High concentrations of selenium in wastewater became even more elevated over time in these wetlands through evaporation in the arid climate of central California, and by 1983 a high incidence of embryo mortality and deformities was noted in wetland birds such as grebes, coots, stilts, and ducks. Because of the severely of these impacts, the ponds were closed and filled. Subsequently, 12 other water reclamation projects in the arid western United States were found to be contaminated with selenium in irrigation drainage water (Seiler et al., 2003).

Although this issue has received attention from planners and researchers, and some remediation has been undertaken, the underlying issues are socioeconomically, politically, and technically complex; thus, progress in finding solutions has been slow.

Radiation/radionuclides. The nuclear age began with the discovery of X-rays in 1895. Fission was first demonstrated in 1938 by German scientists, which led to competition between Nazi Germany and the United States to develop an atomic weapon. The United States developed such a bomb in 1945, and then used two against Japan to end WWII. Nuclear weapons testing and the bombing of Hiroshima and Nagasaki, Japan, during the mid-20th century resulted in the global distribution of radioactive material. The first reactor-produced radioisotopes for medical and industrial use were provided by the Oak Ridge National Laboratory in 1946. Since that time, nuclear medicine has been a boon for the diagnosis and treatment of many diseases. By 1951, electricity was being produced via nuclear fission, and by July 1957 the first civilian reactor for power was on line. However, in October 1957, the first accidental release of radiation from a reactor occurred. Theoretically, nuclear energy can provide "cleaner" and cheaper electricity than burning of fossil fuels. However, in addition to small releases of radioactivity during normal operations and the transportation and storage of nuclear wastes there have been accidental discharges of large amounts of radiation, such as the incidents at Chernobyl, Ukraine, and Three Mile Island, Pennsylvania, USA. In 1986, the explosion and fire at a poorly designed reactor at Chernobyl led to the hospitalization of 203 people and the death of 47 due to radiation poisoning. Also, as of 2004, 9 children from the area had died of thyroid cancer. Fallout containing shortand long-lived radioisotopes covered a wide area of northeastern Europe, and radiation from this incident ultimately was transported throughout the northern hemisphere. Residents of a 30 km radius were evacuated and few have returned.

One source estimated that there would be an eventual 4000 additional human deaths, driven by a 2% increase in cancer rates. In spite of somatic and germ-cell mutations in birds, small mammals, and fish resulting in increased prevalence of aberrant phenotypic traits, native plants and wildlife have flourished, whereas invasive exotic species have declined, in the 30-km exclusion zone around Chernobyl. This has been touted as "proof" that effects of radiation from this nuclear disaster were not as long-lasting as feared. However, these animals could represent the immigration of dispersing animals from surrounding areas into unoccupied habitat. Few detailed, longitudinal studies of wildlife populations and their age structures have been conducted. Perhaps best studied in the Chernobyl area have been barn swallows, which are migratory insectivorous birds with a high degree of philopatry. Fourteen years after the incident, swallows nesting in the area around Chernobyl had high prevalence of abnormal sperm and reductions in antioxidant levels, compared to swallows from a reference area. The exposed birds also had a sustained high rate of partial albinism, and germline mutation rates that were 2- to 10-fold higher than distant reference sites.

Immune suppression and reduction of carotenoid-based sexual coloration were also noted in this population. Mutations produced phenotypic changes in secondary sexual traits (e.g. plumage coloration, feather lengths) that may influence reproductive success. Small mammal populations in some areas near Chernobyl were reduced by as much as 90% following the reactor incident. Increases in populations in the following spring were attributed to immigration from lesscontaminated adjacent areas and the cessation of tillage and crop harvest. Voles collected from the Chernobyl area had the highest radiocesium burdens and dose rates ever recorded for mammals. Researchers have reported high genetic diversity in voles from the highly contaminated zone, which could be indicative of high mutation rates and/or immigration from less-contaminated areas. Cytogenetic damage was documented in voles in Belarus and Sweden, both of which were in the direction of prevailing winds in the first few days following the disaster.

Environmental fate of chemicals in the environment

Our environmental contamination problems are complex but controllable, by limiting environmental contamination so that concentrations are not high enough to harm prokaryotic and eukaryotic life forms (e.g. beyond the intended target site, such as an agricultural plot selected for treatment because an exotic insect pest has been recognized). The assimilative capacity of most natural environments is tremendous; however, it is essential to ensure that pollutants are not produced and used at rates that undermine the capacity of the purposefully exposed site to detoxify it self. Environmental contaminants are degraded, or transformed into simpler, often (though not always) less toxic forms, in several ways. Photodegradation involves transformation of chemicals often involving oxidation to reduce toxicity. Many manmade chemicals are rapidly photodegraded by sunlight. Some chemicals may become more toxic to organisms through the process of photoactivation. For example, dechlorination of hexachlordibenzop- dioxin may produce lower chlorinated but more toxic dioxin analogs. Biotransformation involves the modification of chemicals by the physiological processes (e.g. enzymes in the liver that catalyze hydrolysis, oxidation, reduction, or conjugation) in living organisms – usually in an effort to detoxify and eliminate them. But biotransformation also occurs in soil microbes such as bacteria and fungi under aerobic or anerobic conditions. Plants and microbes within digestive tracts are also important in the biotransformation of environmental contaminants. Of course, metabolic changes can also bioactivate chemicals to produce (by definition) metabolites that are more toxic than their parent compound. For example, the organochlorine pesticide DDT is not itself highly toxic to birds. However, its metabolite p,p'-DDE can cause thinning of eggshells due to disruption of calcium metabolism. Also, during microbial metabolism, inorganic mercury is methylated, producing a potent neurotoxic substance responsible for cognitive, fine motor, and visual-spatial disabilities, especially in developing organisms.

Efforts by organisms to detoxify exogenous compounds can sometimes produce temporary bioactivation manifested in reactive species such as singlet oxygen or hydroxy radicals. These free-radicals or oxidants can cause oxidative stress, damaging cells and may promote tumor formation. More research on fate and toxicity of metabolites and environmental degradation products is needed.

Availability of chemicals in the environment

Because a contaminant is present in the environment does not mean that an organism will have contact with it. For example, surface-dwelling animals may not have direct access to a contaminant buried under of many centimeters of topsoil. However, they may gain exposure by consuming earthworms or plants that bring contaminants to the surface. Thus, contaminants and receptors (i.e. organisms of interest) must overlap in both time and space for there to be a potential for exposure. Should pollutants and receptors co-occur, there are a number of potential pathways of exposure: dermal, oral (dietary and grooming), inhalation, via gills and rarely, injection. The bioavailability of pollutants in the environment refers to the proportion of a substance that is absorbed across the gut, skin, or other portals to enter the bloodstream and other tissues where it can cause a physiological reaction in an organism. Bioavailability can be determined by the amount of organic matter, pH, and cation-exchange capacity of soil or sediment, the presence of antagonistic elements of chemicals, and the nutritional status of an organism.

The future of ecotoxicology

The science of ecotoxicology should exist within a balanced ecological and biomedical context. In the view of the authors, ecotoxicology needs to become more of a preventive science and less of a remediative enterprise. Although diagnostic laboratories must become far more involved in forensic ecotoxicology, simply waiting on animals to come to diagnostic laboratories is insufficient to protect wildlife and ecosystems from chemically induced damage. Thus, ecotoxicology must also be pro-active, involving trips to the field for monitoring of exposures and impacts, and trips to the laboratory for exploratory and confirmatory research. In the authors' view, there needs to be better testing of products before marketing to insure not only the health of humans, but also of pets, livestock, wildlife, plants, and the environment as a whole. For most toxicants, detailed and comprehensive ecotoxicological information exists for very few plant and animal species. Risk assessment therefore often relies upon the use of a few surrogate species, with extrapolation to a huge array of organisms that will likely be exposed when a product is used or an effluent is released. Of concern is that much information from industry on testing of products is buried in reports to the US Environmental Protection Agency and difficult to obtain.

We believe that, upon granting of patents, industry scientists should be required to publish results of toxicity testing in the scientific literature and via the Internet so that the information is fully available to scientists and the interested public. Stakeholders should share in decision-making with regard to risk management and longer-term product registration. Finally, we are of the view that considerable effort in regulatory ecotoxicology should be devoted to removal of problematic compounds and formulations, with their replacement, when warranted, with products unlikely to present undue ecological harm. Ecotoxicology initially was focused largely on determining chemical residues in the environment and not enough on mechanisms and effects, though this has begun to change. Ecotoxicological research typically extends beyond the health effects on individuals, to examine how these changes manifest as population-, community-, and ecosystem- level impacts. However, if environmental mismanagement places greater numbers of species at risk for extinction consideration of toxic effects on individuals will be increasingly necessary.

Far greater attention needs to be devoted to complex mixtures of chemicals, indirect effects of contaminants, fate and toxicity of chemical degradation products, development of more endpoints of exposure and their role in disease, and the validation of risk assessments. There is a continuing need for more integrative ecotoxicology research, i.e. comprehensive, large-scale field studies examining interactions among biotic and abiotic factors, complemented by microcosm and mesocosm studies. More research incorporating reproductive toxicology including multi- generational studies is needed. More studies should examine how contaminant exposure affects animal behavior, since subtle changes in behaviour can have immense negative impacts on an animal's survival and reproductive fitness.

There is a need for more studies designed to identify cause and effect, including assessments of clinical signs, histological lesions, and residues over time at relevant exposures, to enhance diagnostic and forensic capabilities. Animals subjected to toxicity testing to determine LC50 or LD50 or minimum toxic dose or maximum tolerated dose should consistently be evaluated with clinical pathology assays, gross and histological pathology studies, and analyses to determine residues. All of these endpoints should be evaluated at doses with no impact, threshold toxic reactions, marked toxic reactions, and overt lethal toxicity. "Veterinary ecotoxicologists" can participate in meeting critical data gaps by specializing in a variety of disciplines with regard to environmental contaminants, including: epidemiology, pathology, immunology, environmental risk assessment, environmental chemistry/fate, environmental law, ecological rehabilitation, and others. Veterinary ecotoxicologists can also specialize in a particular toxicant group (e.g. metals, hazardous wastes), animal species group (marine mammals, birds, invertebrates), given habitat type or biome (aquatic, terrestrial, desert, forest, estuaries, coral reefs), or given region (midwestern United States, Illinois, north Africa, tropics, polar). The future of a healthy environment for humans, domestic animals, and wild biota depends to a large extent on the degree to which we learn to use and control naturally occurring and synthetic chemicals. Although there have been successes, and some environments are cleaner now than in the recent past, the continually expanding number of chemicals released into the environment increasing free nutrients that prompt biotoxin production and the adverse impacts noted in the environment demonstrate the striking need for vigilance and accountability through research, education, environmental law and enforcement, and development of prudent science- driven environmental policy.

BIBLIOGRAPHY

- Adams, H. Richard (2001), Veterinary pharmacology and therapeutics, Wiley-Blackwell, ISBN 978-0-8138-1743-9
- Anthony, David; (1984), Man and animals: living, working, and changing together : in celebration of the 100th anniversary, the School of Veterinary Medicine, University of Pennsylvania, UPenn Museum of Archaeology, ISBN 978-0-934718-68-4
- Anthony, David; Zoetis (2014), VeritasDVM, Dr. Donal Smith, ISBN 978-0-934718-68-4, archived from the original on 8 February 2019, retrieved 8 March 2022
- Aspinall, Victoria; Cappello, Melanie; Bowden, Sally (2009), Introduction to Veterinary Anatomy and Physiology Textbook, Jeffery, Andrea (forward), Elsevier Health Sciences, ISBN 978-0-7020-2938-7
- Boden, Edward; West, Geoffrey Philip (1998), Black's veterinary dictionary, Rowman & Littlefield, ISBN 978-0-389-21017-7
- Bryant, Susan (2010), Anesthesia for Veterinary Technicians, John Wiley and Sons, ISBN 978-0-8138-0586-3
- Cannon, Marthaxcx; Hijfte, Myra Forster-van (2006), Feline medicine: a practical guide for veterinary nurses and technicians, Elsevier Health Sciences, ISBN 978-0-7506-8827-7
- Catanzaro, Thomas E. (1998), Building the Successful Veterinary Practice: Innovation and creativity, Wiley-Blackwell, ISBN 978-0-8138-2984-5
- Crispin, Sheila M. (2005), Notes on veterinary ophthalmology, Wiley-Blackwell, ISBN 978-0-632-06416-8

- Done, Stanley H. (1996), Color atlas of veterinary anatomy: The dog & cat, Elsevier Health Sciences, ISBN 978-0-7234-2441-3
- Dunlop, Robert H.; Malbert, Charles-Henri (2004), Veterinary pathophysiology, Wiley-Blackwell, ISBN 978-0-8138-2826-8
- Dyce, Keith M.; Sack, Wolfgang O.; Wensing, Cornelis Johannes Gerardus (2010), Textbook of veterinary anatomy, Saunders, ISBN 978-1-4160-6607-1
- Eurell, Jo Ann Coers; Eurell, Jo Ann; Frappier, Brian L.; Dellman, Horst-Dieter (25 May 2006), Dellmann's textbook of veterinary histology, Wiley-Blackwell, ISBN 978-0-7817-4148-4
- Fenner, William R. (2000), Quick reference to veterinary medicine, John Wiley & Sons, ISBN 978-0-397-51608-7
- Finger, Stanley (2001), Origins of neuroscience: a history of explorations into brain function, Oxford University Press, ISBN 978-0-19-514694-3
- Foreyt, William J.; Foreyt, Bill (2001), Veterinary parasitology reference manual, Wiley-Blackwell, ISBN 978-0-8138-2419-2
- Gelatt, Kirk N. (2000), Essentials of veterinary ophthalmology, Wiley-Blackwell, ISBN 978-0-683-30077-2
- Gupta, Ramesh Chandra (2007), Veterinary toxicology: basic and clinical principles, Academic Press, ISBN 978-0-12-370467-2
- Hirsh, Dwight C.; Maclachlan, Nigel James; Walker, Richard L. (2004), Veterinary microbiology, Wiley-Blackwell, ISBN 978-0-8138-0379-1
- Hunter, Pamela (2004), Veterinary Medicine: A Guide to Historical Sources, Ashgate Publishing, Ltd., p. 611, ISBN 978-0-7546-4053-0
- Kahn, Cynthia M., ed. (2010), The Merck Veterinary Manual, Whitehouse Station, N.J., Merck, ISBN 978-0-911910-93-3
- Lane, D. R.; Cooper, B. (2003), Veterinary nursing, Elsevier Health Sciences, ISBN 978-0-7506-5525-5
- Lawhead, James B.; Baker, MeeCee (2009), Introduction to veterinary science, Cengage Learning, ISBN 978-1-4283-1225-8
- Merck, Melinda D. (2007), Veterinary forensics: animal cruelty investigations, John Wiley & Sons, ISBN 978-0-8138-1501-5
- Murphy, Frederick A. (1999), Veterinary virology, Academic Press, ISBN 978-0-12-511340-3
- Nicholas, Frank W. (2009), Introduction to Veterinary Genetics, John Wiley and Sons, ISBN 978-1-4051-6832-8

- Pattengale, Paula (2004), Tasks for the veterinary assistant, Wiley-Blackwell, ISBN 978-0-7817-4243-6
- Pfeiffer, Dirk (2009), Veterinary Epidemiology: An Introduction, John Wiley and Sons, ISBN 978-1-4051-7694-1
- Radostits, O. M.; Gay, C. C.; Blood, D. C.; Arundel, J. H.; Hinchcliff, Kenneth W (2000), Veterinary Medicine: A Textbook of the Diseases of Cattle, Sheep, Pigs, Goats and Horses (9th ed.), Elsevier Health Sciences, p. 1877, ISBN 978-0-7020-2604-1
- Riviere, Jim E.; Papich, Mark G. (2009), Veterinary pharmacology and therapeutics, John Wiley and Sons, ISBN 978-0-8138-2061-3
- Robinson, Wayne F.; Huxtable, Clive R. R. (2004), Clinicopathologic Principles for Veterinary Medicine, Cambridge University Press, p. 452, ISBN 978-0-521-54813-7
- Rollin, Bernard E. (2006), An introduction to veterinary medical ethics: theory and cases, Wiley-Blackwell, ISBN 978-0-8138-0399-9
- Sherman, David M. (2002), Tending animals in the global village: a guide to international veterinary medicine, John Wiley & Sons, ISBN 978-0-683-18051-0
- Shilcock, Maggie; Stutchfield, Georgina (2003), Veterinary practice management: a practical guide, Elsevier Health Sciences, ISBN 978-0-7020-2696-6
- Slatter, Douglas H. (2002), Textbook of small animal surgery, Elsevier Health Sciences, ISBN 978-0-7216-8607-3
- Smith, Gary; Kelly, Alan M., eds. (2008), Food Security in a Global Economy: Veterinary Medicine and Public Health, Phila.: University of Pennsylvania Press, p. 176, ISBN 978-0-8122-2044-5
- Swabe, Joanna (1999), Animals, Disease, and Human Society: Human-animal Relations and the Rise of Veterinary Medicine, Routledge, p. 243, ISBN 978-0-415-18193-8
- Swabe, Joanna (1999), Veterinary Courses and CE., p. 244, ISBN 978-0-415-18193-8, archived from the original on 8 February 2019, retrieved 8 March 2022
- Swope, Robert E.; Rigby, Julie (2001), Opportunities in veterinary medicine careers, McGraw-Hill Professional, p. 151, ISBN 978-0-658-01055-2
- Thrusfield, Michael (2007), Veterinary epidemiology, Wiley-Blackwell, ISBN 978-1-4051-5627-1
- Zajac, A.; Conboy, Gary A.; American Association of Veterinary Parasitologists (2006), Veterinary clinical parasitology, John Wiley & Sons, ISBN 978-0-8138-1734-7

INDEX

www.ingramcontent.com/pod-product-compliance
Ingram Content Group UK Ltd.
Pitfield, Milton Keynes, MK11 3LW, UK
UKHW021328190726
13845UKWH00004B/75

9 789349 964860